Mutation, Randomness, and Evolution

Mutation, Randomness, and Evolution

Arlin Stoltzfus

Fellow, Institute for Bioscience and Biotechnology Research,
Rockville, Maryland, USA

OXFORD
UNIVERSITY PRESS

Great Clarendon Street, Oxford, OX2 6DP,
United Kingdom

Oxford University Press is a department of the University of Oxford.
It furthers the University's objective of excellence in research, scholarship,
and education by publishing worldwide. Oxford is a registered trade mark of
Oxford University Press in the UK and in certain other countries

© Arlin Stoltzfus 2021

The moral rights of the author have been asserted

First Edition published in 2021
Impression: 1

Published in the United States of America by Oxford University Press
198 Madison Avenue, New York, NY 10016, United States of America

British Library Cataloguing in Publication Data
Data available

Library of Congress Control Number: 2020952665

ISBN 978–0–19–884445–7

DOI: 10.1093/oso/9780198844457.001.0001

Printed and bound by
CPI Group (UK) Ltd, Croydon, CR0 4YY

Acknowledgments

I am indebted to many individuals—scientists, historians, and philosophers—for discussions about topics covered in this book, especially David McCandlish, Pablo Razeto-Barry, Sahotra Sarkar, Davide Vecchi, John Beatty, Joanna Masel, Ingo Brigandt, Eric Haag, Arnaud Martin, Lee Altenberg, and Wes Anderson the Philosopher; also Wallace Arthur, Lindley Darden, Jean-Baptiste Grodwohl, Astrid Haase, Philippe Hunemann, Yogi Jäger, Tom Jones, Rees Kassen, Eugene Koonin, Francesca Merlin, Kira Makarova, Roberta Millstein, Stuart Newman, Josh Payne, Josh Plotkin, Justin Pritchard, Premal Shah, David Stern, Jay Storz, Jake Weissman, and Julian Xue. I thank Andrei Chabes, Alex Couce, Ron Ellis, Darin Rokyta, Gloria Rudenko, and Premal Shah for sharing data or figures. I thank Ian Sherman for encouragement and advice, and Charles Bath and the staff at Oxford for production expertise. For camaraderie during thousands of hours of writing in the local coffee shop, I thank Ramyar, Cary, Arcely, Elaine, Lauren, Lucas, Melanie, Nada, Ben, Peter, Valerie, and especially Isabel.

This book surely includes some mistakes, which are my own. For topics that are not in my narrow area of expertise, I have cited relevant work without always knowing the most recent or most relevant work. I apologize to those authors whose relevant work was not highlighted.

I thank the developers of freely available software products involved in the creation of this book: LaTeX (TeXShop, MacTeX, BibTeX), R, RStudio, TextWrangler (BBEdit), and GraphViz.

I am particularly indebted to David McCandlish. Every week for the past eight years, David and I have had a regularly scheduled one-hour conversation, typically about collaborative projects, but also about this book and any other aspects of evolutionary biology. On innumerable occasions, he has helped me to clarify my thinking, offered incisive interpretations of the theoretical literature, and suggested how to explain difficult concepts. This book is partly an expression of our ongoing dialog.

Contents

List of figures

List of tables and boxes

Tables

Boxes

CHAPTER 1

Introduction: a curious disconnect

To see what is in front of one's nose needs a constant struggle
George Orwell

1.1 Mutational origination as an evolutionary cause

MacLean et al. (2010) evolved resistance to the antibiotic rifampicin in many replicate cultures of the bacterium *Pseudomonas aeruginosa*. Resistant strains typically have mutations in the *rpoB* gene, encoding RNA polymerase. For 11 resistant mutants in *rpoB*, MacLean et al. (2010) measured the mutation rate, the effect on fitness, and the frequency evolved. The inter-relationships of these three variables are shown in Fig. 1.1.

What do the results show? The frequency with which an outcome evolves is strongly correlated with the rate of mutation (middle panel). That is, the higher the rate at which a resistant variant emerges by mutation, the more likely its appearance in replicate cultures, reaching a high frequency in each population. This effect does not occur because mutation rates are correlated with mutant fitness (right), which has little effect (left), given that all the mutations are strongly beneficial, with a similarly large chance of fixation.

This immediately prompts questions about the *causes* of the 50-fold range in mutation rates evident in Fig. 1.1—all for single-nucleotide mutations. That is, when faced with evidence that the course of evolution reflects what is mutationally likely, one is naturally concerned with the causes of those tendencies. Yet, in the case of MacLean et al. (2010), no attempt was made to predict or understand such effects.

In Chapter 9, we review additional evidence—both from experimental evolution, and from nat-ural evolution in diverse taxa (including animals and plants)—that biases in mutation influence the course of adaptation. Such results suggest a cause–effect relationship between the rate at which variants are introduced by mutation, and their chances of being manifested as evolutionary changes. As we will see, this effect can be large, in the sense that a several-fold effect in evolution is a large effect. *Tendencies of mutation may impose predictable biases on evolution.* Conversely, patterns in evolution may have mutational explanations, not merely in the sense that mutation is necessary for change, but in the sense that mutation is a difference-maker, a dispositional factor that causes one kind of change to happen more often than another.

Evolutionary biology is the science of evolutionary causes and effects. What are the terms that evolutionary biologists use to describe this causal relationship? What is the theory for its operation? How widespread are its effects? What notable evolutionary thinkers have espoused its importance? The inter-relationships of the three variables measured by MacLean et al. (2010)—rate of mutational origination, fitness effect, and frequency evolved—seem very fundamental. What is known about these relationships, generally?

According to standard accounts in the evolutionary literature, Darwin discovered the principle of selection, and by 1930, Fisher, Haldane, and Wright combined this principle with genetics to yield a general framework for understanding evolutionary causes. Thus, if we were to guess at the answers to the questions in the previous paragraph, we would begin by supposing that Haldane, Fisher, and

Mutation, Randomness, and Evolution. Arlin Stoltzfus, Oxford University Press (2021). © Arlin Stoltzfus. DOI: 10.1093/oso/9780198844457.003.0001

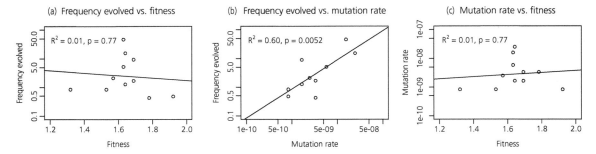

Figure 1.1 Inter-relations of mutation rate, fitness effect, and frequency of evolution for 11 rifampicin-resistant variants (data from MacLean et al., 2010).

Wright discovered the impact of biases in variation and developed the theory for it.

In fact, this is not the case at all. As we will discover, Fisher, Haldane, and Wright each argued against the idea that the course of evolution might reflect internal tendencies of variation, on the grounds that mutation rates are too small to have an appreciable effect. Classical population-genetics theory lacks an account of the causes and consequences of biases in the introduction process, as do the classic works of Mayr, Simpson, Stebbins, Dobzhansky, Huxley, and others. The kind of causation in which a bias in mutational origination imposes a bias on evolution was formally described just two decades ago.

How could such a basic idea escape recognition for so long? As will become clear, the effects of the introduction (origination) process fall within a blind spot or gap in evolutionary thinking. This blind spot is not accidental: it represents the shadow cast by a dense nexus of arguments constructed deliberately to promote and defend the belief that variation is merely a source of random raw materials for selection—supporting a high-level view of selection as the potter and variation as the clay—and to discourage alternatives, particularly the idea that evolution might exhibit internal tendencies due to the way that variation emerges.

A main task of this book is to document this nexus of arguments thoroughly, and then to deconstruct it. Because this nexus was built deliberately, over decades, it cannot simply be excised or deflated with a single precise stroke: it must be attacked vigorously in all its manifestations, like a cancer, until

its power over our imagination has withered. Once this task is completed, we can begin to rebuild our understanding of the role of variation in evolution, drawing on some available theory and data.

The results of MacLean et al. (2010), and those of the more well known study of parallelism by Rokyta et al. (2005), roughly fit the behavior expected from origin-fixation models depicting evolution as a simple two-step process, with a rate determined by multiplying the rate of mutational introduction of new alleles by their probability of fixation. For instance, for the 11 mutations from MacLean et al. (2010), the total rate of evolution can be represented as a sum over 11 origin-fixation rates

$$r = \sum_{k=1}^{11} \mu_{ik} N \pi_{ik} \tag{1.1}$$

denoting the parental genotype as i and the alternatives with k. Here, μN is a rate of origin (mutation rate times population size), and π is a probability of fixation. The chance that a particular mutation j will be observed in a given replicate is a fraction of that sum:

$$p_j = \frac{\mu_{ij} N \pi_{ij}}{\sum_k \mu_{ik} N \pi_{ik}} = \frac{\mu_{ij} \pi_{ij}}{\sum_k \mu_{ik} \pi_{ik}} \tag{1.2}$$

This probability is linearly dependent on both the rate of mutational introduction, and the probability of fixation. Thus, for sites evolving according to origin-fixation dynamics, the 50-fold range of rates of mutation evident in Fig. 1.1 has an expected 50-fold effect on the chance of evolving. The 11 mutational changes have measured fitness effects ranging from $s = 0.3$ to $s = 0.9$, thus the probability of fixation, using Kimura's formula $\pi = \frac{1-e^{-2s}}{1-e^{-2Ns}}$, ranges less than 2-fold from 0.45 to 0.83 (assuming $N = 10^6$). Under these conditions, the difference-making power of mutation is strong and that of selection is modest (with more data, one expects to see a correlation between frequency evolved and fitness).

By contrast, an influence of mutation bias on the outcome of evolutionary adaptation is not expected under the theory that adaptation results from shifting the frequencies of alleles already present in an initial population. For instance, if all 11 variant *rpoB* alleles are present initially, the outcome of evolution will be essentially deterministic, with the fittest allele prevailing every time. Mutational shifts among pre-existing alleles may occur, but because mutation rates are so small, their influence will be negligible compared to the much larger shifts caused by selection.

The notion that natural populations maintain an abundant stock of variation, so that selection never has to wait for a new mutation—the "gene pool" theory—was a key conceptual innovation of the Modern Synthesis, providing the rationale to redefine evolution as a selection-driven shift in the frequencies of alleles.

Yet, in the 1960s, as soon as this shifting-gene-frequencies view became an established orthodoxy, molecular sequence comparisons prompted biochemists to depict evolution quite differently, as a process of the acceptance or rejection of individual mutations. This led King and Jukes (1969) and Kimura and Maruyama (1969) to propose origin-fixation models relating the rate of evolution directly to the rate of mutational introduction (see McCandlish and Stoltzfus, 2014). For years, the significance of this development was muted by the sense that "molecular evolution" is a special case, an isolated world of neutrality irrelevant to adaptation and other major issues. Yet, eventually, theoreticians interested in adaptation and other major issues began to remark on the inadequacy of treating evolution as a process of shifting allele frequencies, arguing instead that the dynamics of long-term evolution depend on discrete events by which mutation introduces new types (Hartl and Taubes, 1998; Eshel and Feldman, 2001; Yedid and Bell, 2002).

That is, some evolutionary geneticists have rejected the idea that all of evolution, including long-term and large-scale changes, follows from the kind of selection-driven process of shifting allele frequencies that one sees in experimental populations of animals and plants over small numbers of generations. This extrapolationist doctrine is usually expressed by stating that macroevolution follows from microevolution (see Dobzhansky, 1937, p. 12).

The implications of this foundational shift are still emerging, and they have not penetrated very far into evolutionary thinking. The orthodoxy that emerged in the mid-twentieth century depends for its validity on rejecting the view that the outcome of evolution reflects the timing and character of individual mutational events, and instead, embracing the view that evolution is a higher-level mass-action process of shifting allele frequencies in a population. Over the years, this position has been quietly abandoned, and is no longer a foundation of evolutionary genetics. *As a result, there is a curious disconnect between what formal models imply about the possible dispositional role of variation in evolution, and the guidance provided by familiar verbal theories that depict mutation as merely the ultimate source of raw materials for selection, or as a weak force incapable of influencing the outcome of evolution.* Of particular interest here is that, when the dynamics of evolution depend on the dynamics of the introduction process, evolutionary change is *biased* by mutation, in the sense that, if the rate of mutation for $A \rightarrow B$ is greater than for $A \rightarrow C$, this elevates the chances of evolving from A to B relative to A to C. That is, a bias in mutational introduction is a prior bias on evolution.

Importantly, nonuniformities in individual mutation rates, as in MacLean et al. (2010), *are not the only sources of biases in the introduction of variation.* For instance, in the study of laboratory adaptation of bacteriophages by Rokyta et al. (2005) (see Ch. 9), a *Met → Ile* change is observed repeatedly, due to mutations at position 3850. The genetic code dictates that this particular amino acid change can take place by three different single-nucleotide mutations: *ATG → ATT, ATG → ATC,* and *ATG → ATA.* By contrast, other amino acid pairs typically have only one or two possible mutational paths interconnecting them (e.g., *Met → Val, ATG → GTG*). Because of this, the *Met* → *Ile* change has a kinetic advantage. Even if all rates of mutation from one codon to another are the same, this amino acid change is probabilized by having multiple mutational routes of origination. In fact, two of the three different mutational pathways at position 3580 are seen by Rokyta et al. (2005).

This is not a bias in the mechanism of mutation itself, but a bias in how mutations are expressed, formally identical to a kind of developmental bias suggested repeatedly in the evo-devo literature (e.g., Emlen, 2000), by which some phenotypes are more likely to arise by mutation-and-altered-development because they have more mutationally accessible genotypes. Molecular and morphological versions of this proposal are united by the concept of a genotype-phenotype (GP) map. The genetic code is the GP map that relates triplet genotypes to amino acid phenotypes. Disrupting the ATG genotype underlying the methionine phenotype is more likely to result in the isoleucine phenotype, accessible via three mutations, than valine, accessible by one mutation.

Likewise, if we generalize further on this effect, we can understand the basis of arguments in the literature of self-organization and evolvability about the differential accessibility of alternative phenotypes in genotype-space (Fontana, 2002). Concepts like genetic proximity or mutational accessibility are relevant to evolution precisely because the evolutionary exploration of genetic spaces is subject to kinetic biases due to mutation.

That is, whereas the patterns of mutation-biased adaptation in MacLean et al. (2010) are specifically (1) an effect of asymmetries in mutation rates, and *do not represent* (2) asymmetries in developmental responses to perturbation, *nor* (3) asymmetries in the density and accessibility of genotype networks, *all three kinds of asymmetries are potential sources of biases in the introduction of variation*. If, starting from state *A*, the rate of introduction of *B* is three times higher than *C*, the expected evolutionary consequences of this 3-fold bias are the same whether (1) *B* and *C* are unique genetic alternatives, and the differential effect reflects mutation bias per se, (2) *B* and *C* are alternative phenotypic states, and (aggregating effects over the entire mutation spectrum) *B* is three times more accessible via mutation-

and-altered-development, or (3) *B* and *C* are phenotypically defined networks in genotype-space, and states in *B* are 3 times more accessible from the parental network *A*.

Note that, when we invoked this kind of causation to interpret the results of MacLean et al. (2010), we did so within a distinctive explanatory paradigm. Confronted with a spectrum of changes, each with an observed frequency (Fig. 1.1), our focus was on explaining why evolution has the tendencies that it has. Molecular evolutionists are familiar with this paradigm—as old as Dayhoff's empirical matrices of amino acid replacement—in which evolution is a process with propensities, and our job as evolutionists is to predict and explain these propensities.

The traditional approach, by contrast, is to establish a plausible narrative for *a unique evolutionary outcome that is considered as a conclusion or end-point*, i.e., what Sober (1984) calls "equilibrium explanation." As Reeve and Sherman (1993) put it, the focus of the adaptationist research program is on explaining "phenotype existence," e.g., why the zebra has its stripes. The paradigm of an explanation is to show that the observed phenotype is the most fit possible, given the various constraints at work. This paradigm has had a profound influence on conceptions of explanation and causation.

To apply the traditional paradigm to MacLean et al. (2010), one would begin with the observation of a single instance, a token outcome, as if we have only a single rifampicin-resistant strain, e.g., the one with arginine at site 518. The existence of this amino acid would be credited to the role of selection shifting the Arg518 allele deterministically from a low frequency to a high frequency, e.g., we could use the measured selection coefficient to model the dynamics of allele replacement, and we could test this experimentally by examining how selection increases the frequency of the Arg518 allele from low to high.

By contrast to this deterministic appeal to selection, the involvement of the Arg518 allele would be described (in the traditional paradigm) by reference to a mutation that occurred *by chance* at some time in the past. Evolution was "contingent" on the presence of this allele in the gene pool of the starting population, so that selection could raise its frequency deterministically from low to high. If some other allele, e.g., leucine 531, provides greater rifampicin resistance, then our account of Arg518 would stress "constraints" or "limits" on the "power of selection" to achieve perfection (Barton and Partridge, 2000). Our explanation would note that the Arg518 allele was present in the gene pool, and the Leu531 allele (for instance) was absent, by chance.

When we look at the traditional paradigm in this way, we can see that it relies on the idealization that selection deterministically ensures that the fittest survives. As this idealistic paradigm has broken down, some evolutionary thinkers have turned to denoting exceptions by appealing to chance, contingency, and constraints (e.g., Futuyma, 2010).

These concepts make it possible to rescue the traditional paradigm, but only superficially. Chance, constraints, and contingency do not denote causal theories to account for evolutionary dispositions or preferences. Chance is not a cause of anything. Contingency is not a causal force. Constraints do not cause outcomes, but explain nonoutcomes, and in the case of mutation biases, they fail even to be explanatory (there is no constraint preventing mutations that occur at lower rates—they simply occur at lower rates). That is, these concepts are explanatory rather than causal: they refer not to pistons or levers, but to excuses. They are not alternative theories, but verbal markers indicating departures from ideals of determinacy and equilibration. Their use in the evolutionary literature represents the unstable transition-state between a failed paradigm and some alternative.

That is, these concepts fail to satisfy the scientific imperative to explore causal theories in which quantities such as mutation rates are used to make precise predictions. However, as demonstrated earlier, a different and more suitable explanatory paradigm exists that allows us to treat variational causes as causes, rather than as limits to the imaginary omnipotence of selection.

1.2 What this book is about

The recognition that biases in the introduction of variation are a cause of evolutionary bias has far-reaching implications for evolutionary research and for evolutionary theory. Biases in the introduction

process have biological causes and evolutionary consequences. Their causes reside in properties of the mutation-generating system, properties of development, and broad features of the architecture of genetic spaces. The *consequences* of biases in the introduction process are in the province of population genetics. Expanding our understanding of population-genetic causation' to include the consequences of biases in the introduction of variation provides a cohesive and previously unrecognized basis for addressing key concerns of scientists interested in neutral evolution, evo-devo, evolvability, and self-organization.

Re-thinking the evolutionary role of variation along these lines is a major challenge for twenty-first-century evolutionary biology. A broad approach to this challenge would require a book much larger than this one, and it would necessarily leave many issues unresolved, dependent on experiments that have not yet been designed or carried out. To respond adequately to this challenge will require the work of many scientists over many years.

This book focuses more narrowly on the development and application of basic principles, relying on molecular examples to establish key arguments. Rather than fully answering the grand challenge of rethinking the role of variation, it aims to prepare the ground by (1) illustrating some of the complexity of mutational processes and using that information to explain why biases are inevitable, and why various precise senses of randomness fail to apply, (2) showing that the mutation-is-random claim common in evolutionary discourse is best understood in terms of a historically popular view in which the internal details of mutation and development are irrelevant to how evolution turns out, and (3) articulating an alternative view based on exploring propensities of variation as causes of evolutionary propensities, and summarizing the current theoretical and empirical support for this connection.

Mutation is often mischaracterized and misunderstood. For instance, the adjective "accidental" or "spontaneous" is often applied, but mutation is not an accident like a branch falling on the roof of your home during a storm and leaving a hole. Damage is not mutation. When cells leave unrepaired holes in their DNA, the typical outcome is cellular death, not mutation. Instead, mutation is like a branch falling on a house and leaving a hole in the roof, followed by a repair-bot detecting the damage and then shingling over the hole using the wrong color of shingles. The mutation is not the hole (which has been repaired), but the slight depression in the roof and the patch of differently colored shingles.

As will become apparent, mutations emerge by complex pathways, often stimulated by damage, yet carried out by enzymes. No unified theory of mutation exists. To learn about mutation as a biological process is to learn about a vast assortment of different processes that contribute to the emergence of mutations. However, because teaching about mutation is not the main purpose of this book, no chapters are devoted specifically to mutation. Instead, this book approaches the topic indirectly, describing mutational processes in order to consider the implications of randomness, and to illustrate key concepts (e.g., mutation spectrum) important for conceptualizing mutation.

The doctrine that mutation is random is not what it seems. Ordinary scientific claims typically are justified by appeal to (1) first principles, which in this case might prohibit nonrandomness, or (2) systematic observations or experiments, which in this case might have shown again and again, under a variety of conditions, that mutation is random. The first option is ruled out because the randomness doctrine long preceded any detailed knowledge of how mutation works—indeed, the claim is made by evolutionary biologists, not by mutation researchers. The second option seems unlikely because sources that promote the randomness doctrine do not cite a large body of systematic evidence, that is, one does *not* see statements like "mutation has been shown to be completely random in a variety of species (Smith, 1939; Johnson, 1944; Jones, 1951)."

"Randomness" has many possible connotations, and the literature of evolution draws on them in diverse ways. The situation has become so confusing that, in the past 25 years, multiple authors—including professional philosophers—have waded into the confusion, aiming to sort out what evolutionary biologists really mean to say. The typical approach is to assume that a special "evolutionary" meaning of randomness can be found

that justifies the doctrine retroactively. We will devote considerable space to exploring these ideas, including the possible conditional independence of mutations and fitness effects—conditional upon a common environment. Yet, conditional independence and other attempts to make sense of the randomness doctrine either fail to match the facts of mutation, or fail to match what evolutionary biologists have been saying.

Other cases exist in which scientists have used a concept for many years without a clear consensus on its meaning. Indeed, the history of science repeatedly shows that confusion or conflicts over the meaning of key concepts is normal and may persist for generations. Scientists took centuries to settle on a common understanding of "heat." The concept of "probability" has proven enormously useful in spite of two centuries of debate over what it means, whether it refers to something real, and whether it can be derived from first principles. Many practically useful concepts, e.g., entropy, are subject to rarefied disputes.

Yet probability, heat, and entropy are *undoubtedly useful concepts*: their use supports precise calculations that help us to understand the world better. What is the purpose of the randomness doctrine? What precise and useful calculations does it support? What logical conclusions depend on it? What advantage would be lost by rejecting it? One often finds evolutionary thinkers urging caution, warning us that "random" is a problematic word, yet if the subject requires caution, *why take a position at all?* These same authors do not feel obliged to stake out a carefully worded position on the randomness of anything else. Why did it become so important to associate mutation with randomness?

A closer look at the classic Synthesis literature reveals that randomness attributions co-occur with other arguments that minimize the importance of variation in evolution, and that these arguments *all refer to the same contrast-case: selection*. Mutation is random, selection is not; mutation is weak, selection is not; selection acts at the (right) population level, mutation acts at the (wrong) individual level; selection provides direction, variation merely provides raw materials. In this context, the randomness doctrine proclaims a deeper mutation-is-unimportant doctrine in which variation is made subservient to selection, supporting the assignment of roles that one finds in the topic article on "natural selection" (Ridley, 2002) in the *Oxford Encyclopedia of Evolutionary Biology*:

In evolution by natural selection, the processes generating variation are not the same as the process that directs evolutionary change . . . What matters is that the mutations are undirected relative to the direction of evolution. Evolution is directed by the selective process (p. 799).

In this way, the randomness doctrine provides a metaphysical guarantee of the classic dichotomy of selection and variation as the potter and the clay, that is, it differentiates selection, the source of order, shape, and direction, from mutation—not the source of those things, because it's "random." Patterns and interesting features and other orderly outcomes in evolution may be safely attributed to selection, because mutation is random. No ordinary definition of randomness cleanly distinguishes one biological process from another, thus a special "evolutionary" definition is developed to fill this gap. *The nature of the required "evolutionary" definition is now obvious: it must apply to mutation but not to selection*, e.g., the randomness claim is interpreted to mean that mutation, unlike selection, does not invariably increase fitness (Section 4.2; see Eble, 1999).

The final aim of this book is to articulate an alternative view, and to make a case for its importance. Evolution is change. The goal of evolutionary research, in my opinion, is to understand change, and particularly to understand why some types of changes happen more often than others. The early geneticists, who first contemplated the role of mutation in Mendelian populations, saw mutation as a difference-maker, as a potentially important source of initiative, creativity, direction, and discontinuity in evolution (Stoltzfus and Cable, 2014). This position was rejected by the architects of the Modern Synthesis, who argued that mutation is merely a random source of raw materials, and further claimed that the view held by early geneticists—in which the course of evolution may reflect the timing and character of individual mutations—was incompatible with the genetics of natural populations, which always have abundant standing variation to fuel evolution.

As we will discover, the shifting-gene-frequencies theory underlying the sweeping claims made by the architects of the Modern Synthesis (see Box 6.2) does not correspond to the open-ended framework that some scientists imagine today, when they refer to a mid-century "Synthesis." To justify a neo-Darwinian division of labor between variation (source of raw materials) and selection (source of creativity, direction, etc), the shifting-gene-frequencies theory of evolution posits a "buffet" view of population genetics, designed precisely to circumscribe the role of mutation, in which everything selection needs to adapt to current circumstances is already present in the gene pool, in carefully maintained abundance. Contemporary thinking still relies on terms and concepts from this narrow theory, including a "forces" theory that seems to rule out a dispositional role of mutation (on the grounds of being a "weak force" easily overcome by selection), even though such a role is perfectly compatible with a broader conception of evolutionary genetics.

The successful promotion of this theory by Ernst Mayr and his cohort of influencers left a blind spot in the development of evolutionary thinking about mutation. Where one might expect to find a description of the propensities of mutation (variation), and the causal principles linking them to propensities of evolution—what we will call the *source laws* and *consequence laws* of variation—one instead finds *a denial that any such principles exist*. Theoretical and empirical results have been encroaching on this blind spot for decades, with little overt recognition.

In this book, we consider some simple theoretical results that are fundamental to understanding the role of mutation and development in evolution, but which are not explained in textbooks. We discover that the introduction of variant forms by mutation-and-altered-development is a predictable cause of nonrandomness in evolutionary change. We find that this kind of cause can be modeled using principles of population genetics, and its effects can be studied using conventional scientific methods. *Patterns of change can have mutational causes, not simply in the sense that mutation is materially necessary for evolution, but in the sense that tendencies of mutation can be difference-makers, causing one kind of thing to happen more often than another.*

Box 1.1 Theory$_A$ and theory$_C$

The concept of a theory plays an important role in evolutionary discourse. The term "theory" has been used in the sense of a grand conjecture for hundreds of years, and continues to be used this way in evolutionary biology, e.g., Kimura's "Neutral Theory" is the conjecture that most changes at the molecular level result from random fixation of selectively neutral alleles; the "Exon Theory of Genes" (Gilbert, 1987) proposes that genes evolved from combining exon-minigenes; the endosymbiotic theory holds that mitochondria and chloroplasts arose from primordial bacterial endosymbionts.

However, the word "theory" does not always mean this. Population genetics theory, for instance, is not the conjecture that populations have genetics, nor is music theory the conjecture that music exists. Instead, these are bodies of abstract principles. Indeed, scientific writings use the same term for both (1) theory$_C$ (concrete, conjectural), a grand conjecture or major hypothesis to account for some set of observed phenomena, as in the "continental drift theory" or "Lamarck's theory of evolution," and (2) theory$_A$ (abstract, analytical), a body of abstract principles relevant to some discipline, methodology, or problem area, as in "music theory," "quantum field theory," or "population genetics theory."

Usually one must rely on context to determine which meaning applies. For instance, a white paper on "The Role of Theory in Advancing 21st Century Biology" (National Academy of Sciences, 2007) emphasizes the development of formalisms rather than conjectures, and says that "a useful way to define theory [note the abstract noun] in biology is as a collection of models," clearly a reference to theory$_A$ (the report also refers to a few theories$_C$).

The two types of theory are evaluated in different ways. The standard of truth for a theory$_C$ is verisimilitude—how well does it match the actual world? A theory$_C$ takes risks, and can be refuted by facts. Because a theory$_C$ is a conjecture—not necessarily a correct one—it still is called a "theory," even by those who doubt its verisimilitude. By contrast, to evaluate a statement in theory$_A$, one does not consider any facts about the world, but only whether the statement is correctly derived from its assumptions. Once a piece of theory$_A$ is valid (correctly derived), it remains valid forever, even if it relies on imaginary things such as infinite populations.

To derive expectations for some possibilities, prior to having any reason to prefer one over another, one requires some theory$_A$. Fisher (1930b) wrote that "No practical biologist interested in sexual reproduction would be led to work out the detailed consequences experienced by organisms having three or more sexes; yet what else should he do if he wishes to understand why the sexes are, in fact, always two?" (in fact, the sexes are not always two, but this is not relevant here). The collection of all the models for different numbers of mating types would be part of the theory$_A$ of sexes. A concrete theory$_C$ of sexes would be something quite different, e.g., it might propose a causal explanation for the actual historic phenomenon of the origin and maintenance of sexual reproduction in animals.

Although the distinction between theory$_A$ and theory$_C$ is not always clear, applying the distinction remains a useful exercise. For instance, the theory of kin selection ("kin selection theory" or "inclusive fitness theory") is frequently described as a set of "tools" (Michod, 1982), implying theory$_A$. Yet, the context for the use of these tools is the broad conjecture of Hamilton that kin selection is crucial to account for the evolution of social behavior in animals. Most of the disputes about kin selection theory are theory$_A$ disputes among mathematicians and philosophers about such things as whether the assumptions underlying certain formulations of Hamilton's rule are correctly described, or whether kin selection theory$_A$ is equivalent with group selection theory$_A$ (Birch and Okasha, 2014).

Obviously, there is a connection between the two types of theories, in that abstract principles of theory$_A$, rendered concrete with observed or conjectured values, can provide the basis of a theory$_C$. In the case of Kimura's Neutral Theory (Kimura, 1983), the theory$_C$—the conjecture that most changes at the molecular level result from the random fixation of selectively neutral alleles—and theory$_A$ were developed somewhat separately. The definition of effectively neutral alleles (perpetually misunderstood by critics) and the probability of fixation under pure drift had been known for decades (see Wright, 1931; Fisher, 1930b, Ch. IV; Haldane, 1932, Appendix). Kimura combined mostly pre-existing theory$_A$ (including plausibility arguments based on the "cost" of selection) with the concrete assertion that the values of certain quantities (relating to population sizes and mutant effects) were such that, for DNA and protein sequences, neutral evolution by mutation and random fixation would be far more common than anyone had imagined previously.

Opponents of the Neutral Theory, who deny the truth of the theory$_C$, are nonetheless quite happy to make use of its theoretical$_A$ infrastructure in efforts to reject neutral models, as in the review by Kreitman (1996). That is, valid theory$_A$ is required to carry out *modus tollens* reasoning, in which X is rejected based on its implication $X \implies Y$ and the observation that Y is absent. In the case of neutral models, $X = neutrality$, and Y is some expectation about rates or patterns. To reject neutrality based on evidence requires a correctly derived model of an imaginary abstract world in which neutrality is true, i.e., the theory$_A$ statement $X \implies Y$ must be abstractly true. The paradox in Kreitman's title "The neutral theory is dead. Long live the neutral theory" is perfectly resolved by the fact that it refers first to theory$_C$, and then to theory$_A$.

In the remainder of this book, the distinction between theory$_A$ and theory$_C$ is not made explicit except in a few cases, so that the reader may apply the distinction and assess its utility, without suffering the annoyance of being forced to do so.

1.3 Who this book is for

In the process of untangling the conceptual mess at the intersection of mutation, randomness, and evolution, this book addresses various topics that are timely and of central importance to evolutionary biology. Nevertheless, the way that the topics are framed may be unfamiliar, even to professional scientists. I would not expect readers to have a clear sense of whether or not they might be interested in reading this book, based solely on the title, or based on a phrase such as "biases in the introduction of variation," or even based on the earlier reference to a conceptual mess. Therefore, it may help to explain how the main arguments relate to several issues that are more familiar.

First, in the past two decades, interest has emerged in taking a more detailed look at the role of variation in evolution, including an interest in the "arrival of the fittest" rather than the "survival of the fittest," apparent in discussions of developmental constraints, evolvability, robustness, facilitated variation, and so on. Most of the previous interest in this issue has been focused on development of visible phenotypes of macroscopic organisms, whereas my background, training, and inclinations are much more molecular than organismal. This book presents a nontraditional view of the role of variation that is firmly grounded in theory and in empirical data, in which tendencies of evolution are related to tendencies of variation. It builds a foundation of basic concepts, which it then uses to address, and sometimes resolve, long-standing problems.

For instance, for decades, even as evo-devo has gained in popularity, others have appealed to population genetics—widely regarded as the language of causation in evolution—to argue that evo-devo has not contributed any new principles or causes to evolutionary thinking. In the absence of a general-cause claim, attempts to justify evo-devo turned to fuzzy claims about alternative "narratives" and "explanatory paradigms." In this book we discover that classic arguments about causation used against evo-devo, such as Mayr's "proximate cause" objection, Dobzhansky's "wrong level" argument, or the weak-pressure objection (e.g., Reeve and Sherman, 1993), are inadequate, and this is not a matter of paradigms or reductionism, but a matter of making the wrong assumptions about population genetics, and failing to recognize the introduction process as an evolutionary cause with distinctive implications. The ability of generative processes to impose a bias on the outcome of evolutionary change is the first-order cause implicitly at work in various evo-devo ideas about higher-order effects.

Second, this book is tailor-made for those in molecular evolution, microbiology, and comparative genomics who feel that evolutionary biology has been stubbornly resistant to the lessons of the molecular era, which has only partially shifted our views of evolution when it should have transformed them. In this regard, several highly original books have appeared recently, including Nei (2013) (mutation, not selection, drives evolution), Lynch (2007b) (nonadaptive mechanisms make genomes complex), Shapiro (2011) (engineering, not accident, provides innovation), and Koonin (2011) (after Darwinism, things get complicated).

The present book shares something with each of these, though the overlap in content is small. Like Lynch, I am concerned to justify a novel position on the causes of observed evolutionary patterns in terms of theoretical implications of population genetics. Like Shapiro, Koonin, and Nei, I believe that evolutionary thinking is deeply shaped by vestiges of a neo-Darwinian view—selection and variation as the potter and the clay—that is broadly incompatible with the facts of molecular evolution. More than any of these authors, I believe that progress depends on conceptual and cultural reform, to include re-evaluating core concepts ("raw materials," "forces"), developing new concepts and metaphors to guide our thinking, changing what is taught to students, and reforming the distorted historiography of our field.

Third, every scientist, philosopher, or lay-person interested in evolution has surely encountered the idea that, beginning in the 1980s, a high-level dispute has been simmering, regarding the adequacy of a mid-twentieth-century orthodoxy formerly known as the "Modern Synthesis." In recent years, this dispute has taken the form of a call for an "Extended Evolutionary Synthesis." This book explains and thoroughly documents the commitment of the architects of the Modern Synthesis to a mistaken conception of the role of variation that remains deeply embedded in evolutionary thinking, e.g., in the "raw materials" doctrine, the "mutation is random" doctrine, and the "forces" theory.

These historic commitments have been obscured by a subsequent process of normality-drift that has perpetually redefined the "Synthesis" (and, to some extent, "neo-Darwinism"), creating a false impression of permanence. We will see exactly how the meanings of key terms have changed, e.g., "raw materials" used to be an analogy to raw materials, and now is merely a synonym for "variation," used even for mutations such as gene duplications *that are in no way analogous to raw materials*. Originally, the term "adaptation" did not implicate the lucky mutant view (the adventitious fixation of a beneficial mutant), which was called "pre-adaptation" and associated with pre-Synthesis

geneticists. Today theoreticians who make models of this process call it "Darwinian adaptation" or "Darwinian evolution," while Darwin and Fisher roll over in their graves. In molecular evolution and evolutionary genetics, scientists frequently assume a form of mutation-limited dynamics that directly contradicts the theory of shifting gene frequencies underlying the original Modern Synthesis.

This has an important implication for high-level debates on evolutionary theory. Most of these debates seem to assume that evolutionary biology must have a Grand Unifying Theory of Evolution covering all of biology. A shape-shifting "Synthesis" is then put forward, typically defined, not as genuine scientific theory, but as a tradition consisting of people and their changing distribution of beliefs: this puts an impossible burden on would-be reformers, who must come up with an alternative that is (1) equally comprehensive, and yet (2) obviously distinct from the shape-shifting "Synthesis." This debate has muddied the waters and prevented reform for decades. In reality, the actual historic Modern Synthesis ceased to be a valid universal theory some time in the 1970s: it was not useful for understanding long-term sequence divergence, nor did its theory of recombination-fueled change in the diploid sexual "gene pool" apply usefully to the asexual prokaryotic organisms that have dominated the planet through most of its history. Evolutionary biology has not had a Grand Unified Theory for nearly 50 years. Apparently, one is not needed.

Finally, many scientific readers may be drawn to this book, due specifically to their interest in the "directed mutation" or "adaptive mutation" controversy that erupted in 1988 based on experiments by Cairns, Foster, Hall, and Shapiro, and their skepticism of the "mutation is random" claim. In brief, Cairns and others suggested that cells have evolved ways of (probabilistically) generating situation-appropriate mutations, an argument that sets traditional thinking against itself, pitting a hopeful belief in the adaptation of mutational mechanisms against the simplifying explanatory paradigm in which selection receives all the credit for the hard work of evolution, on the grounds that mutation merely supplies random raw materials.

The sustained hostile reaction to this idea indicates *the willingness of evolutionists to sacrifice a belief in the pervasiveness of adaptation so as to preserve the neo-Darwinian explanatory dichotomy in which variation plays a strictly passive material role.*

Readers who (like myself) followed the directed mutations controversy will be interested in several relevant points. First, CRISPR-Cas is unarguably an evolved system that generates situation-appropriate mutations, and which contradicts conventional wisdom (e.g., Luria–Delbrück randomness). Second, many systems for generating situation-appropriate mutations have been reported in microbial pathogens, though they rarely receive attention from evolutionary thinkers, e.g., diversity-generating retro-elements, elaborate cassette-switching schemes, phase variation systems, and multiple-inversion "shufflons." Such systems often play a critical role in immune evasion or host-phage arms races. Third, these systems *do not* generally exhibit the kind of nonrandomness in which the chances of mutations are influenced by their incipient fitness effects, as was the case for some models proposed for "directed mutations" (e.g., Cairns's generalized reverse transcription model). Understanding the origin and maintenance of these specialized mutation-generating systems is a major challenge for scientists interested in the evolution of evolvability.

Thus, the random mutation doctrine breaks down in two distinct ways, based on two distinct meanings of the doctrine: independence from fitness, and explanatory irrelevance. Microbial pathogens have specialized systems of mutation that enhance the chances of mutations useful for immune evasion, contrary to the doctrine that mutations occur statistically independently of fitness. However, these contradictions will probably be viewed as exceptions, and they will not win any arguments against the broader doctrine of chance variation, which really means that variation cannot be a dispositional factor in evolution. This version of the randomness doctrine also breaks down, much more broadly, not just in microbes. As we will see, this breakdown can be documented by reference to ordinary mutation biases such as transition-transversion bias, without ever considering specialized mutation systems.

1.4 How the argument unfolds

The focus of this book shifts significantly through its three main parts. The first part, comprising Chapter 2, Chapter 3, and Appendices A and B, addresses how well the biological process of mutation is described by some of the ordinary meanings of "chance" or "randomness" in science: lack of purpose or foresight, uniformity (homogeneity), stochasticity, indeterminacy, unpredictability, spontaneity, and independence (chance).

Throughout this part, I refer to four pathways of mutation described in Appendix A, which explains the process by which a $T \rightarrow C$ mutation (1) arises from an error during genome replication, (2) arises from error-prone repair of damage, (3) emerges symbolically in a computer-generated Monte Carlo simulation, and (4) is engineered in a gene using human technology. Nearly all readers will benefit from reading Appendix A: the examples are provided specifically to build a foundation for explaining and evaluating concepts of randomness. Computer-simulated mutations, for instance, are fully deterministic and predictable (thus not "random" in those senses), but they can be uniform and their use in a simulation can represent "chance." Human-engineered mutations typically are not fully deterministic, but can be highly predictable, nonuniform, and nonindependent.

Some basic concepts of randomness are reconsidered in a more practical context in Chapter 3. This chapter introduces the idea that some ways of thinking about mutation are useful, even if they are only approximately correct. Approximations come at a cost, and thus the practical use of an approximation, e.g., the assumption that mutation is uniform when it really is not, is a matter of weighing costs and benefits. Chapter 3 also introduces the idea that the application of probabilistic reasoning to problems of mutation may be understood as an extension of logic that does not rely on any concept of "randomness." In this context, references to "chance" or "randomness" as something that exists in the physical world, rather than in our minds, represent what E.T. Jaynes called a "mind projection fallacy."

The second part of the book, comprising Chapters 4, 5, and 6, along with Appendices C and D, focuses on what "mutation is random" means, or could mean, in evolutionary discourse. The evolutionary biologists and philosophers who have addressed the issue frequently argue that the "mutation is random" claim does not refer to the ordinary meanings of randomness, but to a special "evolutionary" sense, by which (1) the process of mutation is independent of selection, evolution, or adaptation; or (2) the outcome of mutation is observed to be statistically uncorrelated with selection, evolution, or adaptation. Appendix C provides a catalog of statements by evolutionary thinkers referring to mutation and randomness.

Chapter 4 reviews these ideas and shows that nearly all of them fail for relatively obvious reasons. For instance, adaptation depends on mutation and so is not independent of it; when selection and mutation occur in the same milieu, they are subject to common factors and thus not independent. Only two of these "evolutionary" ideas survive initial scrutiny. One of them is that "mutation is random" signifies the rejection of a view in which the process of variation is pervasively biased toward adaptation, regularly generating situation-appropriate mutations. Another promising idea is a recent proposal that refers to the independence of selection and mutation given (i.e., conditional on) a common environment.

In Chapter 5, we consider some specialized mutation systems in microbes, in relation to the randomness doctrine and the "evolvability" literature. The evolvability research front tends to focus on three types of claims concerning (1) the efficacy of internal factors to shape evolution, (2) the structural principles that account for differences in evolvability, and (3) the emergence of evolved systems with enhanced evolvability. In the microbiological literature, CRISPR-Cas spacer acquisition, cassette shuffling, diversity-generating retroelements, and multiple inversion shufflons are all depicted as evolvability features, i.e., mutation strategies that facilitate long-term survival, e.g., enhancing immune evasion by antigenic variation. In a number of cases, specialized mutation systems have been shown experimentally to enhance long-term survival, thus they represent legitimate examples of the evolution of evolvability.

In Chapter 6 we consider the randomness doctrine in the context of a broader view of the irrelevance of mutation, supported by a catalog of quotations in Appendix D. In this context, the "mutation is random" claim goes along with "raw materials," "weak force," and so on, as one of a variety of claims to the effect that mutation, however necessary, is inconsequential or uninteresting. Sometimes these other arguments are implicated *as the explanation* for the randomness claim. Furthermore, every type of argument about the unimportance of mutation has the same *contrast case*: selection. The arguments all support a dichotomy in which selection is creative, influential, important, predictable, and so on, whereas mutation is none of those things.

Chapter 7 is an interlude in which we consider the problem of variation in evolutionary theory more broadly, before proceeding to devote two chapters to a specific theory about biases in variation. Under the classical neo-Darwinian view that variation merely plays the role of supplying random infinitesimal raw materials, with no determinative impact on the course of evolution, a substantive theory of form and its variation is not required to specify a complete theory of evolution. Alas, reality is not so simple. This view has been breaking down from the moment it was proposed, and is now seriously challenged by results from evo-devo, comparative genomics, molecular evolution, and quantitative genetics. To address a dispositional role of variation in evolution will require, at minimum, an understanding of tendencies of variation (source laws), and an understanding of how those tendencies affect evolution (consequence laws).

In Chapter 8, we explore the theoretical possibility that variational tendencies are consequential in evolution by way of biases in the introduction of variation. In the classic view of evolutionary genetics, mutation cannot be a source of directionality in evolution because mutation rates are too small to overcome the opposing pressure of selection: for mutational tendencies to influence evolution would require either the absence of selection (i.e., neutral evolution), or unusually high mutation rates. However, subsequent theoretical work shows that this way of thinking is based on assuming a "buffet" regime in which the alleles relevant to the outcome of evolution are already abundantly present in an initial population. Instead, one may imagine a "sushi conveyor" regime in which evolution is a two-step proposal-acceptance process, i.e., a process that follows the classic mutationist mantra, "mutation proposes, selection disposes." When the outcome of evolution depends on the introduction of novel forms, mutational and developmental biases may impose biases on evolution, and this does not require neutral evolution, absolute constraints, or high mutation rates.

In Chapter 9, we shift from theory to evidence. Readers familiar with molecular evolution will be aware that genome composition and other aggregate properties reflect effects of mutation bias. Claims of mutational effects have been featured in the literature of molecular evolution for 50 years, usually in association with claims about neutral evolution. In this chapter, however, we are only interested in changes associated with adaptation. We examine cases in which adaptive changes have been traced to the molecular level, and the evidence shows that the mutations involved in adaptation are enriched for kinetically favored mutations, that is, the ones with higher rates of occurrence. The ability of ordinary mutation biases to influence adaptive evolution—without requiring neutrality, absolute constraints, or high mutation rates—confirms a key prediction of the theory developed in Chapter 8.

In Chapter 10, we end by reviewing what has been learned and considering it in the context of broader issues in evolutionary biology. Looking back over the previous chapters, it will now be possible to make some clearer statements about mutation and the idea of randomness itself, randomness and its role in evolutionary theories, and the role of mutation in evolution. The material in the first nine chapters also provides a basis to reflect on the nature of explanations (explanatory paradigms), the role of verbal theories in accounts of causation, and the importance of discerning theories and distinguishing them from people and traditions.

1.5 Synopsis

The following represents a brief guide to the book, which has three parts that correspond roughly to the three combinations of terms

- mutation and randomness—how does the biological process of mutation seem random, or not?
- randomness and evolution—what role does the concept of "random variation" play in evolutionary thinking?
- mutation and evolution—what is the role of mutation in evolution, and particularly, the role of mutational tendencies?

The main goal of the first part is to build a foundation of knowledge and understanding about the nature of mutation, and apply that to evaluating the merit of common statements about the randomness of mutation.

- In **Ordinary randomness** (Ch. 2), we draw on mutation research and the exemplars in Appendix A to probe ordinary ideas of randomness such as uniformity, independence, spontaneity, indeterminacy, and so on.
- In **Practical randomness** (Ch. 3), we briefly draw some practical conclusions, and also consider (1) randomness as a strategy for generating null hypotheses, and (2) randomness as a mind projection fallacy that is not necessary for reasoning under uncertainty.

Whereas the first part considers the relation of mutation and randomness as a question of biology and plausible reasoning, without much attention to evolution, the second part is specifically about the role that the randomness doctrine plays in evolutionary thinking and research.

- In **Evolutionary randomness** (Ch. 4), we consider the idea that randomness has a special "evolutionary" meaning. Most versions of this idea cannot be justified by logic or evidence; and some of the justifiable ideas (e.g., mutation is not caused by future events) are not particularly useful for reasoning.
- **Mutation systems and evolvability** (Ch. 5) probes several more challenging topics, including evolved systems that generate situation-

appropriate mutations, and their relationship to the topic of *evolvability*.

- In **Randomness as irrelevance** (Ch. 6) we consider a variety of metaphors, analogies, mechanistic arguments, empirical arguments, explanatory arguments, and methodological arguments, all to the effect that mutation is just not worth your time. Appendix D is a catalog of supporting quotations.

The final part of the book draws heavily on recent work that is familiar only to a small fraction of evolutionary biologists. This part focuses on building the conceptual, theoretical, and empirical basis of an alternative theory of variation as a dispositional factor in evolution.

- We begin by defining **The problem of variation** (Ch. 7): what happens after we abandon the premise that we can build an evolutionary theory without a substantive theory of form and its variation? What is the current basis for abandoning this premise?

- In **Climbing Mount Probable** (Ch. 8), we consider the conceptual and population-genetic basis for understanding the introduction of variation by mutation (or, for phenotypes, by mutation-and-altered-development) as a source of initiative and direction in evolution.

- In **the Revolt of the clay** (Ch. 9), we review results, largely from recent work, that provide support for previously unacknowledged roles of mutation in evolution, finding strong evidence that mutation biases are important in determining the course of adaptive evolution.

The final chapter, **Moving on** (Ch. 10), includes a summary of key points, and reflections on theories and explanatory agendas.

CHAPTER 2

Ordinary randomness

R.A. Fisher (1930) has assumed that mutation occurs in every conceivable direction; and some of the most important of his conclusions regarding the outcome of natural selection, and the degree of adaptation of a species at any moment, depend upon the correctness of that assumption. There are many things, however, which indicate that in the deal out of mutations the cards are stacked.

A. Franklin Shull (1936)

2.1 Introduction

A convenient simplification, pervasive in the research literature, is to divide mutations into (1) one set representing common mutations, typically comprising all single-nucleotide substitutions or point mutations (Section B.3), which typically are assumed to occur uniformly and independently, and (2) other mutations, which are ignored. This simplification is especially helpful if one has no interest in understanding the role of mutation in evolution, but must include it as a nuisance factor, i.e., a factor that is required to make a model work, but whose effects are not relevant to testing hypotheses or explaining interesting behavior.

The focus of this book, however, is the evolutionary role of the process of mutation (and of variation, more broadly). The goal is not to marginalize mutation, but to make it a focus of research, perhaps even the focus of a research program motivated by the potential breadth and power of mutational explanations for evolutionary outcomes.

Thus, in the book, we take steps to confront the complexity of mutation, and to explore ways to integrate that complexity into evolutionary thinking.

One aspect of this complexity is that the typically ignored class of "other" mutations—including insertions, deletions, inversions, transpositions, lateral transfers, and compound mutations—actually represents the vast majority of possible mutations. Appendix B establishes this point by mapping out the universe of mutations, providing formulas to estimate the number of mutations in each category. This appendix is also useful for explaining what is meant by point mutations, insertions, deletions, inversions, transpositions (translocations), lateral transfers, and compound mutations.

A second aspect of complexity is that, for any specific type of mutation, the mutations may emerge by distinct mechanisms, and their occurrence typically violates the usual simplifying assumptions about uniformity, homogeneity, and independence.

This chapter focuses on the latter aspect of complexity, mostly in regard to nucleotide substitution mutations. The material is structured by reference to various connotations of "randomness" or "chance" in science, including lack of purpose or foresight, uniformity or homogeneity (lack of bias or structure), stochasticity, indeterminacy, unpredictability, spontaneity, and independence (chance). This chapter addresses each concept in turn, considers what it means, and how it might apply to mutation (in subsequent chapters, we consider the notion of a special "evolutionary" meaning of randomness). To supplement this material, Appendix A provides four exemplars of a $T \rightarrow C$ nucleotide substitution mutation: an error during genome replication; error-prone repair of DNA induced by damage; a symbolic mutation in

Mutation, Randomness, and Evolution. Arlin Stoltzfus, Oxford University Press (2021). © Arlin Stoltzfus. DOI: 10.1093/oso/9780198844457.003.0002

a computer simulation; and a human-engineered mutation.

2.2 Lacking in foresight

An obvious meaning of chance or randomness in evolutionary writings is that mutations must reflect a nonintelligent, mechanistic process that does not have a purpose, move toward a goal, or anticipate future needs (e.g., Dickinson and Seger, 1999). Note that this is a negative definition: it refers to the absence of something. A variety of statements by evolutionary biologists (see Appendix C) can be interpreted as suggesting a lack of foresight or future causation, e.g., "Mutation is random in that the chance that a specific mutation will occur is not affected by how useful that mutation would be" (Futuyma, 1979, p. 249). Sober (1984) says that "The defensible idea in the claim that mutation is random is simply that mutations do not occur because they would be beneficial" (p. 105). Here, "would be" indicates the conditional future, i.e., the future that *might* happen.

One way to interpret such claims is that they are rejections of backwards causation, from the future to the present, e.g., this is the most obvious interpretation of the words "because they would be beneficial." This concept of randomness makes the statement "mutation is random" a *metaphysical* claim: once one accepts a universe without future-causation, one may deduce that mutation—along with disease, foul weather, and every other process—is "random." As Sober (1984) says, "One might just as well say that the *weather* occurs at random, since rain doesn't fall because it would be beneficial" (p. 105). This meaning would make it problematic when evolutionary biologists say (as they often do) that mutation is random but selection is not.

Another interpretation of such claims is that they are based on a dualist philosophy in which one may attribute a capacity for foresight to minds, but not to mutation systems or other things that are not minds, i.e., "because it would be beneficial" means "because it is judged (predicted, expected) to be beneficial." This is different from future causation. Human foresight does not require future causation, but only some ability to leverage past experience so

as to learn patterns, or to apply an internal model of the world that can be used for generalized, adaptive learning. If this is what evolutionary biologists mean by the randomness claim, they are saying that natural mutation systems cannot have evolved in such a way as to leverage past experience, so as to learn valuable patterns of mutation. Mutation researchers (as opposed to evolutionary biologists) have often endorsed precisely this idea (e.g., Rosenberg and Hastings, 2004; Shapiro, 2011). Box 2.2 introduces this notion using the example of shuffling of antigen genes in *Trypanosoma brucei*. The topic is addressed in greater detail in Chapter 5.

Note that, although mechanistic processes do not have goals, a dynamical system may show behavior in which the system predictably moves toward an "attractor" state from a large area of state-space (the attractor's "basin of attraction"). A simple one-dimensional example would be a deterministic population-genetic model of alleles $\underline{A}$ and $\underline{a}$, in which the frequency of $\underline{A}$ converges to a particular value f (the attractor), as in the classic mutation-selection balance equation $f = \mu/s$ (Crow and Kimura, 1970). We can imagine a variety of population-genetic models with this property, including a pure mutation model (e.g., a balance of forward and reverse mutation rates), thus the capacity of a dynamic system to converge (exhibiting what seems like goal-directed behavior) under the influence of some process does not distinguish mutation from other processes.

2.3 Uniformity or lack of pattern

Modern probability theory had its origins in the seventeenth century, in the context of the need for rational actions or decisions in the face of uncertainty (Gigerenzer et al., 1989). Given that the outcome of a possible investment is uncertain, how does a rational investor behave? Given that the testimony of witnesses is uncertain, how may a rational judge decide? Given that the chances of a gamble are uncertain, what is the rational basis of making a bet?

Once one replaces certainty with uncertainty, the first challenge in formulating a quantitative theory of action or decision-making is to define what is meant by the probability of an outcome. As Gigerenzer et al. (1989) explain, the mathematicians who set out to define probabilities came up with three approaches that we still use today: "equal possibilities based on physical symmetry; observed frequencies of events; and degrees of subjective certainty or belief" (p. 5). Laplace developed a system based on the first case, in which the possibilities are all equal, as in the rolling of a fair die or the drawing of a card from a well-shuffled deck. More generally, the distribution of outcomes is said to be uniform when a process (a trial or experiment) has many possible outcomes, each with equal probability.

This corresponds to one meaning of randomness as uniformity or homogeneity, i.e., sampling from the universe of possibilities without preference. Whereas this concept of randomness is easily stated, generating uniform random outcomes requires an actual procedure, and finding an efficient procedure is challenging. The table of random digits that one finds in the back of a classic statistics textbook typically was produced by rolling dice. The challenge of generating random numbers has led to the development of numerous computational random number generators, and to the development of numerous tests for nonuniformities and serial correlations (see Appendix A).

Random number generators, or more properly, pseudo-random number generators (PRNGs), are engineered to be superficially random in the sense of producing sequences of uniform uncorrelated numbers. A PRNG generates an endless (but recurring) series of numbers from a formula of the form $x_{i+1} = f(x_i)$, carefully chosen so that statistical tests fail to uncover deviations from an equal chance of drawing a number from any equally sized interval (Brent, 2007). Appendix A (Section A.3) provides an example formula, along with a more detailed description of PRNGs and their uses.

When "random" refers to uniformity, "mutation is random" means that the process of mutation ensures that probabilities are uniformly distributed over the space of possible outcomes. For instance, a common assumption is that mutation has a uniform probability per genomic site, so that the rate of mutation for a genomic segment, such as a gene, may be computed from its length alone (e.g., Houle, 1998). Alternatively, the space of mutational outcomes could be defined as the set of paths from the current genotype to mutants, so that

uniformity would be defined relative to this total set of paths.

Typically, mutations are identified by the topology and scale of the change, e.g., a 2-bp deletion, a 10-kb inversion, a single-nucleotide substitution, and so on. The rates of these types of mutation are not all the same: they occur at widely varying rates, as illustrated in Fig. 2.1, which shows a range of ten orders of magnitude in the average rates for various types of mutations in humans. Short tandem repeats or STRs, like the repeats of CAG found in some genes, typically mutate at rates on the order of 10^{-3} per generation, with most mutations adding or subtracting a single repeat unit (Weber and Wong, 1993). Presumably the rates for some epigenetic mutations, like changes in methylation states, are even higher.

Small insertions occur less often than small deletions, and for either type of mutation, the rate decreases with length, by a power law. The rates in Fig. 2.1 are based on the power-law relationship in Lynch (2010), although a nearly identical relationship is apparent in Fig. 2.2 based on the accumulation of indels in pseudogenes (see Yampolsky and Stoltzfus, 2008). Deletions (triangles) accumulate 4-fold faster than insertions (pluses), but the power-law relationship for both is similar in slope. Because these changes take place in pseudogenes, their rates are assumed to reflect

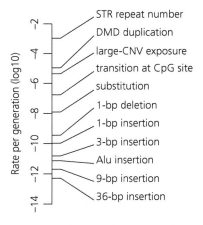

Figure 2.1 Some rates of mutation that span 10 orders of magnitude (see text for sources).

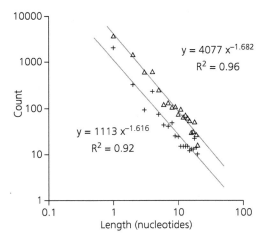

Figure 2.2 The length distribution of short insertions and deletions in pseudogene divergence (see Yampolsky and Stoltzfus, 2008).

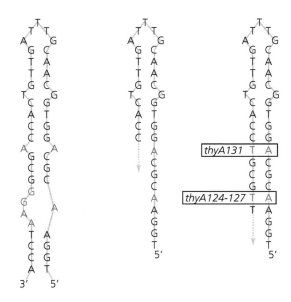

Figure 2.3 Model of mutations templated by an imperfect palindrome (quasipalindrome) in the antisense strand of *E. coli thyA* (after Fig. 1a of Dutra and Lovett, 2006). If the hairpin (left) begins to form during an interruption in DNA replication (center), continued strand synthesis (grey dotted arrow) may result in two different mutations (right), the thyA131 mutation from A to T (T to A on the sense strand), and the thyA124-127 mutation from GGAA to T (TTCC to A on the sense strand).

the mutation rate (without any differential effect of selection).

Robust estimates are not available for the average rate of per-locus duplication in humans. Instead, Fig. 2.1 shows the rate of copy-number variation (CNV) *exposure* of 5×10^{-6}, which is the chance that a base-pair will be included in a large duplication (Campbell and Eichler, 2013). This per-bp rate will be less than the per-locus rate, but will be very close when loci are much smaller than duplications. An example of a rate of duplications for a single locus would be the rate of 1×10^{-5} reported by van Ommen (2005) for the aggregate rate of duplications of the Duchenne muscular dystrophy (DMD) locus.

Even within a single class of mutations, rates are not uniform, as emphasized in recent analyses (Hodgkinson and Eyre-Walker, 2011; Johnson and Hellmann, 2011; Nevarez et al., 2010; Cooper et al., 2011). For instance, we previously learned that MacLean et al. (2010) found a 50-fold range of mutation rates among just 11 single-nucleotide changes that lead to rifampicin resistance (see Fig. 1.1).

In some cases, the reasons for context-dependent nonuniformity are understood. For instance, Dutra and Lovett (2006) show that a mutational hotspot in the *thyA* gene of *Escherichia coli*, responsible for most of the mutations in the gene, is associated with the quasipalindromic sequence shown in Fig. 2.3.

Due to the quasipalindrome, a nascent strand may fold back on itself during DNA synthesis, so that mutations are templated by base-pairing. Mutations of this type have been implicated in a number of human diseases (Bissler, 1998). This process shows a strand-asymmetry due to differences in replication of leading and lagging strands (Kim et al., 2013). Golding (1987) looked for, and found, evidence that such templated mutations are important in natural divergence.

Likewise, the endpoints of rearrangements are not uniformly distributed among all sites in the genome, but frequently map to repeated sequences. This applies both to large rearrangements (Love et al., 1991; Morris and Thacker, 1993; Kidd et al., 2010; Conrad et al., 2010) and to short insertions and deletions (Ball et al., 2005; Kondrashov and Rogozin, 2004), which are often strongly associated in genomic location with short "homopolymeric" sequences, i.e., repeats of the same nucleotide (Schroeder et al., 2016). In the mutation research literature, such patterns typically are not described as mutation biases, because the focus typically is on dissecting the mechanism for a specific mutation, not on providing modelers with generalizations about rates. However, descriptions of patterns are readily translated into generalized statements about rates, e.g., to say that repeats typically are found at the endpoints of deletions is to say that the rate

of the deletion between points A and B is higher to the extent that A and B have a matching sequence context.

Among nucleotide substitution mutations, transitions ($A \leftrightarrow G$ or $C \leftrightarrow T$) generally occur at a rate several-fold higher than transversions (Stoltzfus and Norris, 2016; Katju and Bergthorsson, 2019). Among the many known effects of context on

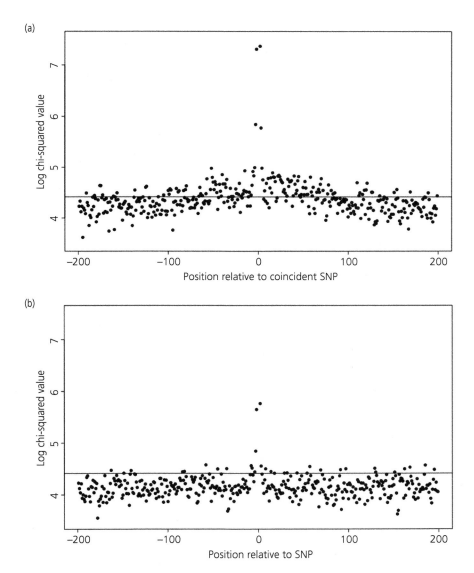

Figure 2.4 Sitewise heterogeneity of triplet frequencies at sites where humans share (above) or do not share (below) a SNP with chimpanzees (see text for explanation). Figure from Hodgkinson et al. (2009), reproduced under the Creative Commons Attribution 4.0 International (CC BY 4.0) license.

substitution mutations, one may include the higher mutability of CpG sites (Boland and Christman, 2008; Cooper et al., 2011; Shen et al., 1994; see Section 9.4 for a brief explanation), effects of other immediately upstream and downstream nucleotides (Blake et al., 1992; Krawczak et al., 1998), and heightened rates in runs of the same nucleotide (e.g., TTTTT), and in the vicinity of direct or inverted repeats (Gordenin et al., 1993; Wojcik et al., 2012).

Some heterogeneity is not attributable to the few sites immediately upstream and downstream of a mutated site. To assess the amount of cryptic variability in mutation rates along the genome, Hodgkinson et al. (2009) identified cases in which a SNP occurs at precisely the same site in human and chimp populations. They found twice as many matching SNPs as expected under a triplet model, i.e., a model that addresses only effects of neighboring upstream and downstream sites, including the CpG effect. They estimate that "the variance due to simple context, including CpGs, is 0.59, whereas the variance due to cryptic variation at non-CpG sites is 1.05." Because the vast majority of the coincident SNPs are not in coding regions, and because the SNPs are not actually clustered in local regions (as might be expected under a model of locally variable effects of selection), they argue that selection is not a likely explanation for this excess variability in the sitewise chance of a SNP.

The excess variability may be due to long-range context effects suggested in Fig. 2.4 (from Hodgkinson et al., 2009). The upper figure is based on 11,000 alignments in which humans and chimps have a coincident SNP at position 0, and the lower figure is computed from 300,000 alignments where the chimp SNP is at position 0, and there is a human SNP within 200 sites, but not at the same site. The vertical position of a dot (for each of the 200 sequence positions upstream or downstream) represents the heterogeneity of triplet frequencies; the horizontal line is drawn such that one expects 5 % of the dots to fall above the line under a null model of no effect. When the SNPs are not coincident (lower panel), the heterogeneity in triplet frequencies is not different from the null expectation, except for the focal site and the nearest neighboring positions, yet for coincident SNPs

(upper panel), heterogeneity is highly unusual over a surprisingly long distance.

2.4 Stochastic or probabilistic

In mathematics, a "random" or "stochastic" variable does not have a unique value, but may take on many values, each of which may have an associated probability, i.e., $Pr(X = x_i) = p_i$. Stochasticity has nothing to do with independence, e.g., we could define a variable x that is simply the sum of the values of all other variables in a system of interest, plus a stochastic error term, making x a stochastic variable whose value is dependent on, and correlated with, all other system variables.

This mathematical concept has no consistent physical meaning: we may find it convenient to use a stochastic variable x to represent the outcome of any process, including a fully deterministic one that is subjectively unpredictable (see Section 2.6), or when the causes of variability in x are not of interest, relative to something else that depends on the variability of x. For instance, in population genetics, selection coefficients may be treated as random variables (e.g., Park et al., 2010; Waxman, 2011).

In short, the concept of "stochastic" is not biological. One may say that, within a particular model, selection (or mutation or recombination) is treated as a stochastic variable, but to attribute a property of "stochasticity" to a physical process would be to make a category error. The same point is made by Rosenberg (2001) when he warns against confusing models with reality, saying that "a theory can be statistical even though the world is deterministic, and vice versa." The inclination of some authors to declare that an aspect of the physical world is stochastic may indicate certain beliefs about its origins (e.g., that it cannot be controlled) or its consequences (e.g., that it has no important consequences).

Thus, where "random" refers to stochasticity, the correct translation of "mutation is random" is "in a formal model, I have chosen to represent the outcome of mutation as a variable that may take on many values." In Sections 3.6 and 3.7, we reconsider the value of this concept from a more practical perspective.

Box 2.1   Counting the universe of mutations

When one encounters a mutation rate in the evolutionary literature, typically the implication is that there is a class of relevant mutations, each member of which occurs at the same rate u. Actually, many types or classes of mutations are possible, and they tend to occur at widely varying rates. Figure 2.1 shows a small fraction of this diversity.

To understand the depth of nonuniformity in mutation rates, it helps to count the universe of mutations, i.e., the possibility-space for mutations. One way to define the possibility-space is to imagine that a genome of length n can mutate to any of $4^n - 1$ genomes of the same length. This number is enormous, e.g., a genome size of $n = 2 \times 10^6$ bp (typical for a prokaryote) would imply about $10^{million}$ possible mutations.

What if we assume instead that mutations only take the familiar forms of point mutations, insertions, deletions, inversions, and transpositions, each with sizes in the range we observe for natural mutations? This question is addressed in Appendix B, which provides formulas for the numbers of each type of mutation. For instance, the numbers of inversions, deletions, and duplications are each rn, and the number of transpositions is $\approx 4rn^2$, where n is the genome length, and r is the maximum size of a rearranged segment.

The table below depicts the universe of mutations given $r = 10^4$ bp ($r = 10$ for insertions) and assuming genome sizes of 2×10^6 bp and 10^9 bp for prokaryotes and eukaryotes, respectively (other assumptions are explained in Appendix B). The universe of mutations is probably dominated by compound events like kataegis, but we know so little about these events that no calculation is possible. If we ignore compound events, the universe of mutations is dominated by lateral gene transfers (horizontal gene transfers), with 8×10^{28} or 4×10^{31} possibilities for prokaryotic and eukaryotic recipients, respectively. If we ignore lateral gene transfers, the universe of mutations is dominated by transpositions. In any case, the universe of mutations is far, far larger than the set of point mutations, which is only on the order of 10^7 or 10^{10} for prokaryotes and eukaryotes, respectively.

This universe is so large that, if all mutations were to occur at the same rate, it becomes unlikely for us to ever see the same mutation twice. According to the calculation provided in Section B.10, if we could monitor every mutation occurring in a prokaryotic population of 10^{10} cells, we would have to wait for 500 years for it to become likely to see the same mutation twice under the assumption of uniformity. If we aim to consider parallel adaptation in eukaryotes, finding an exact parallelism would require us to track changes in thousands of eukaryotic species for millions of years. That is, *if we ever see the same mutation twice under ordinary circumstances, we know that mutation is nonuniform.*

Type	Formula	Prokaryotic	Eukaryotic
Point mutation	$6.7n$	1.3×10^7	6.7×10^9
Deletion	rn	2×10^{10}	10^{13}
Tandem duplication	rn	2×10^{10}	10^{13}
Inversion	rn	2×10^{10}	10^{13}
Insertion	$(4^{r+1}/3)n$	2.8×10^{12}	1.4×10^{15}
Transposition	$4rn^2$	1.6×10^{17}	4×10^{22}
Lateral gene transfer	$2rn_d n_r D$	8×10^{28}	4×10^{31}
Compound	$\binom{d}{e}t^e n/d$	?	?

2.5 Indeterminate

A trial or experiment is said to be indeterminate if the outcome is not determined by initial conditions, i.e., if different outcomes may emerge from exactly the same conditions. Whereas unpredictability may be subjective (see Section 2.6), physical indeterminacy is an objective distinction, i.e., it refers to the actual world, not merely our sensation of it. If "mutation is random" refers to indeterminacy, it means that the outcome of the process of mutation cannot be predicted even with perfect knowledge.

This is clearly not the case for the artificial mutations in a computer simulation of evolution (see Appendix A.3). The simulations used in computer modeling are deterministic, just like the series of "random" numbers computed by a PRNG (pseudo-random number generator): given the same initial seed, every simulation gives the same outcome. In a computer simulation of evolution, mutation is not "random" in the sense of indeterminate.

Is there indeterminacy in the biological world? Organisms are composed ultimately of fundamental particles, and phenomena at the scale of fundamental particles are subject to physical indeterminacy. In our everyday "macroscopic" experiences, we are "screened off" from (i.e., isolated from, buffered from) microscopic indeterminacy, because we experience the world through the collective behavior of huge ensembles of microscopic events, whereas individual microscopic events, like the behavior of a single subatomic particle, are beyond our direct experience. Schrödinger (1944), in his influential book *What is Life?*, was perhaps the first to suggest that mutations represent quantum fluctuations. Authors such as Monod (1972) or Wicken (1987, p. 89) assert (without explanation) that mutation is indeterminate due to quantum-mechanical indeterminacy.

For the biologist, the physicist's terminology of "macroscopic" is misleading: if we look at a living cell with an ordinary microscope, we see only "macroscopic" processes (e.g., the movement of cells, vesicles, or chromosomes). Indeed, even ordinary biochemical and enzymatic processes, such as the diffusion of a substrate to the active site of an enzyme, are predominantly macroscopic, though there is some room for debate about the extent to which quantum events "filter up" and affect these processes. For instance, quantum tunneling may affect the thermal motion and the solvent characteristics of water (Kolesnikov et al., 2016), which would mean that diffusion in an aqueous solution is indeterminate; the active sites of some natural enzymes appear to be sites of quantum-mechanical effects (e.g., Sutcliffe and Scrutton, 2000).

However, we do not need to resolve such subtle issues to arrive at an intermediate answer, because mutational processes clearly show *some* physical indeterminacy. For instance, UVB-induced DNA damage leading to mutation is extremely common, and the source of UVB photons is ultimately nuclear fission reactions in the sun. Thus, whether a particular TT dinucleotide (for instance) is going to be hit by a photon, causing it to dimerize into a TT dimer that may (via error-prone repair) result in a $T \to C$ mutation, is indeterminate. An example of indeterminacy arising internally would be when damage is caused (directly or indirectly) by the spontaneous radioactive decay of some atom within the cell. Whether or not subsequent events are determinate, the initiating event is indeterminate, therefore the entire causal sequence is indeterminate.

The argument applies more broadly to the entire class of mutations that result when electromagnetic radiation damages DNA, and the (possibly larger) class of events where radiation acts indirectly by damaging precursors incorporated into DNA. Indeed, the three main sources of nucleotide mutations identified by Maki (2002) are replication errors, direct damage to DNA, and incorporation of damaged precursors. Quantum indeterminacy clearly affects the latter two categories, and may affect the first, as argued by Stamos (2010).

Given that (1) the occurrence of *some* events of mutation is indeterminate, and (2) evolutionary change *sometimes* begins with an event of mutation (Blount et al., 2008; Rokyta et al., 2005), the conclusion that evolution sometimes depends rather directly on indeterminate events is secure.

To summarize, some aspects of mutagenesis (the mutation process) are physically indeterminate: they cannot be predicted, even with perfect knowledge, because they rely on events affected by quantum-mechanical indeterminacy. In turn, evolutionary outcomes that depend on the timing of the resulting mutations are also physically indeterminate.

2.6 Subjectively unpredictable

Laplace developed a probability theory to deal with events whose outcomes were not predictable, yet he believed that the world is deterministic. He imagined a superior intellect able to know all the positions of natural things and all of the forces acting on them, and able to analyze all of these data in a single formula; for such an intellect "nothing would be uncertain, and the future, like the past, would be present to its eyes" (Laplace, 1951, p. 4). Subjective unpredictability is due to ignorance, whereas objective unpredictability arises from the kind of quantum-mechanical indeterminacy addressed in Section 2.5.

The symbolic mutation process in a computer simulation illustrates Laplace's concept. A superior intellect, knowing the seed number and the generating formula $x_{i+1} = f(x_i)$, could compute any x_i. However, without this knowledge, we cannot predict the next number in the series.

We can imagine predictable, deterministic mutation in the laboratory, although this is typically a scattershot approach (see Appendix A). For instance, microsurgery could be used to remove a chromosome or a mitochondrion from a cell, changing its genotype deterministically. Microsurgery has been used to alter the ciliate cortex—a nongenic mediator of heredity—resulting in patterns that propagate from one generation to the next (Shi et al., 1991).

Do highly predictable mutation processes exist? The answer is positive. Yeast mating-type switching (see Section 5.3.6), for instance, occurs reliably when spores germinate and begin to bud. Many examples could be given of mutational changes to genomes that take place reliably during development, often called "programmed DNA rearrangements" (for review, see Zufall et al., 2005). Such changes often occur by complex, multi-step pathways, as in the case of chromatin diminution in somatic cells of nematodes (Muller and Tobler, 2000), the generation of macronuclear chromosomes in ciliates (Noto and Mochizuki, 2017), or the maturation of antibody genes by shuffling and targeted hypermutation (Jung and Alt, 2004), explained further in Section 5.2.

Because most of these are somatic (not germline) mutational phenomena, they typically are not discussed in regard to the role of mutation in evolution, i.e., they are a separate topic from the main focus of this book. However, the existence of specialized systems of somatic mutation demonstrates conclusively that (1) the rules of biology do not preclude the existence of mutation-generating systems with highly predictable effects, and (2) the rules of

evolution do not preclude the emergence of such systems by natural processes. Indeed, in Chapter 5 we consider the evolution of specialized systems of *germline* mutation.

Nevertheless, most mutations are not very predictable. Why? Could they be predictable in the future, given advances in technology?

A few considerations of basic physics suggest that, for ordinary mutations, prediction methods will never achieve high accuracy, due to subjective unpredictability. Consider the scenario in which an energetic particle is absorbed somewhere in a cell, creating a free radical that may cause genomic damage, which is likely to be repaired, possibly inducing a mutation. The challenge is to predict what will happen in a specific cell in the interval of time from the generation of the free radical to the resolution of any damage. To solve this kind of problem requires us to track the identities and movements of a large number of molecules—all of the molecules in a living cell that might affect a mutational outcome positively or negatively. This represents the dual challenge in Laplace's scenario of omniscience above: (1) knowing the *in vivo* locations and trajectories of all relevant molecules in the cell, and (2) applying a physical model of forces to compute the future state of the system.

How could we map out all the molecules in a cell? Many methods of three-dimensional imaging use visible light and other forms of electromagnetic radiation. The effective diameter of a small molecule such as water or potassium ion ($\sim$ 0.3 nm) is 100-fold below the resolution limit for optical microscopy. Decreasing the wavelength used in microscopy 100-fold to the range of X-rays would improve resolution, but bombarding a cell with X-rays will destroy it, and we have not even come to the question of measuring momenta. Nuclear magnetic resonance (NMR)-based methods of imaging depend on dynamic reorientation of molecules to an external magnetic field, i.e., NMR is an intrusive or disruptive method with respect to detecting the momenta of molecules.

In short, assigning identities, locations, and momenta to all molecules in a cell at the requisite scale of precision, without changing the state of the cell, is apparently impossible, due to what is called the "observer effect" in physics.

Furthermore, the forward simulation problem is many orders of magnitude beyond current technology. Fine-grained computer molecular dynamics (MD) simulations may follow up to $\sim 10^6$ atoms for a timescale of many microseconds (Aminpour et al., 2019), and these simulations are not complete physical models, but include shortcuts (e.g., ignoring higher-order interactions, using a bulk-solvent model). By contrast, the timescale for a peroxide radical or a repair enzyme to diffuse a substantial distance in a cell is on the order of milliseconds or seconds (e.g., see Elowitz et al., 1999), i.e., the timescale is 10^2 to 10^5 times larger.

Also, the number of atoms to represent in a whole-cell simulation is far more than 10^6. If we think of an *E. coli* cell as a 1-micron by 3-micron cylinder, with a density several times that of water, a single cell weighs $\sim 10^{-11}$ grams, which is $\sim 10^{13}$ daltons, or $\sim 10^{12}$ atoms, given that living matter is predominantly hydrogen (1 dalton), carbon (12), nitrogen (14), oxygen (16), and phosphorous (31). That is, bacterial cells have 10^6 times more atoms than a large fine-grained MD simulation. Merely storing the three-dimensional coordinates of 10^{12} atoms for one time-point would require 10^{13} bytes, i.e., 10 terabytes, and the time-points of MD simulations are typically femtoseconds (10^{-15} seconds), thus 10^{12} time-points per millisecond (Aminpour et al., 2019).

If the problem is $\sim 10^4$ times longer in time and 10^6 times larger in number of atoms, then (given that computational intensity increases greater than linearly with number of atoms), the forward MD simulation to predict a mutation is much more than 10^{10} times harder than current fine-grained MD targets.

Such considerations suggest that subjective unpredictability of mutation is unavoidable.

2.7 Spontaneous

When biologists use the word "spontaneous," the connotations are the same as for the ordinary English adjective that may be applied to persons or events. A spontaneous event (or person) proceeds from an internal cause (or impulse), without external impetus or constraint. When we refer to

Figure 2.5 Spontaneous combustion. Due to the capacity to trap heat from decomposition, large compost piles can catch fire. Image by Ramiro Barreiro reproduced under the Creative Commons Attribution Share-alike (CC-BY-SA-3.0) Unported license.

"spontaneous combustion," e.g., of a compost heap (Fig. 2.5), we mean that a substance ignites from an internal heat-generating process, without an external spark. The widely used term "spontaneous mutation" implies that mutations emerge from internal processes, without external impetus or constraint.

Within this broad definition, the spontaneousness of mutation could mean several different things.

Sometimes "spontaneous mutation" is applied in the mutation research literature with a narrow technical meaning of "natural" or "not artificially enhanced," as a way of indicating that the process of mutation under investigation is that of a wild-type cell under ordinary conditions, without any attempts to artificially enhance mutation rates. The context of this usage is that, until recently, mutation researchers nearly always increased the rate of mutation artificially, to make the process of mutation easier to study. Most of the mutation research literature is based on experiments with repair mutants or mutagens, so that the resulting rates and spectra of mutations are of little use to the evolutionary biologist. Better sources of information on evolutionarily relevant mutation rates include mutation-accumulation experiments (Halligan and Keightley, 2009), as in Katju and Bergthorsson (2019), and analyses of the presumptively neutral divergence of sequences such as pseudogenes (e.g., Hwang and Green, 2004).

An additional meaning of "spontaneous" (e.g., Schroeder et al., 2018) is suggested by a series of classic experiments on mutation reviewed in depth by Sarkar (1991). The most famous of these is Luria and Delbrück's "fluctuation test" based on mutations in *E. coli* conferring resistance to bacteriophage T1. When a suspension of bacterial cells is plated (i.e., spread on a petri dish filled with growth medium), each cell grows up into a colony that is easily visible to the naked eye. If the growth medium contains bacteriophage T1, the cells will die unless they happen to contain a mutation that confers resistance.

The protocol of the fluctuation test is simply to repeat this procedure using separate cultures, each grown from a single cell, and to count the resistant colonies on each plate. The distributions of resistant colonies will differ depending on whether mutations occur prior to plating, or are induced by exposure to T1 phage, as illustrated in Fig. 2.6. If rare resistant mutations are induced by exposure

to T1, then the number of resistant colonies will follow a Poisson distribution, with a variance equal to the mean. If the mutations are not induced, but may occur at any time during growth, they will follow what became known as the Luria–Delbrück distribution.

The key feature of this distribution is that the variance is much higher than the mean, due to "jackpot" cultures in which the mutation occurs early in the process of growth, leaving many resistant descendants that grow up into many separate colonies.

In their experiment with *E. coli* and T1 phage, Luria and Delbrück found that the ratio of the variance to the mean was much higher than 1, and consistent with the results being due primarily to mutations that occurred prior to selection, thus uninduced (though the experimental design is not sensitive enough to exclude the possibility of a small minority of induced mutations). To say that mutation is "random" or "spontaneous," in the sense defined by the fluctuation test, is to say that the process, *because it begins and ends prior to the onset of selection*, is neither induced nor directly shaped by the current environment.

As Sarkar (1991) relates, this experimental protocol was repeated in the 1940s with the same biological system, and with other systems. The results tended to support the interpretation of predominantly noninduced mutations, with some deviations that were attributed tentatively to other effects, such as phenotypic lag. Phenotypic lag refers to the fact that the expression of the phenotype that survives selection does not necessarily emerge immediately after the mutation, but may take time to build up. For instance, if the resistance phenotype involves the loss of a surface protein that a phage uses to infect a cell, the mutation does not take effect immediately because the membranes of daughter cells still carry unmutated proteins synthesized prior to the mutation: if turnover is slow, the shift to complete resistance may take several generations. That is, the experiments become difficult to interpret when *the phenotypic realization of the mutation, i.e., its developmental expression, is not complete prior to the onset of selection,*

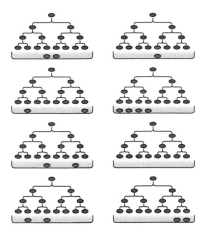

Figure 2.6 The difference between induced (left) and spontaneous (right) mutations in the Luria–Delbrück experiment (see text for explanation). Wikipedia image by MadPrime, reproduced under the Creative Commons CC0 1.0 Universal Public Domain Dedication.

an issue to bear in mind for the discussion in Chapter 4.

Let us consider a more stringent sense of "spontaneous," in which the process of mutation is entirely internal to cells, uninfluenced by any external conditions. To say that a mutation is spontaneous in this sense is to say that it arises entirely as the result of internal processes, without outside influences. Some kinds of mutations seem fully spontaneous in this sense. For instance, a replication error that takes place with a template and precursors undamaged by external radiation seems to reflect events that happen entirely inside the cell (see Appendix A). One also could imagine the case where an atom inside a cell spontaneously decays, releasing an energetic particle that causes damage to DNA, which is then repaired in an error-prone way that introduces a mutation.

Teaching resources and textbooks frequently equate mutations with copying errors, but this is misleading. Even for the most prevalent mutations, nucleotide substitutions, the relative contributions of the three categories of Maki (2002)—replication errors, incorporation of damaged precursors, and direct damage (see Appendix A)—are not known for *any* organism. Several lines of evidence indicate that the contribution of damage to ordinary nucleotide mutations is substantial. For instance, new mutations in humans identified by genomic sequencing show an effect of parental age that suggests a major contribution of damage-induced mutations (Gao et al., 2019). Context-dependent patterns of mutation inferred for mammalian genomes show a strong correlation with context-dependent patterns of susceptibility to damage by reactive oxygen species (Stoltzfus, 2008).

When a pathway of mutation begins with damage, this is often damage due to radiation (or to free radicals that result from radiation), repaired in an error-prone way by the cell's internal DNA repair systems. For instance, ultraviolet radiation may generate a TT dimer, an external influence that triggers damage repair and possible mutation (see Appendix A.2). The radiation comes from outside of the cell, thus the mutation is not spontaneous in the sense of being entirely internal.

In short, the widely held notion that mutations are normally spontaneous, in the sense of normally arising from internal processes such as replication errors, is not a firmly established conclusion based on empirical findings or on basic principles.

Finally, some kinds of mutations are strongly non-spontaneous in the sense of being strongly shaped by external inputs. For instance, a strong form of nonspontaneity applies when hereditary substances are transferred into a cell from the outside. Not only does this happen every day in the laboratory, when biologists introduce human-modified genes into cells (see Section A.4), it also happens every day in the wild, when cells take up foreign matter—nucleic acids, viral particles, and even cells—containing hereditary material (Mell and Redfield, 2014). For instance, the insertion of a transposable element or a bacteriophage genome into a prokaryotic genome is an event of mutation that—considering the enormous numbers of prokaryotic cells on the planet—must occur enormous numbers of times every minute. These nonspontaneous processes may lead to evolutionary events of lateral gene transfer (Smith et al., 1992; Koonin and Wolf, 2008). The CRISPR-Cas system of acquired bacterial immunity that we discuss later (5.3.3) is based on the nonspontaneous incorporation of fragments of foreign DNA.

To summarize, the "spontaneity" of mutation might mean several things. In the research literature, it sometimes has a technical meaning, referring to natural mutation under ordinary conditions, rather than laboratory mutation using mutant strains or chemical mutagens. Mutations are spontaneous in the Luria–Delbrück sense if they emerge prior to a condition of interest (e.g., the presence of toxins or pesticides), rather than being induced by the condition. Most mutations are not spontaneous in the sense of reflecting an entirely internal process. Instead, many kinds of mutations are influenced by external factors, and some are shaped strongly by external inputs.

Box 2.2 Specialized mutation systems

Most mutations can be understood as side effects of mechanisms of DNA replication and repair, but this is not always the case. For instance, mutations that result from mobile elements reflect an ongoing battle between hosts and their highly evolved genomic parasites, e.g., the transposase enzyme encoded by a transposon catalyzes the movement of genomic segments in a predictable way that reflects the transposon lifestyle: typically very specific for the source segment (the transposon, as distinct from other segments of DNA), but typically indiscriminate in regard to the point of insertion.

More interestingly, some mutation-generating systems are specialized so as to enhance specific types of useful mutations or to target critical genes underlying adaptation. This phenomenon is most familiar in the context of microbes that evade host immune responses by *antigenic variation*, i.e., switching of antigen proteins. That is, mutational mechanisms of antigenic variation in microbial pathogens (see Ch. 5) are routinely described in the microbiological literature as evolved "strategies" or "mechanisms" for survival or adaptation or persistence by immune evasion, in discussions of the causative agents of trypanosomiasis (Barry et al., 2012; Hall et al., 2013; Horn, 2014), Lyme disease (Bankhead and Chaconas, 2007; McDowell et al., 2002), relapsing fever (Dai et al., 2006; Norris, 2006), and malaria (Recker et al., 2011; Guizetti and Scherf, 2013), and in broad reviews of multiple systems (Foley, 2015; Vink et al., 2012; Palmer et al., 2016).

The context for the evolutionary emergence of mutational strategies of immune evasion is the vertebrate adaptive immune system, which has the capacity to generate new antibodies that target previously unknown microbial pathogens by recognizing specific surface proteins. When this system is successful on a pathogen in the bloodstream, for instance, the pathogen is "cleared" by the immune response. However, some pathogens cause relaps-ing infections characterized by repeated cycles of decline and regrowth in which the dominant antigen "is replaced by another of sufficient antigenic distance that current antibodies are ineffective against cells expressing the second antigen" (Dai et al., 2006). This cycle can go on indefinitely, so long as the pathogen continues to vary its antigens and the host continues to produce new antibodies to target them.

The system of antigenic variation in *Trypanosoma brucei* (a single-celled eukaryote), the causative agent of trypanosomiasis, provides a dramatic example. *Trypanosoma* species express a surface glycoprotein in enormous amounts, as a kind of external shield. Though only a single variable surface glycoprotein (VSG) gene is expressed at any given time, the genome of *T. brucei* has about 15 possible expression sites (see Mugnier et al., 2016), and about 2,000 silent VSG genes, most of which are partial (Glover et al., 2013; Horn, 2014). Key aspects of the genomic organization are shown schematically in Fig. 2.7, which illustrates arrays of VSG genes in subtelomeric regions, and telomeric expression sites, including one active bloodstream expression site. The active expression site (ES) corresponds to a single physical location within the nucleus, analogous to a nucleolus.

Variability in the VSG coat protein is the result of several distinct mechanisms: (1) transcriptional switching that turns off expression of one ES and turns on another, (2) ectopic recombination that swaps segments between an inactive and active ES, and (3) gene conversion events that transfer information from the thousands of silent VSG genes into the ES (Barry et al., 2012; Hall et al., 2013; Horn, 2014). In addition, the silent VSG genes are located in subtelomeric regions subject to higher rates of mutation than the rest of the genome (Barry et al., 2012). The rate of antigenic switching from all of these effects is about 10^{-3} per cell per generation (Barry et al., 2012).

Figure 2.7 A genome-wide system of surface antigen switching with thousands of cassettes in *T. brucei*. Figure from Glover et al. (2013), reproduced under the Creative Commons Attribution 4.0 International (CC BY 4.0) license.

Continued

Box 2.2 *Continued*

To understand the significance of this rate, note that, in the context of an infection, a single individual expands into a clan, and success depends on the capacity of the clan, in aggregate, to survive host defenses. As Barry et al. (2012) point out, "The first wave of parasites in a cow can peak with a total in excess of 10^{11} trypanosomes, therefore $> 10^8$ will have switched to a different variant, while a mouse can support $> 10^8$ parasites with $> 10^5$ emerging variants." That is, a mutation that is a remote possibility in a single individual may become the guaranteed mass-action response of a clan.

The trypanosome system exemplifies a larger class of systems. Palmer et al. (2016) describe three such systems in bacterial species of the genera *Borrelia* (causing relapsing fever and Lyme disease), *Neisseria* (gonorrhea, meningitis), and *Anaplasma* (anaplasmosis in livestock). In generic terms, these systems all have a recipient locus or expression site subject to overwriting (by gene conversion or ectopic recombination) with sequence information from multiple silent donor loci, typically nonfunctional partial genes. Such systems typically allow, not simply an *n*-fold switch among types determined by the *n* donor loci, but $> n$-fold diversity due to the potential for mosaics (with segments from multiple donor loci) and hypermutation, as is the case for *T. brucei* (Hall et al., 2013).

Furthermore, such schemes of cassette shuffling are not the only kinds of specialized mutational systems described

in the microbiological literature. The five additional types of schemes listed below are described in more detail in Chapter 5.

- Diversity-generating retroelements (DGRs) represent a widely distributed (Paul et al., 2017; Wu et al., 2018) scheme of targeted hypermutation mediated by error-prone reverse-transcription (Guo et al., 2014), typically inducing variability of C-type lectin domains.
- Inversion systems like the R64 "shufflon" have a small number of unique DNA segments interspersed with repeated motifs that serve as endpoints for inversion mutations, resulting in combinatorial shuffling by inversion (Komano, 1999; Sekizuka et al., 2017).
- Mating-type switching is a regular part of the life history of several yeast species, including *S. cerevisiae*, which switches by a gene-conversion event that is programmed in regard to timing and outcome (Hicks and Herskowitz, 1977; Hanson and Wolfe, 2017).
- The widespread phenomenon of phase variation, known for nearly a century, refers to switching between expression states, often by mutation but also by epigenetic mechanisms (van der Woude and Bäumler, 2004).
- In CRISPR-Cas systems of phage defense used in prokaryotes, pieces of phage DNA are added to spacer arrays that are used to encode site-specific DNA shears (Koonin, 2017).

2.8 Independent (part 1)

According to a definition often attributed to Aristotle, a "chance" event is the intersection of independent causal sequences (e.g., Eble, 1999). An example would be tossing two dice and finding the same number on both of them (Fig. 2.8). The two tosses are independent causal sequences: whether they converge on the same number is a matter of chance.

Importantly, this sense of randomness refers to a relation rather than to an intrinsic property—a matter of one thing relative to another, e.g., the chance of a mutation relative to the current temperature. This notion of causal independence invites the use of the corresponding concept of independence from probability theory, allowing joint probabilities to be determined simply by multiplication, e.g., the joint probability $\Pr(A, B)$ is simply the product of

$\Pr(A)$ and $\Pr(B)$ when they are (unconditionally) independent. For instance, in the case of the six-sided uniform dice, the chance of getting 5 on the first roll, and 5 on the second, is $(1/6)^2 = 1/36$, if the two rolls are independent causal sequences.

Our main interest in regard to independence is to ask whether mutation is independent from biotic or environmental conditions, e.g., temperature or physiological states. That issue is addressed in the next section. This section is devoted to the issue of whether one mutation is independent from another.

Some examples of this type of independence or nonindependence were encountered earlier. For instance, in the analysis depicted previously in Fig. 2.4, which involved a comparison of SNPs between humans and chimps, it was assumed that each coincident SNP arose independently

Figure 2.8 The chance conjunction of numbers on two dice is an example of independence. Image by Steaphan Greene, reproduced under Creative Commons Attribution Share-alike (CC-BY-SA-3.0) license.

by mutation, in the sense that the mutation that occurred in the chimp lineage was independent of that in the human lineage. This assumption seems safe for most mutations, though not for lateral gene transfers, because the frequency of a donor sequence in a particular environment affects the rate of acquisition in all recipient genomes in that environment.

However, when we are considering the emergence of different mutations in the same cell, or the same lineage, the assumption of independence is not safe. Double-nucleotide changes at adjacent sites are rare, but not nearly as rare as one would predict from independence (Hill et al., 2003; Reid and Loeb, 1993). Compound events involving multiple changes in the same vicinity are seen in laboratory mutation assays (e.g., Lang and Murray, 2008; Phear et al., 1987; Drobetsky et al., 1987) and mutation-accumulation experiments (e.g., Keith et al., 2016), and appear regularly in clinical studies (e.g., Cordella et al., 2006), e.g., large clinically important rearrangement mutations often include additional changes near the endpoints (Conrad et al., 2010). Occasionally, evolutionary comparisons suggest nonindependence, e.g., between indels and nearby nucleotide substitutions (Jovelin and Cutter, 2013). Recently, Venkat et al. (2018) argued that phylogenetic tests for positive selection often produce misleading results due to compound mutations affecting nearby sites. Drake et al. (2005) review results from diverse taxa suggesting that

clusters of mutations occur more often than would be expected if mutations were independent of each other.

Though the exact causes are unknown in most cases, one obvious explanation for clusters of linked mutations is that DNA repair stimulated by damage or recombination frequently involves excision and error-prone re-synthesis of a strand over many nucleotides, not just a single site (see Section A.2). Appendix A.2 also notes that a single particle of ionizing radiation can produce a burst of free radicals, which can then go on to create a localized burst of damage. Clusters of nucleotide mutations that occur in the same generation have been observed in a variety of organisms, on length-scales ranging from one to hundreds of kb (Chan and Gordenin, 2015).

The extreme case of such clustering is the recently named phenomenon of *kataegis* (Greek, "thunder"), which refers to regional domains of hypermutation that may extend over many thousands of sites. Alexandrov et al. (2013) studied this phenomenon in human cancer cells, with results shown graphically in the "rainfall" plots in Fig. 2.9. The upper cloud of points between 10^5 and 10^6 bp indicates that a mutation is typically far from the next mutation. However, the vertical streaks each represent clusters of many mutations that are much more closely spaced, on the scale of 10^2 or 10^3 bp.

Another possible mechanism for an excess of multiple mutations, discussed by Drake et al. (2005), is transient hypermutation, when some fraction of cells enters a transient state with a greatly increased rate of mutation.

A final possible mechanism of nonindependence would cause mutations to occur preferentially near the sites of previous mutations. This idea is based on observations from yeast indicating that gene conversion and DNA repair events tend to be induced by the presence of sequence differences between chromosomes. Thus, the chance of mutagenic gene conversion and repair events is increased to the extent that the inheritance of past mutational changes makes homologous alleles differ. Data on clustering of variations for human SNPs and microsatellite variants provide some support for this hypothesis (Amos, 2010).

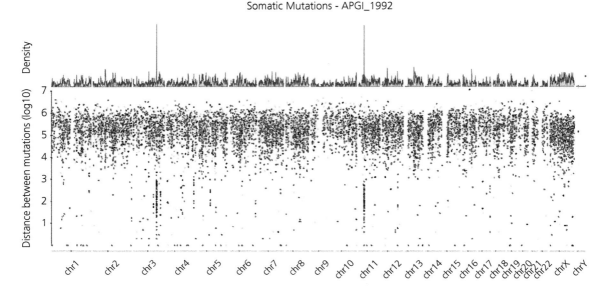

Figure 2.9 Rainfall plot showing kataegis in a pancreatic cancer line (data from Alexandrov et al., 2013, plotted using code modified from Bernat Gel at https://bernatgel.github.io). The horizontal axis is genomic position, covering the entire genome of 23 chromosomes. The vertical axis is distance to the next mutation on a log-scale. In this case of kataegis, mutations are heavily biased toward $C \rightarrow G$ and $C \rightarrow T$ mutations (black and red dots, respectively).

2.9 Independent (part 2)

Having addressed (in Section 2.8) whether mutation events are independent of each other, we now may address whether mutation is an independent or chance process relative to events or conditions other than mutation.

The use of a PRNG in an evolutionary simulation illustrates how this kind of randomness (independence) differs from other concepts (see Appendix A.3). I once used a flawed PRNG in a computer simulation of evolution that sequentially generated (1) a number to decide whether a mutation would occur, and then (2) a second number to decide the type of mutation. In the first step, the method of implementing the decision was to ask if $m < \mu$, where μ is the mutation rate for a gene (a very small number) and m is a number between 0 and 1. However, the particular flaw of this PRNG was that an extremely small number was followed, not by a uniformly distributed number, but by a smallish number. Because the first step in the mutation algorithm depended on testing whether $m < \mu$, where μ is an extremely small number, the next number

was always a smallish number, which meant that the next step in the simulation algorithm—choosing the type of mutation—was biased.

How does this case distinguish chance from predictability, uniformity, and so on? The sequence of numbers from the PRNG is deterministic, uniform, and—if one knows the internal state—fully predictable. The flawed PRNG produces a sequence of outputs that have serial correlations (i.e., nonindependent), which means that, even without knowing the internal state, it exhibits some strongly predictable behavior, because a very small number is followed reliably by another small number. However, if one does not know the internal state, most of the behavior of the PRNG is very unpredictable.

Importantly, the relation of the flawed PRNG to the simulation remains a matter of "chance" (independence) even though it produces a highly predictable effect based on the nonindependence of one number from the next. The relationship still meets Aristotle's "independent causal sequence" condition because the computer code for the PRNG is a separate function, walled off from the rest of the simulation program, operating independently—the same deterministic sequence of numbers would be generated regardless of how the simulation program uses the numbers.

If genes are "sealed off from the outside world" (Dawkins, 1976, p. 21), e.g., if mutation were a kind of immaterial process distinct from the rest of biology, mutation would occur independently of biology, and the effect of mutation on evolution (regardless of any intrinsic biases, and their consequent effects on outcomes) would be guaranteed to be a matter of chance in this narrow sense.

We can imagine an artificial case in which mutation is independent of conditions. Suppose that we have six different growth conditions for *E. coli*, and six engineered genes of types A, A, A, B, C, and D. For each growth condition, we can pick one of the six genes blindly, then introduce it using genetic engineering. In such a case, the mutations are introduced independently of growth conditions: the chance that a given culture gets a particular gene is $Pr(A) = 3/6$ for A genes, and $Pr = 1/6$ for the other types, regardless of conditions.

Yet, mutation is a biological process taking place inside a cell that is also the locale for a multitude of other biological processes. It cannot possibly be independent.

Some concrete examples may be helpful to illustrate how mutation processes are affected by other states and processes inside the cell. For instance, DNA synthesis, though mediated by enzymes, is a chemical reaction whose outcome depends on the concentration of reactants, including the pool of precursors (see Section A.1). Changing the composition of dNTPs has a predictable effect on mutation: adding more dCTP favors mutations to C, adding more dTTP favors mutations to T, and so on.

Kumar et al. (2011), along with Phear et al. (1987), provide evidence both for this effect, and for an indirect effect in which proofreading is inhibited when a favored nucleotide is inserted downstream (see figure legend for explanation). The leftmost pathway in Fig. 2.10 illustrates both effects at once, and is based on the conditions explored by

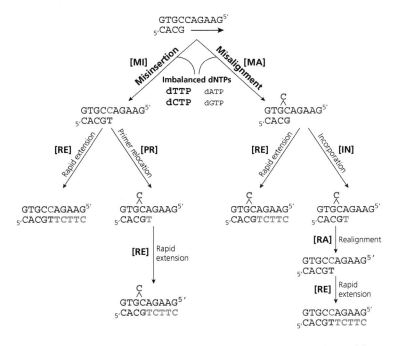

Figure 2.10 How dNTP concentrations influence mutation. In a mutant yeast strain that over-produces dTTP and dCTP, certain paths of mutation are favored. In particular, the excess of dTTP increases the chance that T will be misinserted (left path, bold red "T"). When this occurs at a site normally followed by a string of Ts and Cs (as for the bold green "TCTTC" sequence), the excess of dTTP and dCTP favors rapid extension ("RE" in the figure) rather than proofreading ("PR"), with the result that misinsertion mutations are enhanced at such sites. Figure kindly provided by Dr. Andrei Chabes.

Kumar et al. (2011) using a mutant yeast strain that over-produces dTTP and dCTP. Kenigsberg et al. (2016) present evidence that changes in nucleotide precursor pools during the mammalian cell cycle contribute to differences in nucleotide composition between early- and late-replicating regions of the genome, i.e., they have presented an evolutionary hypothesis in which the dependence of mutation on a cellular condition that varies predictably (precursor pool composition) has a predictable effect on the composition of the genome.

Whereas precursor pools exert their influence on mutation via diffusion and chemical mass-action, other effects on mutation are local, based on processes happening in the physical locale of a possible site of mutation, e.g., processes such as chromatin formation, or gene expression. Friedberg et al. (2006) devote an entire chapter with 226 references (their Ch. 10) to heterogeneity in nucleotide excision repair relating to transcription, methylation, and nucleosome structure. For instance, an enormous amount could be said about chromatin structure, which not only affects the physical distribution of damage, e.g., hydroxyl-radical-mediated damage (Hayes et al., 1990), but also affects the efficiency of repair (for discussion, see Odell et al., 2013; Burgess et al., 2012).

For such reasons, mutation at a given site is affected by whether or not the site is chromatinized, and where it lies within (or outside of) a nucleosome. Chromatin, in turn, is affected by all sorts of other conditions in the cell, such as replication and transcription. The organization of chromatin in humans and other animals differs systematically in different cell types (e.g., Valouev et al., 2011), which of course express different sets of genes with systematically different metabolic capabilities. Thus one expects that the pattern of mutation in different cell types will be nonindependent of gene expression and metabolic capabilities across cell types.

Furthermore, one expects that transcription of a gene (even in organisms without chromatin) will increase mutation, because transcription involves temporary unpairing of the strands of the helix, which leaves the bases much more exposed to damage. In fact, an affect of transcription on the rate and spectrum of mutations has been shown repeatedly (Datta and Jinks-Robertson, 1995; Beletskii and Bhagwat, 1996; Klapacz and Bhagwat, 2002; Kim et al., 2013; for review, see Kim and Jinks-Robertson, 2012). This is a second way, distinct from the effect of chromatin just discussed, in which mutation is not independent of gene expression or metabolic capabilities, e.g., if we survey mutation in a cell as it shifts from one condition to another, we expect changes in the pattern of mutation to correlate with changes in metabolism, because they are not independent.

Recently, Krašovec et al. (2017) summarized data from 67 studies of mutation rates measured in microbial cells grown in culture, showing an inverse correlation with density over many orders of magnitude, a stunning effect shown in Fig. 2.11. They then generated more robust evidence for this correlation in *E. coli* and *S. cerevisiae* by conducting their own experiments designed to reduce the influence of possible artefacts. Finally, they showed that, in both species, the dependence of mutation rate on density could be removed by eliminating a particular damage-avoidance pathway (MutT) that protects cells from the mutagenic effects of 8-hydroxyguanosine.

Thus, the notion that mutation is walled off from biology, though it may be a useful idealization in some cases, is not a biological principle and is at odds with well-established facts about mutation. Mutation is a biological process, and a biological process taking place inside a cell is not independent of the other processes taking place inside the same cell, which are not independent of what is happening outside the cell.

Whereas the evolutionary literature sometimes gives the impression that the independence of mutation follows from a mechanistic or materialistic view, and that anything else is an appeal to mysticism, precisely the opposite is true: dependence is expected, whereas independence would suggest supernatural causes. Of course, one may simply assume that all such dependencies are trivial, and treat mutation as being independent (even when it is not), reaping the benefits that come from applying the statistical concept of independence. We address this possibility in Chapter 3.

However, our purpose here is to ask whether the independence of mutation from biology is true, not to ask whether it may be a useful approximation in

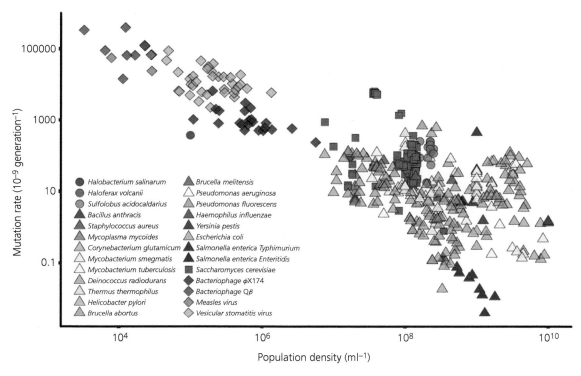

Figure 2.11 Negative correlation of mutation rate and cell density for diverse microbes. Image from Krašovec et al. (2017), reproduced under the Creative Commons Attribution 4.0 International (CC BY 4.0) license.

some cases. The answer to the former question is clearly negative for the case of independence from gene expression, metabolic states, and so on. Evolutionary researchers sometimes recognize this fact, e.g., in their experiments on laboratory adaptation of yeast to increased copper concentrations, Gerstein et al. (2015) carefully consider potential effects of copper on the mutation rate and spectrum, rather than assuming that such an effect is prohibited by the randomness doctrine.

2.10 Synopsis

This chapter introduced some conventions for addressing mutation without much explanation. The possibility-space for mutation is the universe of mutations. Appendix B maps out a clearly defined version of this possibility-space. A process of mutation or a pathway of mutagenesis represents a way of sampling from this universe that may be represented by a vector of rates (or frequencies) **u** the same size as the space of possibilities.

This is what is meant by a mutation spectrum. Practical uses of the concept of a mutation spectrum nearly always refer to a model of the mutation spectrum in which distinct mutations are assigned to classes (e.g., treat all $A \rightarrow G$ changes the same), but references to mutation spectra also may be reduced by referring to only part of the genome, or by ignoring mutations other than nucleotide substitution mutations.

We began this chapter with the idea that "random" may mean many different things. Having addressed these various meanings, the problem can be simplified:

- Mutation is not uniform or homogeneous, e.g., mutation is nonuniform in timing, location, and effect.
- Mutation lacks foresight. However, this statement is negative and relatively empty. The nonmagic nature of mutation tells us nothing important or distinctive about mutation as a biological process.

Table 2.1 Concepts of randomness summarized

Concept	Application to mutation	Validity
Uniformity	Mutations are all equally likely.	Wholly invalid: contradicted by readily available facts of genetics.
Unpredictability	The next mutation cannot be predicted.	The predictability of mutations is typically many orders of magnitude less than 1.
Stochasticity	Mutation is a stochastic process.	(A category error, not a claim whose validity can be evaluated.)
Indeterminacy	The outcome of mutation is not fully determined by conditions.	At least partly true: many mutation pathways have an indeterminate step.
Lack of foresight	Mutations proceed from a mechanism: they are unplanned accidents without design or purpose	(A metaphysical position, not an evidence-based claim.)
Spontaneity	Mutation is an internal event not instigated or influenced by external factors.	Steps in mutation are mostly internal, but also reflect external influences directly (rarely) or indirectly (often).
Independence (chance)	The emergence of a mutation is independent from (1) other mutations, and (2) other processes in the cell.	Invalid due to shared biological factors.

- To say that "mutation is stochastic" would be to mix up aspects of reality with the variables used to represent those aspects in mathematical models.
- Mutation is *often* affected by physical indeterminacy. Whether mutation is *pervasively* affected by indeterminacy is an unresolved issue.
- Mutation is not fully predictable, due mostly to subjective unpredictability, but also to physical indeterminacy.
- Mutation is not spontaneous in the strict sense of emerging entirely internally, without any external impetus or constraint. At least in bacteria, mutations appear to be largely spontaneous in the Luria–Delbrück sense, with some important exceptions.
- Mutation is not independent of the rest of biology, nor is it independent of prevailing conditions. Mutations are not always independent of each other.

The conclusions suggested here on various ideas of randomness are summarized in Table 2.1.

CHAPTER 3

Practical Randomness

The true logic for this world is the calculus of probabilities
James Clerk Maxwell

3.1 Introduction

In Chapter 2 we dismissed a variety of concepts without considering their practical utility, because we were confronting the randomness doctrine as a potentially valid universal principle, rather than as a rough guess or working assumption. For instance, we have rejected the idea that mutation is uniform. Nonetheless, we might be able to justify limited assumptions of approximate uniformity, e.g., based on averaging over many sites or paths. Likewise, we must reject an unqualified claim of causal independence, yet a narrow assumption of independence may be justifiable in a particular context of modeling and analysis.

Note that some of the concepts we rejected previously do not warrant any further attention.

- The "lacking in foresight" concept tells us nothing interesting.
- In regard to indeterminacy, improving our understanding of the extent to which mutation is physically indeterminate is possible, but seems to have few practical implications. The practical issue is predictability, but subjective unpredictability seems to play a larger role in reducing the predictability of mutation.
- The concept of spontaneity has some practical implications, but they are all subsumed under independence. That is, the Luria–Delbrück fluctuation test probes whether mutations are spontaneous by probing whether their occurrence is responsive to an external condition controlled by the experimenters.

Below we address some of the remaining concepts with practical relevance—uniformity, predictability, and independence. Then we consider the role played by assumptions of uniformity and independence in practical strategies for hypothesis-testing, or more generally, in approaches to scientific reasoning in the face of uncertainty.

First, however, we must consider why and how scientists employ simplifying assumptions, rather than demanding only well-established universal truths.

3.2 What good is a randomness assumption?

In the introductory chapter (see Section 1.2), we encountered the question of whether the "random mutation" doctrine is useful—in the way that other problematic concepts (e.g., entropy, probability) are useful—, without considering possible answers to this question. Kauffman (1993) reviews the literature of evolutionary biology and concludes that "the initial idealization that variation can occur in any direction, known by all biologists to be literally false, is too useful a simplification to give up" (p. 11). The material in Chapter 2 reveals a biological process of mutation far more complex than what is typically represented in evolutionary models, or in accounts of evolutionary causes.

What makes this simplification useful? What makes it allowable in the context of scientific inquiry? That is, what is the justification for adopting a position that is not known to be true,

Mutation, Randomness, and Evolution. Arlin Stoltzfus, Oxford University Press (2021). © Arlin Stoltzfus. DOI: 10.1093/oso/9780198844457.003.0003

or one that is known to be untrue? Why would scientists consider adopting, for the sake of utility, a proposition that could not be justified without this utility?

One possible high-level answer is that scientists studying evolution gain an enormous explanatory and methodological simplification by assuming that all conceivable variations are abundantly available, and that selection alone chooses among them, so that nothing about the generation of variation affects the course of evolution. Under these conditions, selection is the potter and variation is the clay: to understand the course of evolution, we only need to study the selective forces at work, and there is no need to study the details of variation, which do not matter. This theory that the internal details do not matter to evolution was particularly valuable when no one knew the internal details.

Such a theory provides a basis to ignore the generation of new variation entirely. A quite remarkable fact, for those unfamiliar with theoretical population genetics, is that a large body of classical mathematical work simply does not include terms for mutation (e.g., Edwards, 1977), on the grounds that evolution begins with abundant variation in the gene pool, and subsequent mutation represents a weak force with no important influence on the final equilibrium (see Section 6.4.2 and Box 8.2).

Another practical reason for simplification is that mutation rates are difficult to measure under natural conditions, due to the rareness of mutation. Before DNA sequencing, scientists could not systematically distinguish different mutations at the molecular level, and did not know what (precisely) was in either the numerator or the denominator of a genomic mutation rate. That is, telling different mutants apart quantitatively ultimately requires sequencing, and DNA sequencing was not widely applied until the early 1980s. Scientists make models for the data that they have, and not for the data they do not have. The gaps in our knowledge are filled with assumptions and theories.

For related reasons, the generation of variation, or the stock of available variation, is much more difficult to manipulate than selection; therefore the role of variation is less likely to be the focus of effective interventions, whether to test hypotheses or merely to explore impacts. For instance, the breeder's equation—which depends equally on a term for selection and a term for variation (see Section 7.5.1)—is called "the response to selection" because selection is always the experimentally manipulated variable. The heritable component of variation could be manipulated, in principle, but this is exceedingly difficult. To manipulate the mutation spectrum without large impacts on the total mutation rate remains difficult (e.g., Couce et al., 2015). Likewise, in the context of evo-devo challenges, changing the way that a phenotype is encoded, without changing the phenotype, is exceedingly difficult (for a positive example, see Schaerli et al., 2018). In general, to understand the development of evolutionary thinking, one must recognize an enormous practical and historical difference between outside and inside, between our capacity to measure and to control the external circumstances of organisms (including reproduction and survival) versus their detailed internal behavior.

Of course, many contemporary models include mutation. When mutation is included in models of evolutionary processes, what is the basis for treating it simplistically, e.g., with a single scalar value, when decades of mutation research contradict this assumption?

One obvious answer to this question is that mutation rates are simplified to make theoretical models tractable. Developing theories in which mutation takes a single scalar value is easier than representing a vast spectrum of mutations with rates that cover many orders of magnitude and that are sensitive to conditions. For instance, a substantial body of literature—including a body of literature about the evolution of the mutation rate (e.g., Hua and Bromham, 2017)—refers to mutation via a single scalar parameter for the genomic mutation rate, ignoring the fact that actual rates of specific mutations range over many orders of magnitude (Fig. 2.1), with no apparent lower limit.

A crude approach to addressing this variance in mutation rates is to introduce a binary distinction.

For instance, Fisher and Haldane developed an informal "recurrent mutation" concept to distinguish the class of mutations that occur often enough that the mutant alleles can be assumed to be reliably present in a population, as distinct from rare mutations that may be ignored. They argued that adaptation would take place based on variants already present in the population, rather than by a mutation-limited process that, they assumed, was too slow to account for the facts of evolution (see McCandlish and Stoltzfus, 2014). A similar assumption underlies present-day approaches in which evolution is broken down into steps (e.g., the mutational landscape model in Section 8.6), where the next step is chosen from the set of mutational neighbors, defined as genotypes that can be reached by a single change, typically replacing the nucleotide or amino acid at a single site. Such models divide the universe of mutations into (1) accessible mutations that might contribute to evolution, i.e., the ones whose mutant alleles are considered, and (2) inaccessible ones that are ignored.

More formally, when developing analytical theories, one sometimes asserts that the value of a particular parameter is restricted so as to ensure that some mathematical term may be dropped, on the grounds of not having any large effect on the result. For instance, later we will define an origin-fixation regime that has tractable behavior as $\mu N \rightarrow 0$, which ensures that no more than two alleles segregate in the population at any time, i.e., this restriction allows us to ignore the complications that follow from allowing a second mutation to occur before an earlier mutation has either been lost or gone to fixation.

Indeed, sometimes an unknown factor disappears entirely. This is a highly desired result, i.e., theoreticians deliberately pursue approaches in which unknown values are factored out of the final equation. For instance, one might imagine that the burden of removing deleterious mutations from a population (the mutation load) would depend on knowing *how deleterious the mutations are*, yet in certain well-known conditions, the fitness distribution drops out of the calculation and only the mutation rate is needed (Crow, 1970).

Molecular phylogenetics provides a number of examples showing how and why scientists simplify the world. The "character-state" methods imply a model of evolution in which a species is fully characterized by a list of the states (i.e., $S_1, S_2, S_3, \ldots$) assigned to its n characters, like the nucleotide states (i.e., T, C, A, G) assigned to each position in a sequence of length n. In simple models, an evolutionary lineage is then characterized by a series of transitions between character-states, where the transitions occur one character at a time, independently, and represent a first-order Markov process (i.e., no historical memory) that is homogeneous in time. Any lineage may split stochastically into two lineages, which then change independently.

Clearly, this is a gross simplification of evolution. The model literally lacks populations—each species is treated as a single discrete particle. All of the characters and states are treated the same, with no correlations between characters, no species interactions, and no explicit influence of adaptation or optimality.

Yet, a model that treats evolution in this way—merely a stochastically branching Markov process of generic character-state transitions—is incredibly valuable, and can be applied to any molecular or morphological characters in any species over any time-scale. Methods based on this simplistic model were applied with great success by the early pioneers of molecular phylogenetics. The simplicity of the model facilitates the derivation of mathematical formulas to compute probabilities, e.g., the independence assumptions allow the "pruning algorithm" of Felsenstein (1973), used to compute likelihoods.

In reality, each gene or protein evolves differently. Bloom (2014) has shown that phylogenetic inference using a particular protein can be improved greatly by using deep mutational scanning to develop an empirical model of the fitness effects of amino acid replacements for that specific protein. However, one typically does not have access to the fitness landscape (or the mutation spectrum) for each protein-coding gene. The ability to proceed in ignorance of underlying details is obviously a benefit, whether the details are about selection or mutation.

In general, methods in molecular phylogenetics have become more sophisticated over the past 40 years, allowing for violations of these simplifying assumptions (for a general discussion with a different focus, see Ch. 13 of Felsenstein, 2004). For instance, in the 1980s, it became clear that changes in molecular evolution are biased toward nucleotide transitions relative to transversions (see Section 8.9); in response, in the 1990s, methods of phylogenetic inference were adjusted to account for this bias (e.g., see Wakeley, 1996). Before Kimura (1980) introduced a two-parameter model of nucleotide changes distinguishing transitions and transversions, all changes were treated identically under the one-parameter model of Jukes and Cantor. Today, phylogenetic analyses frequently use the general time-reversible (GTR) model with six rate parameters, one for each forward–reverse pair of nucleotide changes.

Many other examples could be given. Contemporary methods in molecular phylogenetics typically assume that rates of evolutionary change are gamma-distributed across sites, rather than uniformly distributed. The assumption of independence of sites was abandoned in some models 20 years ago, to address paired sites in RNA molecules that switch character-states in coordinated fashion so as to maintain pairing (Tillier and Collins, 1998). Some contemporary models allow different rates of evolution on different branches, nonindependence of branches, fusion of branches (via lateral transfer), and hemiplasy due to ancestral polymorphism (e.g., Wen et al., 2016; Szöllösi et al., 2012; Schrempf et al., 2019). Many contemporary methods draw on an implicit or explicit origin-fixation framework in which the influences of mutation and fixation are parameterized separately (for review, see Rodrigue and Philippe, 2010; McCandlish and Stoltzfus, 2014).

In each of these cases, one could have argued in advance that the original simplifying assumption was unrealistic, i.e., the assumption could have been rejected *a priori* on the basis of known counterexamples. Instead, the original model was used for the sake of simplicity, and the unrealistic simplifying assumptions were abandoned subsequently, not because they were suddenly discovered to be unrealistic, but because it became technically possible and scientifically valuable to implement better models that led to better phylogenies.

In many cases, issues of realism and of underlying evolutionary causation have remained unresolved in spite of advances in technology. For instance, when phylogeneticists confronted evolutionary transition-transversion bias, they acknowledged ignorance of whether it was caused by selection or mutation (Wakeley, 1996; Mooers and Holmes, 2000). For the purposes of phylogeny inference, having a transition-transversion bias parameter is useful, even if one does not know whether its value is determined by mutation or selection. In general, the distributions of inferred values of parameters in phylogenetic models sometimes reflect mutation more strongly than selection (e.g., transition-transversion bias), sometimes reflect selection more strongly than mutation (e.g., shape parameters for gamma-distributed rates across sites), and sometimes reflect biased gene conversion (e.g., GC content effects in outcrossing organisms), which is neither mutation nor selection.

To summarize this section, one may adopt a simplified view of mutation for practical and methodological reasons, without committing to any substantive claim about the uniformity or independence of mutation, and without committing to the theory that the details of mutation are irrelevant to how evolution turns out. The simplified treatment of mutation in molecular phylogenetics is not a special case: in fact, phylogenetic methods were just as likely to ignore complexities of selection, epistasis, ancestral polymorphism, and so on, as complexities of mutation. Over time, more realistic features of selection, mutation, and demographics have been introduced into models of molecular evolution, facilitated by advances in computing techniques and fueled by larger and larger amounts of data, i.e., the conditions that make the development of more complex models both feasible and scientifically valuable.

Box 3.1 Three kinds of adaptationism

Godfrey-Smith (2001) suggests that adaptationism is not a single view, but comes in three different flavors: empirical, methodological, and explanatory.

The most obvious kind of adaptationism is the belief that adaptation is actually pervasive in nature. A scientist might adopt this position of empirical adaptationism on the grounds of *observations* indicating that organisms are pervasively adapted, or perhaps, from a *theoretical* understanding of the mechanism of evolution by which adaptation always prevails. An empirical adaptationist might believe (for instance) that neutral evolution does not happen, and that features that appear to be nonadaptive inevitably will be found to have a cryptic adaptive purpose. Empirical adaptationism is subject to refutation, e.g., evidence that evolved traits are not adaptive is evidence against empirical adaptationism.

Using textual analysis, Godfrey-Smith (2001) shows that not every adaptationist thinks this way. He identifies two other styles of adaptationism, methodological and explanatory, that do not require the assumption that adaptation is pervasive. Methodological adaptationism holds that, whether adaptations are common or rare, one must study them because this is the only effective way to do evolutionary research. In his defense of the adaptationist research program against the famous critique by Gould and Lewontin (1979), Mayr (1983) makes this explicit:

> When one attempts to determine for a given trait whether it is the result of natural selection or of chance (the incidental byproduct of stochastic processes), one is faced by an epistemological dilemma. Almost any change in the course of

evolution might have resulted by chance. Can one ever prove this? Probably never. By contrast, can one deduce the probability of causation by selection? Yes, by showing that possession of the respective feature would be favored by selection. It is this consideration which determines the approach of the evolutionist.

What "determines the approach of the evolutionist" is that selective explanations are accessible to science, and "chance" explanations are not. Interestingly, Bateson expressed the opposite methodological position that selective explanations are a matter of idle speculation, whereas variational explanations might at least draw on observable facts about variation (Bateson, 1894, p. vii).

An *explanatory* adaptationist may allow the possibility that adaptation is not pervasive, and may allow the methodological possibility of studying evolution in other ways (e.g., using neutral models), while still insisting that what matters most is to explain adaptation by natural selection: adaptation (in this view) is the uniquely distinctive *explanandum* (thing to be explained) of biology, and everything else represents noise and irrelevant detail. As Godfrey-Smith (2001) points out, precisely this position is argued by Dawkins (1987) when he writes that "Large quantities of evolutionary change may be nonadaptive, in which case these alternative theories may well be important in parts of evolution, but only in the boring parts" (p. 303). This form of explanatory adaptationism cannot be refuted by any evidence of nonadaptive processes, so long as they are limited to "the boring parts" of evolution. This idea will become important in Chapter 9, when we attempt to construct an empirical argument for the importance of mutation biases.

3.3 Uniformity

What practical conclusions can we draw from the observed heterogeneity of mutation rates?

First, let us consider the large-scale heterogeneity by which the rates for various types of mutation span many orders of magnitude. A conclusion important for practical evolutionary reasoning is that we have no basis to imagine a lower limit on the rate of relevant mutations in any situation. For any mutation with a low rate, e.g., a duplication with a simultaneous base substitution in one copy, we can imagine something less likely, e.g., a duplication with 2 substitutions. In evolutionary modeling, one

is not safe in assuming that all possibly relevant mutations are present in a population, or that all possibly relevant mutations will be encountered in some finite period of time. However, a narrower assumption might be justifiable, e.g., it might be possible to assume that, in a given period of time, the population will experience at least m mutations that affect some trait by a given degree, i.e., this is a form of assumption that one may make without ignoring the exceedingly long left tail in the distribution of low mutation rates.

Mutations do not occur uniformly, but the assumption of uniformity can be treated as a heuristic assumption that may be justifiable depending on reality and our tolerance for error. For instance, ordinary dice show nonuniformities that gamers tolerate but gambling casinos do not. Likewise, in spite of pervasive nonuniformity of mutation, we might be able to justify a kind of uniformity by (1) restricting attention to a particular class of mutation, (2) averaging effects across many sites or paths, and (3) treating these values as estimates with a variance.

On this more modest assumption of uniformity, we might attempt to predict the total mutability of a gene based solely on its length, as in the model of Houle (1998). Under the assumption of strict uniformity, the rates of mutation for genes of lengths $n_A > n_B > n_C > n_D$ bp are $n_A\mu > n_B\mu > n_C\mu > n_D\mu$, respectively, where μ is the rate of mutation per bp. If mutation is not uniform, then this assumption has a cost that depends on the heterogeneity of mutation rates, and the number of values being summed together. For instance, if some genes have rare "hotspot" sites with rates 10^3 times the average, then $n_A\mu > n_B\mu > n_C\mu > n_D\mu$ is still our best model for relative mutability, but this model will fail much more frequently. The lower the heterogeneity and the longer the mutated regions, the more reliable is the estimate.

Finally, note that a pernicious side effect of assumptions of homogeneity is that they bias our estimation of recurrences, consistently resulting in underestimation. For instance, parallel evolution between two independent samples is simply a sum of squares over all the possibilities, and this sum necessarily increases with the variance in individual probabilities (see Section 8.12).

3.4 Independence

As noted earlier, independence is a relation that might apply to mutation relative to other processes. We may wonder, for instance, whether mutations in bacterial cells growing in a flask in a laboratory in Michigan are independent of fluctuations in the stock market in New York. Perhaps they are. However, they clearly are not causally independent of other things going on in the same flask in Michigan.

In the absence of true causal independence, the justifiability of applying a statistical concept of inde-pendence depends on our tolerance for error and on our experimental design, an entirely practical matter. Scientists are concerned with the issue of independence precisely because we wish to imple-ment analytical methods in which the assumption of independence makes a complex system more tractable. This is one answer to the question (posed in Section 1.2) about what makes the randomness doctrine useful.

Whether or not this assumption is safe depends on circumstances. Sometimes the assumption is not safe. For instance, we considered earlier the case of Gerstein et al. (2015), where the authors had prior knowledge that copper affects the mutation rate of microbes, therefore, in analyzing data on the labo-ratory evolution of copper resistance, they did not simply assume that mutation would be unaffected.

Let us consider another practical example. Sup-pose that we are making a model of the emergence of antibiotic resistance among pathogenic microbes, a problem of enormous importance for public health (Marston et al., 2016). The outcome of evolution in our model depends on the occurrence of some enu-merated set of mutations that increase resistance, each with a characteristic rate μ_i and an expected degree of resistance.

Of course, a convenient assumption is that each μ_i represents a characteristic rate independent of everything else in the system, including the concen-tration of antibiotic, $[A]$. Yet, when bacterial cells are exposed to nonlethal doses of some antibiotics, the result is a programmed stress response that increases the rates of many kinds of mutation (Lau-reti et al., 2013). Thus, we may find that the accuracy of our model is substantially increased by abandon-ing the assumption of independence from $[A]$ and instead implementing a functional relationship $\mu_i = f(i, [A])$. The documented fact that sublethal doses of antibiotic may increase the rate of evolution of resistance by virtue of an effect on mutation rates is obviously of practical importance in designing interventions to maintain public health by limiting exposure of bacterial populations to such doses, e.g., Gutierrez et al. (2013) write that

Because this mutagenesis [induced by subinhibitory doses of antibiotics] can generate mutations conferring antibi-otic resistance, it should be taken into consideration for the development of more efficient antimicrobial therapeutic strategies.

In summary, mutation is not independent of anything else in biology, but it may be possible to justify independence from certain other processes in modeling or data analysis. Conversely, knowledge of the dependence of mutation on other factors may be leveraged to improve the accuracy of models.

3.5 Predictability

What about predictability? The predictability of individual mutations is evidently low, unless we place arbitrary constraints on the problem. For instance, consider the highly constrained challenge of predicting whether the next mutation to occur in gene X is a microindel or a base substitution. If base substitutions happen ten times more often than microindels, on average, and we know of this bias, we will predict that the first mutation is a substitution, and (ignoring other kinds of mutations) we will be right 10/11 times (Box 3.2 explains why this is the best strategy of prediction).

Likewise, if we attempt to predict which codon in gene X will be hit with the first mutation, knowing (from a highly accurate model of context effects) that the mostly likely target is codon 126, with a relative probability of 0.016 of being hit, we will predict codon 126, and we will be correct 1.6 % of the time. If, the single most probable mutation occurs with a probability of 10^{-4} per generation, then we can achieve an accuracy of 10^{-4} by predicting that this mutation will happen in the next generation.

What are the practical circumstances that depend on our ability to predict mutations, or to predict the outcome of a process that is dependent on mutations, even though individual mutations are difficult to predict? Actually, the reasons to improve our success in predicting mutations include obvious economic, health, and security considerations. Let us imagine that we have some model to compute the chances of an outcome such as (1) whether a microorganism will evolve to become pathogenic, (2) whether a microorganism will develop drug resistance, (3) the emergence and progression of

a cancerous tumor, or (4) the duration of a viral infection. These are processes that depend on mutations. For instance, we might want to use a probabilistic drug-resistance model to design a drug therapy that addresses the dangers of evolved resistance.

This kind of approach is not a matter of science fiction, but has already been done for a number of years in regard to optimizing HIV treatment (Prosperi et al., 2009). Furthermore, the models do not assume that mutation is uniform, e.g., the model of Prosperi et al. (2009) includes transition-transversion bias. Understanding rates of specific mutations to resistance in HIV (e.g., Schader et al., 2012) is also a part of designing effective therapies. Keulen et al. (1997) argue that a specific pattern in the emergence of lamivudine resistance in HIV is due to one variant tending to appear earlier due to "the mutation bias of the RT [reverse transcriptase] enzyme."

As another example, consider the case of Down syndrome in humans, typically due to an aberration of chromosomal segregation that results in trisomy for chromosome 21. Though one cannot predict with high accuracy when a mutation will occur, knowledge of the tendencies of mutation is valuable. In particular, the chance that an egg carries a mutation giving rise to Down syndrome increases with the expectant mother's age (Allen et al., 2009). This predictable effect can be leveraged in strategies for screening and intervention, by encouraging older expectant mothers to undergo screening.

Note the difference between these two cases. The case of Down syndrome is one in which the target of prediction and intervention is an individual mother–child pair. For mothers under the age of 30, the chance of bearing a child with Down syndrome is less than 1 in a 1000 births, whereas for mothers over 45, the chance increases to 1 in 25 (Hook et al., 1983). By contrast, the utility of predictive knowledge in the case of HIV is not based on predicting one mutational outcome with a low probability in one trial, but on aggregating over possible mutations in the resident population of HIV in a patient, where a billion new virions may be generated in a day, with a mutation rate on the order of 10^{-5} per nucleotide per replication cycle (see Prosperi et al., 2009). In such cases, the numbers may be so large

---

Box 3.2 A simple point about prediction and biases

No scheme of prediction has a higher expectation of success than predicting that the outcome of a trial is the single most likely outcome. If we have 100 independent trials, our best bet is to make the same prediction 100 times, betting on the single most likely outcome each time.

A mathematical proof is as follows. Assume a vector of unique probabilities of mutation **p** that sum to 1, and a corresponding vector of nonnegative prediction weights **q** that sum to 1. In mathematics, these are called stochastic vectors. The overall rate of success is

$$\sum_i p_i q_i \qquad (3.1)$$

If we bet everything on the most likely outcome, i.e., $q_m = 1$ for the maximally likely outcome, and $q_i = 0$ for all other outcomes, then the overall chance of success is simply the chance of the most likely outcome, p_{max}. The proof that no prediction strategy does better than p_{max} is simply that, for any **p** and **q** that are stochastic vectors,

$$\sum_i p_i q_i \leq \sum_i p_{max} q_i = p_{max} \qquad (3.2)$$

In the case of ties, i.e., multiple outcomes with the same maximal probability, every distribution of prediction weights among the maximally probable outcomes gives the same result (and ties among the less probable outcomes are irrelevant).

What if we have reason to believe that some outcomes are more likely than others? Suppose a process has n possible outcomes. With no other knowledge, our chance of success in prediction is $1/n$. Now, suppose we know of a biasing factor that makes some outcomes more likely, such that some outcomes have probability p, and others Bp, where $B > 1$. If we are to make a prediction using this new knowledge, we will of course bet on an outcome favored by the biasing factor B, though we have no basis to prefer any specific outcome

favored by B. How much does this improve our chances of success?

Suppose a fraction f of possible outcomes is favored by B. The chance that the next outcome is one of these is $Bf/(Bf + 1 - f)$; the chance of predicting the right outcome from this set is $1/(fn)$; and our chance of success is the product of these two factors, $B/(n(Bf + 1 - f))$. The ratio of this product to our naive prediction success, $1/n$, is the ratio of improvement, $B/(Bf + 1 - f)$. This relation is plotted for $B = 10$ in Fig. 3.1.

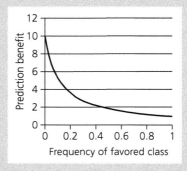

Figure 3.1 Prediction is improved by knowledge of a bias, here a 10-fold bias that favors some class of changes with a variable size.

The rightward convergence on 1 (i.e., no improvement) indicates that, if nearly all outcomes are in the B-favored set, then the knowledge that this set is preferred hardly helps us. By contrast, if almost none are in the B-favored set, then we get the maximum advantage, which is B. The purpose of this simple exercise is to verify our intuitive sense that knowledge of biases yields quantitative gains in prediction, with the largest gains applying to *large* biases acting on *rare possibilities*.

---

that the most likely mutations are highly likely to have occurred multiple times.

Other clinically important lines of argument could be cited in this regard, e.g., Leighow et al. (2020) showed the importance of mutation biases in the emergence of resistance to the drug imatinib in leukemia patients, and argued for including such effects in the design of effective treatments. Cannataro et al. (2018) have shown a very strong

effect of the rate of mutational origination on the prevalence of driver mutations in lung cancer (see their Fig. 2), such that the most clinically prevalent driver mutations are not the fastest-growing ones, but the most mutationally likely.

Once we put the issue of predictability in a practical context, it ceases to be a mysterious property of mutation. Almost nothing in biology is fully predictable. However, the more we understand about the process of mutation (e.g., the effect of local sequence context, the effect of free radicals), the more accurately we can predict rates, and the more accurately we can estimate (for instance) the risk of pathogens evolving resistance to antibiotics. As a practical matter, improving our understanding of the mutation process will lead to better predictive models of cancer, drug resistance, and other processes that depend on mutation.

3.6 The random null hypothesis

Another practical approach is to embed some concept of randomness in a strategy for hypothesis-testing. The concept of a null hypothesis is widely used in biology. Let us begin by considering carefully what researchers are doing when they employ a null hypothesis. Suppose a researcher suspects that birds living at higher altitudes frequently have blood hemoglobins with higher oxygen affinity than those living at low altitudes. A typical null hypothesis in this case would be that the factor of interest, altitude, has no effect. Suppose that data are collected for 20 pairs of bird species, each pair consisting of closely related low-altitude (L) and high-altitude (H) species. For each pair, we determine whether oxygen affinity is higher in the L or the H species.

The null hypothesis is not that the oxygen affinities will be exactly equal for each pair. Instead, there may be other factors that affect affinity, from measurement error to evolutionary factors other than altitude. Because of these other unknown factors, the oxygen affinity for a given species cannot be predicted precisely, and the affinity for two species at the same altitude may differ. Thus, the expectation under the null hypothesis is not that the affinities are exactly equal, but that they are drawn from some unknown distribution (reflecting all relevant factors) that is the same for both H and L. Under this null hypothesis, the expected result is exactly the same as for flipping a coin 20 times: we expect H species to have higher affinity 10 times, i.e., half the time.

Suppose that the actual result is 12 H, 8 L. This is evidence that altitude influences affinity in the manner expected: the H species show higher affinity more often. However, the null hypothesis cannot be rejected on this basis. If we were to flip a coin 20 times, the chance for us to see a deviation as great as 12 to 8 is about 25 %. By contrast, the chance of seeing a deviation at least as great as 15 to 5 is 0.021, and the chance of seeing a deviation at least as great as 16 to 4 is 0.0059. That is, if altitude has no effect on affinity, we might see a result as extreme as 16 H to 4 L, but only very rarely.

With the result of 12 H to 8 L or even 14 H to 6 L, the researcher would not be able to reject the null hypothesis. Any ratio greater than 1 is evidence for an effect of altitude, but an outcome is not considered to be "significant" unless the chance of seeing the outcome under the null hypothesis (of no effect) is below a small number, called a "critical level," which is often 0.05 in biomedical research (where data are often expensive to acquire), and is lower in other fields.

This example illustrates that, not only is the *content* of the null hypothesis chosen in a skeptical way—removing the influence of a factor of interest—the null hypothesis is also *applied* in a way that works against forming conclusions too easily. Thus, we can view the use of null hypotheses as a method that embodies scientific skepticism and caution. The null model is erected deliberately as a roadblock that must be removed *by the force of data* before victory is declared.

Most null models are proposed on a case-by-case basis, as in the bird example. However, sometimes scientists engage in broader discussions over generalized null models, e.g., Duret (2008) describes Kimura's neutral theory as "the null model of molecular evolution." The issue of a generalized null model has arisen at different times in paleontology, biogeography, ecology, and molecular evolution, and in every case the issue of neutrality has arisen, sometimes in confusing ways (for discussion, see Gotelli and McGill, 2006). Why are null models often defined in terms of neutrality? Why is a neutral model often considered the appropriate null model? This tendency follows from the interpretation of null models given above: based on historical patterns of unwarranted credulity, *the things that biologists most need to be skeptical about are selection and adaptation.* Therefore, to choose a neutral null model is to enforce a useful kind of skepticism.

Wagner (2012) suggests that the randomness of mutation refers to a null hypothesis. What does this signify? Actually, Wagner (2012) is an advocate of facilitated variation, e.g., he says that "the phenotypic variation on which natural selection feeds can be viewed as non-random and highly structured. Not only that, it is structured in ways that facilitate evolutionary adaptation and innovation" (p. 96). However, Wagner (2012) says that we can posit the randomness of mutation as a kind of null hypothesis. Once we learn of some effect, such as transition-transversion bias,

we can form new, better-informed expectations. For example, we might call point mutations random if the likelihood that a mutation transforms a base into another base depends only on whether the mutation would be a transition or transversion but would otherwise be independent of the base considered. This expectation might seem quite reasonable, but it also turns out to be violated. For example, mutations are often context dependent. That is, whether, say, a C mutates into a T may depend on whether there is a G next to it . . . Thus, mutation is nonrandom with respect to the expectation above.

As Wagner (2012) explains, this approach makes the idea of nonrandomness "a moving target" (p. 102). Note that this position is valuable for someone who favors the idea that variation is facilitated. That is, we established earlier that the content of a null hypothesis is dictated by skepticism, and specifically by eliminating some effect on which we base our subjective hopes for success. If one is an advocate of facilitated variation, the null hypothesis to reject (by the force of data) is that the novelty-generating capacity of evolved systems is not better than that of unevolved systems.

3.7 Beyond randomness: the principle of indifference

Interestingly, we can retain all of the familiar apparatus of reasoning from probabilities, without resort to "randomness," null hypotheses, and arbitrary procedural assumptions about critical levels. How is this possible? The answer depends on a major argument in statistical theory that has emerged in the past fifty years.

Starting over two centuries ago, intuitively appealing concepts of probability theory appeared, often as as arbitrary rules, or as rules justified by appeals to "randomness," "symmetry," and other concepts (Gigerenzer et al., 1989).

Then in 1946, R.T. Cox made a striking proposal: the classic rules of probability theory emerge, not from some physical notion of chance or randomness that actually exists in the world, but *logically*, from the *desiderata* of plausible reasoning, i.e., from the demands of constructing a logically coherent system to conduct reasoning under conditions of uncertainty. E.T. Jaynes (2003) took this argument further, concluding that "if degrees of plausibility are represented by real numbers, then there is a uniquely determined set of quantitative rules for conducting inference" that validates classical probability theory, and that these rules are "valid principles of logic in general, making no reference to 'chance' or 'random variables'" (p. xx of Jaynes, 2003).

The starting criteria for deriving a system of plausible reasoning are (1) the use of real numbers for degrees of plausibility, (2) qualitative correspondence with common sense (e.g., supporting intuitive forms of reasoning), and (3) internal consistency. Readers who are curious about how these criteria lead to the product rule and the sum rule of probability theory (from which nearly everything else follows) may wish to read Chapters 1 and 2 of the seminal work of Jaynes (2003), *Probability Theory: The Logic of Science*.

From this perspective, the concepts of randomness and probability in conventional uses of probability theory (and sometimes in the evolutionary literature) are what Jaynes calls a "mind projection fallacy," in which the uncertain state of the observer's knowledge is projected onto the world as though it were an aspect of reality. By contrast, the logical justification of probability theory does not ask us to use probabilities on the grounds of corresponding to things in the world: instead, probabilities are chosen as the quantities suitable for calculations about degrees of plausibility or rational belief, within a logically coherent system of plausible reasoning. The coherence of this system, rather than some isolated relationship between probabilities and reality, justifies the use of probabilities as an extension of logic.

In this system of plausible reasoning, "randomness" and various arbitrary procedural assumptions (e.g., null hypotheses) are replaced with the rule that we will treat what we truly do not know with maximum indifference or modesty, consistent with Laplace's suggestion of a principle of indifference. When we are faced with a choice among n possibilities that do not differ in any known way, the principle of indifference means that each possibility will be assigned a prior probability of $1/n$.

The application of this way of thinking means that we may begin an analysis in some specific case with (for instance) a uniform set of probabilities for the outcome of some process, but we will have reached this point, not because of any beliefs about the inherent randomness or uniformity of the process, but because this kind of assumption ensures that we will reason consistently about things that are in the domain of plausible reasoning, i.e., things that are uncertain for any reason including physical indeterminacy and lack of knowledge.

3.8 Synopsis

In Chapter 2, our sole focus was on whether the randomness of mutation is a principle of biology. However, the biological literature sometimes suggests that randomness is a heuristic assumption, e.g., Kauffman (1993) suggests that this assumption is false but "too useful a simplification to give up" (p. 11).

In this chapter, we stepped back to consider (1) the practical implications of randomness and nonrandomness and (2) approaches to scientific discovery that depend on randomness or related concepts. Though mutation cannot be predicted with perfect accuracy, this is not a barrier to making models and designing interventions that depend on the occurrence of mutation; and whatever we can learn about mutation will be helpful in improving such models and interventions.

The randomness doctrine is dispensable, i.e., a rigorous approach to evolutionary analysis does not require us to take a position on the supposedly "random" nature of mutation. As noted, Wagner (2012) suggests a strategy of assuming that whatever we do not know about mutation is maximally homogeneous and independent. This is a potentially useful strategy for finding deviations from randomness. However, taking a specific position about strategies, randomness, or null hypotheses is not necessary. We can assume that what we do not know about mutation is as disordered as possible, merely on the grounds that *this is part of a logically coherent approach to reasoning in the face of uncertainty*.

The topics covered in this chapter may help to define a coherent nonideological approach to mutation in the future, but they do little to make sense of past claims. That is a question we will attempt to answer in the next three chapters.

CHAPTER 4

Evolutionary Randomness

> Whether definite variations are by chance useful, or whether they are purposeful, are the contrasting views of modern speculation. The philosophic zoologist of to-day has made his choice. He has chosen undirected variations as furnishing the materials for natural selection. It gives him a working hypothesis that calls in no unknown agencies; it accords with what he observes in nature; it promises the largest rewards.
>
> **T.H. Morgan (1909)**

4.1 Introduction

In previous chapters, we addressed some conventional meanings of randomness in science, and considered a practical approach that does not require any ontological claims of randomness.

Let us now consider that "mutation is random" might have a special evolutionary meaning, as many have argued (e.g., Eble, 1999; Merlin, 2010; Millstein, 1997). Even prior to evaluating the substance of these claims, the context clearly provides two reasons to believe that the "mutation is random" claim has a special evolutionary significance. First, this claim is defended specifically in the evolutionary literature. There is a voluminous body of research on mutation, but the arguments defending the randomness of mutation emanate, not from scientists who study the mechanistic basis of mutation, but primarily from evolutionary biologists with philosophical or theoretical interests.

Second, within the evolutionary literature, the claim that mutation is random is strongly associated with a particular view of evolution, the neo-Darwinian view that evolution results from an unequal marriage of selection and variation, in which selection does all the important work and receives all the credit. The mutation-is-random doctrine is often listed as a key principle or tenet of neo-Darwinism (e.g., Lenski and Mittler, 1993; Sniegowski and Lenski, 1995). The scientists who

hold this view routinely claim that mutation is random, whereas critics avoid this assertion and often contradict it (e.g., Shapiro, 2011).

Therefore, we have reason to expect a special evolutionary meaning of randomness linked to neo-Darwinism. Appendix C provides a large sample of statements about mutation from the literature, indicating the scope and historical relevance of the "mutation is random" claim.

4.2 Rejection of pervasively directed mutations

In the distant past, over a century ago, various evolutionary thinkers considered the idea that variation has a pervasive tendency to generate beneficial variants, and that such variations, as distinct from "random" ones, have a major and unappreciated role in evolutionary adaptation. For instance, Darwin's colleague Asa Gray accepted the importance of the struggle for life, but also proposed divine intervention to bring about useful variations at the right moments in evolutionary history (see Beatty, 2010). Other theories depended on an internal force expressed through the process of variation, e.g., Berg's 1922 theory of nomogenesis (Berg, 1969). In the contemporary literature, this general idea is sometimes mistakenly labeled "Lamarckian," though Lamarckism is a specific and thoroughly discredited version of this idea in which the

Mutation, Randomness, and Evolution. Arlin Stoltzfus, Oxford University Press (2021). © Arlin Stoltzfus. DOI: 10.1093/oso/9780198844457.003.0004

useful variations are brought about by effort in the form of use or disuse (see Section 4.3).

Morgan (1903, p. 463) presents this idea very explicitly in considering how to account for adaptedness:

I can see but two ways in which to account for this condition, either (1) teleologically, by assuming that only adaptive variations arise, or (2) by the survival of only those mutations that are sufficiently adapted to get a foothold. Against the former view is to be urged that the evidence shows quite clearly that variations (mutations) arise that are not adaptive.

That is, in the history of evolutionary thought, the formal possibility has been considered that mutation (variation) is a dispositional cause of adaptation, by virtue of generating variants in an adaptive direction. This notion was rejected by the architects of the Modern Synthesis, e.g., Simpson (1953) explains that "Calling them [i.e., mutations] 'spontaneous' and 'random' means simply that they are not orderly in origin according to the demands of any one of the discarded theories" (p. 135); Stebbins (1982) explains that mutation is random by saying that "there is no internal force or guiding principle that has directed the course of evolution" (p. 69); Mayr (1959b, p. 4) defines randomness in opposition to "finalistic" theories. These are all explicitly negative claims, i.e., they define randomness as the negation of some theory.

Merlin (2010) reviews many of these same sources (all the quotations she provides are included in Appendix C). She argues ultimately that the consensus position of "biologists of the Modern Synthesis" is defined negatively, and is simply a rejection of pervasively directed mutations: mutations meet the "evolutionary chance" definition if they are not specifically and exclusively adaptive responses to environmental conditions.

However, this definition is so broad that it fails to explain known controversies. In reality, a genuine controversy over "directed" or "Cairnsian" mutation erupted and persisted for years (e.g., Sniegowski and Lenski, 1995; Sarkar, 1991; Foster, 1998; Brisson, 2003) after Cairns et al. (1988) argued that, in *some* cases, "cells may have mechanisms for choosing which mutations will occur." Advocates of this idea did not deny the existence of spontaneous

mutations in the Luria–Delbrück sense, but rather criticized the protocol of Luria and Delbrück—in which nonmutant cells are immediately killed—for precluding the detection of mutations that emerge in response to an environmental stressor.

Thus, whereas mainstream thinking, beginning with the early geneticists, has long rejected an extreme view in which mutations are pervasively directed toward adaptation, the orthodoxy expressed by Sniegowski and Lenski (1995) or Roth et al. (2006) goes considerably beyond this to oppose even the more modest idea of a *limited tendency* to produce situation-appropriate mutations. Indeed, the literature of the Modern Synthesis era clearly includes a sweeping claim to the effect that mutation is in some sense independent, unlinked, or uncorrelated with regard to selection, adaptation, or evolution. Many examples are give in Appendix C.

4.3 Rejection of Lamarckism

As suggested by Merlin (2010) or Sober (1984, p. 105), the insistence that mutation is "random" might be simply a way of rejecting Lamarckism, a historically important theory in which adaptation is caused by the inheritance of adaptive responses that emerge by use and disuse of organs. A century ago, geneticists familiar with research on mutation and inheritance rejected a Lamarckian mechanism thoroughly and immediately. This rejection was not always argued from the lack of a correspondence with expectations of adaptation, e.g., Cuénot (1909) rejects Lamarckism on the grounds that hereditary variants emerge suddenly, rather than being brought on gradually by use and disuse, nor by any kind of effort.

This topic would deserve no further comment, except for the remarkable persistence of anachronistic references to Lamarck's ideas, and to a persistent, false contrast between Lamarck and Darwin. Ghiselin (1994) describes the latter as a "bogus dichotomy," and refers to the Lamarck of textbooks as an imaginary figure "replicated in countless books by successive teams of plagiarists."

Why is Lamarck's theory irrelevant today? Lamarck's theory of adaptation was expressed with admirable clarity in the form of two laws, literally the *Première Loi* and the *Deuxième Loi*. The first law says that the more frequent (or less frequent) use of an organ will strengthen and grow it (or weaken and diminish it). The second law says that these changes will be preserved by reproduction.

The appeal of this kind of theory rested on the fact, well known to every observer of nature, that living organisms show a pervasive tendency to respond advantageously to conditions *within their lifetimes* via bodily or behavioral changes: the blacksmith's arms grow stronger with use, the fox grows thicker fur as winter comes on, the plant grows toward the light and so captures even more of it. Today this kind of adaptive responsiveness is sometimes called "physiological adaptation" as distinct from evolutionary adaptation. If these helpful responses to changing conditions are inherited, the result is *an automatic mode of evolution with very broad significance*—as broad as the innumerable examples of physiological adaptation. Generations of naturalists, including Lamarck and Darwin, assumed that modifications such as enhanced muscles would be inherited, because they held what historians call a *developmental* view of heredity (see p. 114 of Bowler, 1988), in which development, growth, and hereditary are all manifestations of the same growth-directing substances circulating in the body.

This developmental view of heredity was refuted by the discoveries of genetics. Enhanced muscles built up through exercise are not passed on to offspring. Factors that follow Mendelian rules do not mix or change in potency, but either maintain exactly the same state, or undergo rare mutations whose timing and character typically have no obvious connection with external conditions, or with use and disuse.

A second, equally profound change in thinking was that scientists, after centuries of interpreting adaptive responsiveness as a somewhat mysterious innate property that distinguishes living matter from nonliving matter, began to see it as a *heterogeneous collection of evolutionary adaptations*, each with a separate historical explanation, e.g., the capacity to form calluses in response to abrasion was no longer seen as an inherent property of integuments, but was re-interpreted as a specially evolved capacity. This change in thinking is incipient, for instance, in Morgan (1903); in the same

era, Delage and Goldsmith (1913, p. 6) remarked on the difficulty of rethinking biology in terms of direct mechanical causes, after centuries of relying on a life-force to invest living systems with teleological behavior.

The combination of these two developments destroyed the prospects for Lamarckism *thoroughly and permanently*. Genuine Lamarckism cannot be a major mechanism of organic evolution. Genuine Lamarckism does not explain numerous hypothetical examples repeatedly proposed by Lamarck and his followers, e.g., the gradual evolution of webbed feet in aquatic birds due to swimming.

What about Darwin's ideas? In the *Origin of Species*, Darwin relied on Lamarckian adaptation ("use and disuse") as a major mode of evolutionary modification, second only to "natural selection" (see Box 4.1). Darwin believed that all variation is stimulated by altered conditions either directly or indirectly (Winther, 2000). Darwin's conception of hereditary factors as responsive growth-directing substances circulating in the body was not unusual, but reflected common beliefs about heredity, e.g., Bowler (1988, p. 114) refers to the "developmental view-point that Darwin had left largely intact" (see also Winther, 2000).

Thus, the stated scientific views of Darwin and Lamarck do not provide the basis to define contrasting "Darwinian" and "Lamarckian" positions on variation and inheritance, particularly when these are applied to contemporary debates about mutations and their role in evolution.

The reasons to avoid such usages can be summarized as follows. First, Darwin clearly accepted Lamarck's theory, as Ghiselin (1994) stresses in his attack on the "bogus dichotomy" of Darwin vs. Lamarck.

Second, no one today accepts Lamarck's theory. The contemporary scientists who invoke Lamarck's name recite the phrase "inheritance of acquired characteristics" and then point to non-Mendelian hereditary processes, some with environmentally responsive states (e.g., methylation states). However, the inheritance of acquired characteristics is not Lamarckism. Lamarck's second law simply expresses the pre-Mendelian developmental view of heredity accepted by Darwin and others, which was not Lamarck's invention (Ghiselin, 1994). Lamarck's second law does not lead to a general theory of adaptation without the general theory of adaptive responsiveness provided by the first law.

The way to vindicate Lamarck's theory today would be to show that altered methylation states (for instance) allow the blacksmith's stronger arms to be passed on to his children, allow the giraffe's longer neck to be passed to offspring by efforts to forage at greater heights, allow aquatic birds to gradually evolve webbed feet by swimming, and so on. Yet no contemporary scientist is making such an argument: no "Lamarckian" today is actually a Lamarckian. Philosophers of science who have studied scientific debates about mutation, including Sarkar (1991, pp. 239 to 240), Millstein (1997, p. 134), and Merlin (2010, p. 11), all make precisely this same point, in agreement with my own conclusion and that of Ghiselin (1994) or Penny (2015).

Finally, the modern conception of mutation emerged in the early twentieth century, long after Darwin and Lamarck, and is generally inconsistent with the developmental view of heredity that they both held. To the extent that Darwin recognized the occurrence of changes that might be considered mutations—sudden changes inherited discretely— he rejected them as monstrosities unimportant for evolution. The names of Darwin and Lamarck should not be invoked in the same sentence as the word "mutation," except to note the anachronism of doing so.

To summarize, the post-1988 debate on mutation is emphatically not the molecular version of an ancient debate between Lamarck and Darwin. To suggest otherwise not only distorts history and invites confusion, but also robs contemporary scientists of credit for raising important issues, conducting crucial experiments, and defining novel positions on phenomena that fall utterly outside the expertise of long-dead authorities such as Darwin and Lamarck.

Box 4.1 Darwin's theories of evolutionary modification

In the *Origin of Species* (OOS), Darwin relies on 3 major means of modification—natural selection, effects of use and disuse, and direct effects of environment—and two minor means—correlations of growth and single variations. Computer-based searches of electronic texts aid in assessing Darwin's preferences, though one still has to read and interpret each passage. The following results are from parsing the sixth edition of the OOS, digitized by Project Gutenberg (with thanks to Sue Asscher).

Selection Searching for any word that begins with "select" implicates about 523 sentences, almost all of which refer to natural or artificial selection.

Use and disuse (Lamarckism) Searching for the complete words "habit," "habits," "use," or "disuse" implicates about 293 sentences, but most of these are spurious matches on the common word "use." About 97 refer to the effects of use and disuse.

Direct action Searching for the complete word "action" preceded (anywhere in the sentence) by "direct" or "definite" implicates about 28 sentences, about 23 of which refer to direct action of the environment.

Correlations of growth Searching for "correlat" implicates about 46 sentences, many of which refer only to variation (not evolution); about 20 sentences implicate correlated growth as a means of modification.

Single variations Searching for "single" or "spontaneous" before "variation" yields about 14 sentences, a few of which might be affirmative references to modification by single variations, but most of which are ambiguous due to the uncertainty in Darwin's use of "spontaneous."

To understand these ideas, one must begin with Darwin's views of variation. Later in his career, Darwin developed a mechanistic "gemmules" theory of variation that combined the classic Greek idea of pangenesis (hereditary substances flowing throughout the body) with aspects of cell theory (Zirkle, 1935). However, in the OOS, Darwin declares ignorance of the mechanistic basis of variation. His understanding of empirical patterns ("laws") of variation was based on his own voluminous observations, his experiments, the literature, and the dubious testimony of breeders with whom he corresponded.

Based on these sources, Darwin recognized multiple kinds of variation and multiple modes of inheritance. His most firm and repeatedly stated belief about variations in the OOS is that they are stimulated by the effect of "altered conditions of life" on the "sexual organs." Organisms living under domestication, or in new and unusual conditions, exhibit more frequent and more extreme variability. A geneticist today would say that Darwin confused environmental and genetic sources of variability. Darwin assumed that all chains of causation for variability began outside the organism (see Winther, 2000), thus with "absolutely uniform conditions of life, there would be no variability" (Darwin, 1868, Ch. 22).

The second key generalization about variation in the OOS is that, although the emergence of variation is induced or triggered externally, the *character* of the variations reflects both the external conditions and *more importantly* the internal nature of the organism. Variations might be triggered by a change in external conditions, but the variations show patterns, e.g., correlations of growth, characteristic of the organism.

A third important generalization is that variation is sometimes of a "definite" kind, i.e., it takes a specific form or has a particular direction, and in other cases it represents "indefinite" variability:

> We see indefinite variability in the endless slight peculiarities which distinguish the individuals of the same species, and which cannot be accounted for by inheritance from either parent or from some more remote ancestor.

That is, "indefinite variability" (called "fluctuating variability" in his later work on variation) is like noise, and it emerges anew each generation, not by inheritance. The word "definite" is used, with similar connotations, by Bateson (1894, p. 64) or by Eimer (1898, p. 21). Again, indefinite variability (fluctuation) is like noise and is subject to blending, whereas definite variability is not like noise (and may show complex patterns of inheritance).

With this in mind, consider a passage from the first volume on variation by Darwin (1883), reporting on attempts at breeding wild ducks:

> Mr. Hewitt found that his young birds always changed and deteriorated in character in the course of two or three generations; notwithstanding that great care was taken to prevent their crossing with tame ducks. After the third generation his birds lost the elegant carriage of the wild species, and began to acquire the gait of the common duck. They increased in size in each generation, and their legs became less fine. The white collar round the neck of the mallard became broader and less regular, and some of the longer primary wing-feathers became more or less white.

Given his assumption of "inheritance of every character whatever as the rule, and noninheritance as the anomaly," Darwin is here describing what he believes to be heritable variations that emerge and accumulate in a few generations of captive breeding. From this passage, we can see why Darwin believed that "direct action" was an important means of modification when organisms experience a new environment.

Presumably Darwin also believed that some of the changes above were due to the use and disuse of organs, i.e., Darwin was a Lamarckian. He believed that a persistent pattern of use or disuse of organs, which he sometimes called "habit," created heritable changes pervasive enough to modify a species. In the realm of behavior, Darwin believed that learned experiences could be passed on, e.g., Darwin (1873) describes a case in which an antipathy to butchers is passed down to the children and grandchildren of a mastiff mistreated by a butcher.

Another mode of modification involved distinctive single variations that arise and are inherited discretely, rather than being diluted by blending. This sounds like a mutationist theory in which beneficial mutations arise and tend to be preserved, but the text of the OOS does not resolve this issue clearly.

Darwin refers to effects of correlations as a minor means of modification, separate from "natural selection," but it actually represents a separate mode of explanation rather than a separate mechanism or mode of action: Darwin clearly portrays the influence of correlations of growth as *a side-effect of natural or artificial selection*, e.g.,

> It is also necessary to bear in mind that, owing to the law of correlation, when one part varies and the variations are accumulated through natural selection, other modifications, often of the most unexpected nature, will ensue.

Finally, we come to the means of modification for which Darwin is known today, namely "natural selection." The OOS is frequently vague and inconsistent in the use of this term. One must read the OOS as a whole, and consider how Darwin responded to critics such as Fleeming Jenkin, to conclude that "natural selection" is a mode of change combining fluctuation (indefinite variability), blending inheritance, and the struggle for life. Individual members of a species show "endless slight peculiarities" with slight effects on survival and reproduction. These peculiarities are blended together during reproduction, and in the offspring, we see a new suite of endless slight peculiarities, shifted by the struggle for life, continuing the process. Over time, this results in gradual change.

4.4 Independence from adaptation or evolution

Some randomness claims are evolutionary in the sense of asserting a claim of independence (or lack of correlation) in terms of evolutionary concepts—often adaptation or evolution. For instance, Eble (1999) defines evolutionary randomness as "independence from adaptation and the directionality imposed by natural selection" (see Appendix C, item C.2). Similarly, Rokas (2004) characterizes a concept of "developmental bias" as the idea that mutations, "although occurring at random in respect to the trait under selection, are not random in their direction." Lenski and Mittler (1993) suggest that a "random" mutation is one whose initial occurrence is independent of its fitness value. Other examples are given in Appendix C.

Some of these claims are unworkable. In this category, we must include claims that rely on animistic ideas such as "needs" of organisms or "demands" of environments, e.g., in Appendix C, see C.1, C.4, C.8, C.18, C.24, C.28, C.39, or C.42. Interestingly, this kind of claim goes back to Bateson (1909b, p. 289), who writes that

Each new character is formed in some germ-cell of some particular individual, at some point of time. More we cannot assert. That the variations are controlled by physiological law, we have now experimental proof; but that this control is guided ever so little in response to the needs of Adaptation there is not the smallest sign.

Likewise, the notion of randomness with respect to "the trait under selection" (Rokas, 2004) is problematic. Without variation, no traits are "under selection." The subset of potential "traits under selection" *ex posteriori* is not independent of intrinsic tendencies of variation (e.g., consider why one never hears the suggestion that humans are "under selection" for the ability to squirt hot acid out of our backsides into the faces of adversaries, in the manner of the bombardier beetle, even though this might be beneficial). Of course, researchers often have a predetermined focus on a trait of interest such as body size or genome composition, but the generation of variation may exhibit a bias on this focal trait, e.g., body size in *C. elegans* (Azevedo et al., 2002), or genome composition in various organisms (Long et al., 2018).

The claim that mutation occurs independently of the direction of adaptation or evolution is difficult to fathom more generally. Over any significantly long period of time, the process of adaptation or evolution is dependent on the occurrence of mutations. Mutation does not occur before evolution or adaptation—mutation is part of the process. In particular, the course taken by evolution depends on which mutations occur, both (1) in the token-cause sense that a particular episode of adaptation may depend on a particular distinctive mutation whose occurrence could not be taken for granted (Blount et al., 2008; Rokyta et al., 2005), and (2) in the general-cause sense that the course of evolution may be biased systematically by tendencies of mutation, an idea that we will explore in Chapter 8.

To make this point concrete, consider again the case of MacLean et al. (2010), who studied one-step adaptation to rifampicin in replicate cultures of *Pseudomonas aeruginosa*, and counted how many times various resistant forms rose to prominence in replicate populations. They measured fitness effects and mutation rates for 11 resistance-conferring single-nucleotide substitutions in the *rpoB* gene. The left panel of Fig. 4.1 shows the frequency of achieving prominence (in replicate populations) as a function of the rate of mutation, revealing a significant correlation ($R^2 = 0.60$, $p = 0.0052$). As shown in the right panel, this is not because fitter mutants are favored by higher mutation rates: there is no significant correlation between mutation rate and fitness ($R^2 = 0.0097$, $p = 0.77$). Instead, the chance of a particular path of adaptation depends on mutation rates: paths favored by higher rates of mutation contribute to adaptation more often. The course of adaptation to rifampicin resistance is not independent of mutation, but depends sensitively on *rates* of mutation.

Other examples of this kind of effect are given in Ch. 9. The results rule out statistical independence, because the course of adaptation depends on which mutations occur.

Given that this kind of independence claim occurs repeatedly in the evolutionary literature, how could

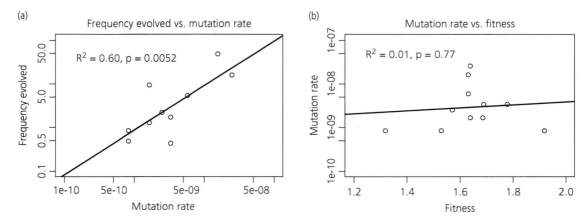

Figure 4.1 Dependence of experimental adaptation on mutation rates. In this case of experimental evolution of resistance (data from MacLean, et al., 2010), the frequency (among replicate populations) that a resistant variant reaches high frequency is correlated with its rate of origin by mutation.

we interpret it more sympathetically? The claim is literally erroneous and should not be repeated, but one must consider the possibility that the authors of this claim are struggling to say something else. One possibility is that this claim of "independence" is really a way to express the claim of explanatory irrelevance or inconsequentiality that we address in Chapter 6.

Another possibility is that the independence-from-evolution claim is a way of invoking the standard metaphysical doctrine that future events cannot cause present ones. This possibility arises because of the peculiar way that scientists under the influence of the mid-twentieth-century "Synthesis" tend to frame "evolution" or "adaptation" as a process that comes *after* the introduction of variation by mutation, in contrast to a more general conception of the evolutionary process that *includes* events of mutational introduction. This narrow framing is formalized in the "shifting gene frequencies" theory of the Modern Synthesis (see Box 6.2), and is apparent in some of the statements in Appendix C, e.g., "A central tenet of evolutionary theory is that mutation is random with respect to its adaptive consequences for individual organisms; that is, the production of variation precedes and does not cause adaptation" (Sniegowski and Lenski, 1995).

4.5 Independence from fitness effects

In Section 4.4, we rejected claims that mutation is independent from evolution or adaptation. However, other claims of independence warrant further attention, namely the ones that refer to mutation being independent from selection or fitness or value (e.g., items C.6 or C.13). Let us begin by expressing this type of claim in a way that is tractable and not obviously wrong. Invoking "natural selection" invites confusion, because it often refers to the long-term accumulation of effects, or implies a compound theory of variation and differential reproduction (e.g., Ridley, 2002; University of California Museum of Paleontology, 2008). Here and in the next chapter, we refer to fitness or "fitness effects" to avoid this kind of ambiguity.

In the simplest case, we want to know whether the chance of a mutation from genotype A to B is uncorrelated with the fitness effect of replacing A with B. More generally, we want to know something about the entire spectrum of mutation. When an event of mutation occurs, the result is a mutant genotype. The observable measure of this outcome is occurrence or frequency (how often the outcome occurs), and the underlying quantity is a rate. That is, a mutational path can be defined from the current genotype to some alternative genotype, and each such mutational path has a characteristic rate under a given set of conditions. The entire process of mutation, then, can be represented by a vector of rates, one rate for each possible mutational change, representing a mutation spectrum. That is what we will mean by mutation, **u**.

The process by which fitness is realized is also complex, involving all of the interactions of individuals with the environment. This process can span from one generation to the next, when a mutation occurs unexpressed in a gamete of a parent. Nevertheless, let us imagine that we can specify the outcome of this process with a vector **w** of fitness values for each possible alternative created by mutation. Now the issue of independence can be expressed in regard to **u** and **w**. For instance, in deriving a mathematical model of parallel evolution in which the chance for a given option i depends on a mutation rate u_i and a selection coefficient s_i, Chevin et al. (2010) assume that **u** and **s** are uncorrelated, which is tantamount to assuming that **u** and **w** are uncorrelated. That is, the claim of independence in this case is a clearly stated position that represents a formal assumption in a theoretical model of evolution.

Is this assumption justified? For each possible alternative to the current state of an evolvable system, we can measure both a mutation rate and a fitness. Are the two uncorrelated?

Let us imagine a tractable experiment to detect correlations. For instance, suppose that we choose a set of five model organisms, e.g., fruit-flies, *E. coli*, maize, *C. elegans*, and mice. For each organism, we will specify an arbitrary set of possible genetic changes, e.g., a set of 20 specific genomic changes, including some nucleotide substitutions, some deletions, some transpositions, and so on. For instance, in the *E. coli* genome, we could change from T to C at the arbitrarily chosen position 1314348, which happens to fall in the *ompW* gene. It does not matter that the set of changes will be different for each model organism. What matters is to specify an arbitrary set of changes in advance. Then, we will measure the mutation rates and the fitness effects for each genomic change, and ask if the two measures are correlated or uncorrelated.

We might want to extend the design of this experiment in two ways. First, if our focus is on the potential to produce substantially beneficial mutations, which may be rare, the earlier experiment is not a very powerful approach, because we may fail to find a sufficient number of beneficial mutations among our small set of 20 × 10 mutations. Instead, we may want to consider a far larger set of mutations, e.g., 200 or 2,000 for each species.

Second, some assertions about randomness and nonrandomness make reference specifically to whether a *change* in conditions induces beneficial mutations (e.g., Sarkar, 1991; Merlin, 2010). We can revise our experimental design easily to address this. For each species, let us designate a set of standard conditions as the "normal" environment, and then specify some number of alternative environments, e.g., three of them. Then we will ask whether the *change* in **u** is correlated with the *change* in **s** when we shift from normal conditions to an alternative condition. This could be a simple comparison of signs: does fitness increase (decrease) when mutation rate increases (decreases)?

The randomness doctrine seems to suggest that such an experiment will reveal no correlations. Is this an empirical generalization based on past observations, or a necessary law, based on principles of biology?

First, let us consider whether we could predict the outcome of the proposed experiment based on biological principles. That is, the randomness-as-independence claim might be a deduction from principles of biology, leading to the conclusion that mutation and fitness must be uncorrelated, on the grounds that no other outcome is biologically possible.

In Chapter 2, we considered the biology of mutation, and found that the chances of mutation reflect a variety of mundane conditions such as gene expression and dNTP concentrations. These same conditions may influence or mediate effects of fitness. That is, the realization of mutations and the realization of fitness involve processes taking place in the same cellular milieu. To the extent that everything happening inside a cell influences everything else, there is no independence. Therefore, the randomness-as-independence claim is not a conclusion from principles of biology.

Now, having rejected the idea that independence is an *a priori* expectation based on correct biological principles, let us address the possibility that independence is an *empirical generalization*—a summary of available evidence. Perhaps biologists have learned from past experiments that mutation and selection are uncorrelated, given the limit of detection in some large and systematic experiments.

Yet, this cannot be the case, because the required experiments simply have not been done. The study by MacLean et al. (2010) is quite distinctive in reporting paired values for both fitness effect and mutation rate for a set of alternative alleles, and this study is very recent and covers just 11 mutants. More systematic evidence of this type simply does not exist, and this is why, when evolutionary biologists refer to the idea that mutation is random, they do not cite large systematic studies showing that mutation rates and fitness effects are uncorrelated.

Let us review exactly what kinds of relevant systematic data are available. Over a century ago, Morgan (1903, p. 463) reasoned that, because "the evidence shows quite clearly that variations (mutations) arise that are not adaptive," it cannot be the case that mutations are pervasively directed toward adaptation. However, these early results did not provide a systematic basis to argue that mutation rates and fitness effects are completely uncorrelated.

As noted previously (Section 2.7), the Luria–Delbrück experiment was repeated in a handful of cases (for review, see Sarkar, 1991). However, this experiment only shows that some mutations are spontaneous in a particular sense: it does not show that all mutations are spontaneous, and it does not address the issue of correlation (because it does not involve the measurement of fitnesses or mutation rates).

What more systematic kinds of studies of mutation have been done, and what do they tell us? Many experiments have been carried out by evolutionary biologists interested in the distribution of mutational effects, e.g., the famous mutation-accumulation (MA) experiment in *Drosophila* by Mukai (1964), which aimed to estimate the rate of deleterious mutations. MA experiments are based on propagating organisms under forgiving conditions, so that they are free to accumulate mutations (Halligan and Keightley, 2009). The decline in fitness over time provides information on the rate of deleterious mutation. The MA approach involves measuring the fitness of lines (each with potentially many mutations) rather than the fitness effects of individual mutations. When the MA approach is combined with sequencing, it can be used to estimate average rates for classes of mutations (see below), but this approach does not involve measuring either fitness effects of individual mutations, or rates of individual mutations.

We may also consider systematic attempts to characterize the entire distribution of fitness effects (DFE) for new mutations (Eyre-Walker and Keightley, 2007; Keightley and Eyre-Walker, 2010). However, these experiments typically focus on aggregate properties, rather than measuring rates and effects for individual mutations (which would be necessary to find correlations). The focus is on fitness and not on mutation. In some cases, individual fitnesses have been measured, but not natural mutation rates. For instance, in a series of experiments, Sanjuan and colleagues characterized the DFE for nucleotide substitution mutations in various viruses (Domingo-Calap et al., 2009; Peris et al., 2010; Sanjuan et al., 2004). The sampling strategy for mutations was to use genetic engineering to create a predetermined list of randomly designed changes: natural mutation rates were not involved in the experiments.

Mutation-scanning studies typically go further by associating specific effects with specific mutations. Mutation-scanning studies (e.g., Cunningham and Wells, 1989; Kleina and Miller, 1990) are used by biochemists to characterize the functional sensitivity of a protein by mutating many sites, typically using either genetic engineering or error-prone PCR, rather than natural mutations. The largest emerging source of systematic information on mutant fitness effects is the new technology called "deep mutational scanning" (Fowler and Fields, 2014). This approach combines genetic engineering, robotics, and deep sequencing to measure fitness values for many thousands of mutants in a single series of experiments. However, natural mutations and mutation rates are not involved.

To summarize, one interpretation of the randomness doctrine is that the chance of mutation to an alternative genotype is uncorrelated with the fitness effect of substituting that alternative genotype. In this section, we considered whether this proposition can be derived (1) from theory, i.e., from biological principles, or (2) from facts, namely from systematic studies of fitness effects and mutation rates. Based on Chapter 2, we know that the process of mutation is not walled off from the rest of biology; therefore we do not find a principled basis for the independence claim. Likewise, we do not find an empirical basis for the claim: available research simply does not include systematic studies of the relationship of mutation rates to fitness effects.

4.6 Exceptions and possible exceptions to independence

Section 4.5 showed that there is no clear argument for the independence claim. This claim is not a correct deduction from principles of biology, nor is it an empirical generalization from systematic studies of the relationship of mutation rates to fitness effects, because such studies do not exist.

What about arguments *against* independence?

Let us first consider empirical arguments, making use of DMS (deep mutational scanning) data. As noted already, these studies only address fitness effects. In the DMS approach, an enormous library of mutants is generated and then expressed in such a way that the propagation of each mutant gene is dependent on the mutant product it encodes. One may even create a library that includes every possible single-codon-replacement, i.e., with each codon separately randomized to cover all 64 possible codons. By using deep sequencing to quantify the mutant sequences before and after a period of growth, the effect of the mutation on fitness can be measured. Though this approach does not include natural mutation rates, we can take advantage of estimates of common mutation biases from MA experiments summarized by Katju and Bergthorsson (2019).

Specifically, let us consider two different types of asymmetry: (1) transitions take place at a higher rate than transversions, and (2) mutations that substitute one nucleotide take place at a higher rate than tandem substitutions that change two or three nucleotides at once (e.g., see Reid and Loeb, 1993) .

In an earlier meta-analysis, Stoltzfus and Norris (2016) addressed the effects of transitions and transversions using fitnesses of 1239 mutants from low-throughput mutation scans, rather than DMS studies (see Box 9.1). Because the scale of fitness differs in each study, fitness values are converted to quantiles between 0 and 1 before combining across studies. The results indicate that a transition has a chance of 53 % (CI, 50 to 56 %) of being more conservative than a transversion, compared to the null expectation of 50 %, a marginally significant result ($P = 0.024$).

At left, Fig. 4.2 shows results of a more recent meta-analysis of DMS studies, using fitness quantiles for 12,386 transitions and transversions (Stoltzfus and McCandlish, in prep). Note that, although the original DMS studies generate nonsense changes, synonymous changes, and multiple-nucleotide changes, the results here include only the singlet changes in which a single-nucleotide mutation results in an amino acid change. The labels in the diagonal indicate both the "from" nucleotide (rows) and the "to" nucleotide (columns), thus the upper right histogram is for $T \rightarrow G$ mutations.

In this large set of protein mutants, the different nucleotide mutation pathways have different distributions of effects. We know from Chapter 2 that different nucleotide mutation pathways have different mean rates. Thus, in any given organism, one expects that there will be correlations between mutation and fitness, though the pattern of correlations may differ in different organisms.

At right, Fig. 4.2 shows the same data grouped into transitions (red) and transversions (blue). We noted in Chapter 2 that transition mutations typically take place at a rate several-fold higher than transversions. When all transitions (red) are combined, and compared with all transversions (blue), there is a slight but significant difference: a transition has a 51 % chance of being more benign than a transversion, a significant difference, but only slightly higher than the null expectation of 50 %. Thus, in this case, the type of mutation that typically

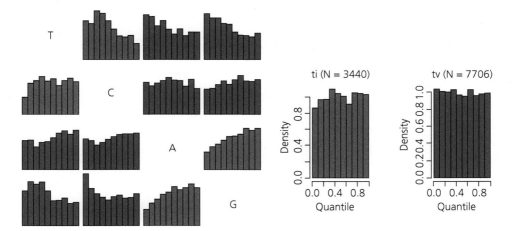

Figure 4.2 Fitness effects (quantiles) for singlet replacements in deep mutational scanning experiments, with transitions in red and transversions in blue. Left: 12 types of mutation categorized by nucleotide-from (row) and nucleotide-to (column). Right, the same data combined by transitions or transversions.

has a higher rate (transitions) is also more benign, though the difference is minuscule.

A second result emerges from comparing the singlet mutations to the doublets and triplets. Given a codon with three sites, each with three alternative nucleotides, the numbers of possible changes are $3 \times 3 = 9$ for single-nucleotide or "singlet" changes, $3^2(3!/2!) = 27$ for "doublet" changes, and $3^3 = 27$ for "triplet" changes. Most of these mutations change from one amino acid to another. Thus, in experiments that randomize a codon completely, we expect a mixture of singlet, doublet, and triplet changes, with considerably more of the latter two categories.

Firnberg et al. (2014) carried out deep mutational scanning of TEM-1 β-lactamase, finding that the median selection coefficients for singlets, doublets, and triplets were -0.36, -0.52, and -0.63, respectively. Figure 4.3 shows a meta-analysis of this singlet-doublet-triplet effect using fitness quantiles for 80,000 mutants from many DMS studies (Stoltzfus and McCandlish, in prep). If the three types were equal in their fitness effects, each distribution would be flat, but the distribution slopes upward for singlets, indicating that they are slightly more benign. In particular, the chance that a singlet mutation is

more fit than a randomly chosen doublet or triplet is 0.55, as compared to the null expectation of 0.5.

Thus, in regard to fitness effects, singlets are less damaging than doublets, which are less damaging than triplets. In regard to mutation rates, the average rate of singlet mutations is much higher than that for doublet mutations, which is, in turn, higher than that for triplet mutations. Though we do not have an *a priori* logical reason to expect that singlets will be more benign, this relationship is suggested by the hypothesis that the genetic code is adapted so as to favor benign mutations, by placing similar amino acids in mutationally closer positions (Sonneborn, 1965; Freeland and Hurst, 1998; Freeland et al., 2003).

In summary, systematic evidence is beginning to appear on the fitness effects of mutations in protein-coding genes. DMS studies provide distributions of fitness effects for different types of mutation that (from other evidence) are known to occur at different rates. Out of the entire spectrum of mutation, we have examined two cases in which different types of mutations have different mean rates, and in both cases, we find that they have different mean effects on fitness. Of course, these are only two cases, and the differences may be utterly unimportant, but

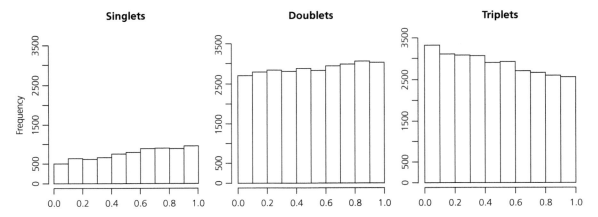

Figure 4.3 Quantile distribution for fitness effects of singlet, doublet, and triplet replacements in deep mutational scanning experiments (Stoltzfus and McCandlish, in prep).

that is not the relevant issue. The relevant issue is whether any systematic evidence weighs against the claim that fitness effects and mutation rates are uncorrelated. In fact, systematic evidence suggests that the claim is unlikely to be true.

What about theoretical arguments based on principles of biology? Does our understanding of mutation from Chapter 2 (and elsewhere) provide arguments *against* the proposition that mutation and fitness are independent?

Let us consider a few cases in which nonindependence is expected from basic mechanistic considerations. Koonin and Wolf (2009) point out that the chance of acquiring a particular gene by lateral gene transfer from another species is obviously dependent on the value of that gene in a shared environment, on the grounds that genes that are more valuable in a given environment have a higher expected frequency. For instance, consider the case of resistance to copper. In an environment contaminated by copper, microbes containing a transposon bearing copper-resistance genes have a survival advantage. This advantage increases the expected frequency of copper resistance genes in the environment. The environment, in turn, provides the pool of potential donor segments for lateral gene transfer to all resident organisms. Therefore, the chance of acquiring a transposon carrying genes conferring copper resistance is increased in environments contaminated by copper.

A similar argument applies to CRISPR spacer acquisition, explained in more detail in Section 5.3.3. Because CRISPR spacers are sequence fragments taken from invading phage genomes and used to create target-specific DNA shears that provide immunity, the chance of acquiring a spacer that immunizes against a particular phage is an increasing function of its concentration in the environment, and likewise, the fitness benefit of acquiring immunity against that same phage is an increasing function of its concentration in the environment.

Other more mundane examples suggesting nonindependence follow from cases we considered previously, e.g., the familiar example of a $T \rightarrow C$ mutation due to error-prone repair of a UVB-induced TT dimer. The higher the level of UVB radiation, the greater the damage to the genome in the form of pyrimidine photo-dimers, creating lesions that lower fitness by disrupting replication and transcription. The effect of mutation is not uncorrelated with this condition, because the net effect of error-prone repair targeted at pyrimidine dimers—separate from considering any effect of fitness—is to reduce the frequency of pyrimidine dimers (see Section A.2). That is, the effect of damage and repair is to reduce the number of targets for UVB damage, which alleviates (to some degree) the reduction in fitness otherwise caused by photo-damage. Other things being equal, this effect would reduce the frequency of pyrimidine dimers in populations especially affected by UVB, e.g., microbes in alpine environments. This correspondence does not require us to posit any specially adapted properties of the mutation process: the correspondence of effects is merely an outcome of the fact that mutation and fitness both occur in the same biological milieu, subject to the same biophysical conditions, leading to dependence.

One may envision various other cases in which damage-induced mutation correlates with fitness effects via the common factor of chemical instability. For instance, A:T base-pairs are less stable than G:C base-pairs. Thus, AT-rich regions of DNA unzip more easily, exposing the nucleobases to damage and mutation. Increased temperature reduces the stability of AT-rich regions differentially, leading to more mutation in AT-rich regions and therefore less AT-richness (Fryxell and Zuckerkandl, 2000), i.e., mutation at AT sites is enhanced under the same condition that reduction in AT sites (increasing stability) is advantageous. Similarly, free radicals affect CpG sites via oxidative deamination, reducing the number of CpG sites, i.e., reducing the possible targets for damage, thus decreasing the overall burden of damage, which is *prima facie* advantageous.

A biochemically distinct example of nonindependence would be the case of DNA precursor pools affecting mutation, discussed in Section 2.9. For instance, a higher concentration of dCTP precursors increases the rate of replication errors toward C, which increases the concordance between DNA composition and nucleotide precursor pools, which (other things being equal) represents a reduced metabolic cost, thus higher fitness.

Of course, these are largely speculative arguments. The proposed mutational effect is documented in some cases, but not the proposed fitness correlation. The point of elaborating such scenarios is not to claim that they are definitive, but to reinforce our previous conclusion that the independence claim is *simply not a valid deduction from principles of biology*. A detailed understanding of the biology of mutation, rather than leading us to reject non-independence as mechanistically impossible, instead suggests multiple hypotheses in which mutation and fitness are nonindependent.

Box 4.2 Aristotle's 4-fold view of causation

Aristotle provided a theory of causes including four types: material, formal, efficient, and final. A concise scholarly description of this 4-fold system is provided by Falcon (2019). The material cause of a thing refers to the matter (substance, material) out of which the thing is made. In Aristotle's example of making a statue, the bronze is melted and poured into a mold, which provides the shape or form. The bronze is the material from which the statue is made, and is responsible for some properties of the statue, such as its heaviness. The shape or form of the statue, by contrast, is not inherent in the bronze, but in the mold. The mold is a formal cause. To cast the bronze into the mold requires the efforts of a craftsmen who sets the work in motion, melting and pouring the bronze. Note that, whereas contemporary thinkers tend to envision the craftsman as the agent or efficient cause, Aristotle thought of the *craft* of statue-making (rather than its embodiment in the craftsman) as the efficient cause.

Finally, the entire process would not have taken place without an end or purpose, which was not merely to melt and pour bronze, but to produce a particular statue (e.g., to honor the gods). This is the final (ultimate) cause.

Aristotle's scheme is nearly always explained with the example of building a house, where the material causes are the bricks, boards, and so on, the efficient cause is the craftsman who lays the bricks and joins the boards, the formal cause is the blueprint or plan for the house, and the final cause of the house (its purpose) is to satisfy the desire for shelter. Darwin's "architect" analogy (see Section 6.6.1) also uses the example of constructing a building.

Aristotle was not the first philosopher to speak of causes or to propose various kinds of causes. He was preceded by materialists who focused on material causes, and idealists who focused on forms. Aristotle's view was distinctive in proposing that, to have complete knowledge of a thing, one must understand its plural causes (its "whys"), and in providing a 4-fold scheme to cover this plurality of causes.

4.7 Conditional independence and related ideas

Recently, Razeto-Barry and Vecchi (2016) proposed to interpret the randomness doctrine as a claim that mutation and selection are independent, except via the effect of shared conditions. They express this idea in terms of the statistical concept of conditional independence, and contrast it with mutations whose chances of occurrence correspond positively with the chances of being beneficial.

This treatment of randomness is useful for at least two reasons. First, it provides a way to recognize a key point from this chapter and Chapter 2, which is that the emergence of mutations, and the realization of fitness, take place in the same milieu, and therefore are not independent. Second, Razeto-Barry and Vecchi (2016) define an alternative to random mutation and provide an example that, they argue, satisfies the definition.

The approach presented in this section is inspired by Razeto-Barry and Vecchi (2016), but the argument is not quite the same. The misnamed "Lamarckian" alternative outlined by Razeto-Barry and Vecchi (2016) requires the extreme condition that nonrandom mutational mechanisms have a net fitness benefit (i.e., the sum of selection coefficients for all the mutations generated by the mechanism is > 0), but this restriction is not required to negate their concept of randomness, nor would it be required in evolutionary models for the emergence of specialized mutation systems, such as those described in Box 2.2. Interested readers are encouraged to refer to Razeto-Barry and Vecchi (2016).

To begin, let us consider a specialized system that generates mutations whose rate of occurrence corresponds positively with the chance of being beneficial. For instance, CRISPR spacer acquisition (for more detail, see Section 5.3.3) is a specific mutational pathway that depends on Cas restriction enzymes, and which inserts short segments of "spacer" DNA into CRISPR arrays. For simplicity, we will consider one type of mutation as the *focal* or *foreground* mutation m_f, with the understanding that this represents one example of a class of similar mutations. For instance, in the case of CRISPR spacer acquisition, m_f represents acquisition of a spacer derived from the genome of a specific phage that poses a threat to fitness. For comparison, we can consider a *background* mutation m_b, e.g., a spacer from a different unrelated phage, or a deleterious self-derived spacer. The mutation rate for mutation m_f is u_f, and its selection coefficient is s_f, and likewise for m_b we can define u_b and s_b.

The key feature of CRISPR spacer acquisition is that the rate of occurrence corresponds positively with the chance of being beneficial. That is, if environments A, B, and C have resident phages ϕ_A, ϕ_B, and ϕ_C, then the chance of acquiring a spacer that protects against ϕ_A is highest in environment A, where it is most beneficial, and the same is true, *mutatis mutandis*, for ϕ_B and ϕ_C. If we define environments e_1 and e_2 such that the phage is absent or present (respectively), then the relationships of u, s, and e are as shown in Fig. 4.4.

The argument of Koonin and Wolf (2009) regarding lateral transfer (Section 4.6) represents a related example of nonindependence. In this case, the focal mutation, m_f, is the integration of a mobile resistance cassette, acquired from the environment and inserted into the genome. In the scenario of copper resistance mentioned previously, the environmental conditions would be the absence (e_1) or presence (e_2) of high levels of copper. The background mutation m_b could be defined as an arbitrary genomic mutation with no particular relationship to

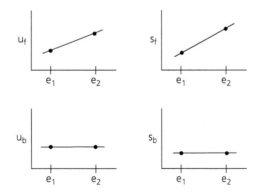

Figure 4.4 A correspondence across environments, helpful for thinking about the possibility of specialized mutation systems, following Fig. 3 of Razeto-Barry and Vecchi (2016). Top row, the focal type of mutation, for which the rate of occurrence u_f corresponds positively with the fitness effect s_f in different environments. Bottom row, a background mutation that does not show this correspondence.

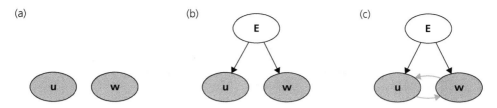

Figure 4.5 Independence (left), conditional independence (center), and co-dependence (right) are illustrated by graphs of causal relations between environment or conditions (**E**), mutation (**u**), and fitness (**w**).

this condition. If so, then the general relationships of variables will be as shown in Fig. 4.4.

This scenario of lateral transfer from Koonin and Wolf (2009) provides a useful example. Clearly, mutation rates and fitness effects are not independent, because there is a positive correlation across environments. However, this correlation is mediated by a single common factor, which is the environmental concentration of the donor sequence. The chance of acquiring a segment by lateral transfer (reflected in u_f) is a positive function of its environmental concentration, and this environmental concentration is positively correlated with fitness (indeed, the total concentration of a sequence might be *equated* with fitness under a genic view of selection).

Figure 4.5 provides a series of three graphs to illustrate different kinds of dependency. In these graphs, the environment (**E**) determines the fitness effects of alternative genotypes (**w**) and also may influence the rates of mutation to alternative genotypes (**u**). To assert conditional independence of mutation and fitness relative to an environment is to assert that, once we know the environmental conditions, this tells us all we need to know about **u**, and no further knowledge is gained by knowing **w**.

To say that **u** and **w** are conditionally independent, i.e., conditional on a common environment **E**, is to say that, relative to Fig. 4.5b, no missing arrows connect **u** and **w**, either directly, or indirectly via some factor other than **E**.

The scenario of lateral transfer seems to be a relatively straightforward example of conditional independence, because the relatedness of mutation rates and fitness effects can be represented by reference to a single environmental factor, namely the concentration of the donor sequence.

What would be an example of *dependence* between mutation and fitness? Obviously, once a mutation has happened, this outcome may influence the fitness of the organism, but that is not the important issue. Instead, the important issue is whether the genesis of a mutation could be influenced in some way by its incipient fitness effects, so that incipient beneficial effects would push the system further along the path to completing the mutation, or incipient deleterious effects would inhibit completion of the mutation.

This is exactly the kind of effect hypothesized by Cairns et al. (1988) in their "non-specific reverse transcription" (NSRT) model. Imagine a starving colony of bacteria in a petri dish containing lactose. The bacteria have exhausted other sources of energy, but their inherited genotype does not support the use of lactose. One cell happens to have an RNA transcript with an error such that proteins synthesized using this transcript are able to metabolize lactose. The more this transcript with the beneficial error is translated, the more protein is made, providing energy to the cell. Other cells without this error are dying, but the cell with the error has an increased chance of surviving. The longer the cell survives, the greater the chance that the transcript with its beneficial error will be reverse-transcribed and incorporated into DNA, resulting in a heritable mutational change.

Thus, in the NSRT model, mutational pathways have intermediate states with fitness effects that influence the chance of becoming a heritable mutation. Specifically, the intermediate state is an RNA transcript with a beneficial transcription error that might be reverse-transcribed back into the genome, resulting in a heritable change. We could illustrate this kind of effect with a diagram like Fig. 4.6, in

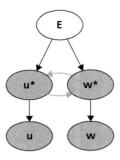

Figure 4.6 Dependence of mutation and fitness via intermediates.

which the transcript is an intermediate realization of mutation ($\mathbf{u}^*$) and this influences an intermediate realization of fitness ($\mathbf{w}^*$) in a way that influences the chances of mutation.

That is, this NSRT mechanism, and some other possible mechanisms discussed by Cairns et al. (1988) in the "directed mutations" controversy (see Sarkar, 1991, pp. 256 to 257), propose feedback loops whereby the influence of a mutational intermediate on (a correlate of) fitness probabilizes the mutation. In their critique, Sniegowski and Lenski (1995) designate this kind of mutation process as unorthodox ("non-Darwinian") and say that it "truly would be directed mutation."

Before proceeding, let us consider more carefully how to diagnose an interdependence of mutation rates with fitness effects. If we were to characterize a biological system with the NSRT mechanism by measuring mutation rates *with and without lactose*, we would get two different rates for the focal mutation, and two similar rates for the background mutation, as in Fig. 4.4. We might say on this basis that condition e_1 determines mutation rates u_{f1} and u_{b1}, and condition e_2 determines mutation rates u_{f2} and u_{b2}. Based on this measurement alone, we would have no basis to suggest that mutation and fitness are entangled, only that different environments result in different mutation rates.

In order to understand the causal relations at work, we must consider the effect of *interventions* that alter the values of variables, or alter their relationships. In the case of a hypothetical scenario, we can interrogate causal relationships using thought experiments. For instance, an obvious intervention is simply to change the degree of phenotypic effect,

e.g., imagine a series of ten different Lac^+ mutants with increasing efficiency in metabolizing lactose. Suppose that everything else about the ten variants is the same (e.g., same propensity to arise by transcription error, same propensity to be reverse-transcribed). The expectation from the NSRT model is that the mutation rate for the ten variants will correlate positively with the efficiency in metabolizing lactose.

Alternatively, we could change relationships qualitatively. For instance, imagine that we engineer the parent strain so that it carries a second-site suppressor of m_f, e.g., an antibody that binds specifically to the m_f mutant protein so that it no longer confers a Lac^+ phenotype. In terms of molecular sequences, the mutation at the site of m_f would be exactly the same, and thus we expect the transcription error to occur at the same rate, but it now has a different phenotypic effect and, because it will not help the cell to recover from starvation, it is not a favored mutation under the NSRT model. That is, if the NSRT model is true, the suppressor will lower the rate of the m_f mutation to background levels, and this will provide evidence that fitness and mutation are entangled.

What about CRISPR spacer acquisition? As a first approximation, this appears to be closely analogous to the Koonin–Wolf argument about lateral transfer, which is not a violation of conditional independence. That is, in the simplest possible CRISPR mechanism, the rate of gaining a spacer from some phage depends on its environmental concentration [ϕ], and the fitness benefit of gaining that spacer also depends on [ϕ], so that the two are independent, conditional on [ϕ]. However, some aspects of CRISPR spacer acquisition—to be discussed in Section 5.3.3—make rates of mutation dependent on the *behavior* of a sequence rather than merely its environmental frequency. For instance, the pathway of spacer acquisition tends to be activated in response to conditions of infection, and in some CRISPR-Cas systems, spacer acquisition shows a preference for rapidly replicating, highly expressed sequences characteristic of more virulent phages.

To the extent that these effects are important, we expect that, if we engineer a series of increasingly aggressive phages and measure the rate of spacer acquisition, we will find that the rate of spacer acquisition mutations is correlated positively with virulence, due to the inter-relationships of virulence with replication and expression rate.

Note that, in such a case, we also can imagine interventions that break the correspondence between mutation rates and fitness effects. For instance, we could engineer a harmless but rapidly replicating phage genome with highly expressed genes that do not damage the host, and this will increase the rate of spacer acquisition while negating any benefit of spacer acquisition. The ability to engineer artificial exceptions does not vitiate the idea of a causal relationship with fitness. Of course, fitness is, in some sense, a high-level statistical abstraction that applies to lineages, and that can never be tracked perfectly by any molecular mechanism. Yet, if we reject attempts to link fitness to mechanisms on the grounds that fitness is a high-level abstraction, this is tantamount to asserting that fitness exists in an immaterial realm, immune from effects of causation. A more practical way to think about the issue is that, when we confront fitness in any specific context, we are confronting the influence of diverse factors, and if something has a causal relationship with some but not all of the factors that determine fitness, then it has a relationship with fitness that must be taken into account.

What about the other cases considered earlier as exceptions—transitions vs. transversions, singlet mutations vs. doublets or triplets, effects of precursor concentrations, and effects of UV damage?

In regard to transitions and transversions, or in regard to singlets vs. nonsinglets, the issue is a general inequality of fitness effects, not a correlation with environments. For instance, let us define m_f as a singlet mutation, and m_b as a doublet mutation. In general, singlet mutations tend to occur at higher rates, and they tend to be more benign. Thus, we are not concerned with any particular difference between environments e_1 and e_2. Instead, $u_f > u_b$ across environments, and $s_f > s_b$ across environments. The cause of this effect of fitness is simply that singlet and doublet mutations each represent a set of amino acid pairs (e.g., Phe to Ile is a singlet, and Phe to Asn is a doublet), and the fitness effects for changing within a singlet pair are more benign on average. If we were to intervene by changing the genetic code, this would have no effect on the

mutation rates because (in the absence of some special mechanism such as NSRT) the mutations are already completed before proteins are synthesized.

In regard to the effect of precursor concentrations, suppose that e_2 represents an excess of dCTP experienced by the site of mutation at the time of replication. Assume that the focal mutation m_f is a $T \rightarrow C$ mutation, whereas m_b is a $T \rightarrow A$ mutation. Due to the law of mass action, m_f has an elevated probability of occurrence under e_2, whereas the chance of the other mutation is (by assertion) unchanged. Generally, given an excess of dXTP, where $X \in \{T, C, A, G\}$, mutations from not-X to X represent a kinetic advantage in replication and an improvement in the match between genomic nucleotide composition and dNTP pools that represent (by our previous assumption) advantages in fitness. Therefore, the relationships of u, s, and e follow Fig. 4.4.

4.8 Mutation and altered development

So far, we have considered mutation narrowly, in genetic terms, consistent with the molecular focus of this book. However, to elaborate a broader evolutionary view, one must also consider mutation as a process that differentially probabilizes the introduction of alternative phenotypes, realized through development. That is, we must consider the implications of mutation-and-altered-development, or what Arthur (2004) calls "developmental reprogramming."

Let us consider again the idea of intermediates to fitness and mutation, where those intermediates are developmental. The diagram is exactly the same as the previous Fig. 4.6, in which some intermediate realization of fitness ($\mathbf{w}^*$) influences an intermediate realization of the chances of mutational outcomes ($\mathbf{u}^*$). We simply interpret $\mathbf{u}^*$ and $\mathbf{w}^*$ as developmental intermediates, such that the fitness of a developmental intermediate affects the realization of an alternative phenotype.

More specifically, consider the case of a developmental "constraint" in centipedes discussed by Arthur and Farrow (1999). In spite of the name, no "centipede" has exactly 100 legs, because their leg-bearing segments each have precisely two legs, and the number of leg-bearing segments is always odd, never an even number such as 50. For instance, geophilomorph centipedes have 15 to 191 leg-bearing segments. Thus, the development of leg-bearing segments seems to dictate that they are gained or lost in pairs. This raises the question of why the total number is odd, not even: the answer is presumably that the single segment bearing the 2 venom claws (which are not counted as legs) originates developmentally as a leg-bearing segment.

This is an issue of development in the sense that we have no reason to doubt that adults with even numbers of leg-bearing segments are physically possible and would be viable. As Arthur and Farrow (1999) point out, juveniles of some species have an even number of leg-bearing pairs, followed by an odd number of segments with rudiments that will develop into legs after molting. Perhaps with sufficient practice, one could use microsurgical tools to remove a segment and create a centipede with an even number of leg-bearing segments.

Arthur and Farrow (1999) argue that this is a developmental constraint, in the strict sense that centipedes with an even number of leg-bearing segments never emerge, even in the earliest stages of development. However, they also note that one cannot reject the possibility that centipedes with an even number of leg-bearing segments could emerge very early in development, but fail to hatch due to "internal scrambling of the developmental process." Thus, although developmental processes can be modified in many ways, in this case either (1) they cannot be modified to generate alternatives with an even number of leg-bearing segments, or (2) they cannot do so without introducing some extremely deleterious intermediate.

The latter type of case is a genuine paradox if we are trying to separate development from fitness. The lack of survival of a scrambled embryo prevents it from developing further. One may describe this outcome by saying that fitness affects the chances of development. At the same time, we could argue that development affects fitness in the sense that the developmental paths leading to a viable form proceed through inviable intermediates.

One could attempt to sort this out in arbitrary ways, e.g., by insisting that one must not use the word "developmental constraint" when fitness

effects are involved, as per Williams (1992). However, this would only change how things are labeled: it would not change the fact that the chances of realization of alternative phenotypes depends on the realization of fitness.

By contrast, if we consider mutation narrowly as a genetic event, this kind of paradox is largely averted by the way that mutation is defined, and by the discrete nature of Mendelian genetics: a change to DNA is not considered a mutation if it cannot be replicated and transmitted, and a change that is transmitted within a Mendelian system is typically fully and completely transmitted.

4.9 Synopsis

Though the randomness doctrine may have begun a century ago as a well-justified rejection of Lamarckism or pervasively directed mutation, the claim that "mutation is random" eventually acquired a much broader significance among evolutionary biologists. The doctrine is understood to exclude not only a view in which mutations are pervasively directed toward adaptation, but also even the more modest idea of a limited tendency to produce situation-appropriate mutations (e.g., Cairnsian "directed" mutation).

This sweeping position can be understood in various ways, some of which we rejected as unworkable. We considered two kinds of traditional claims, one in which tendencies of mutation are independent from tendencies of evolution, and another in which they are independent of fitness effects. Either claim might be an argument from principles of biology, or an empirical argument, based on observations or experiments. Regardless, the conclusion of our analysis is the same: the randomness doctrine does not have a firm empirical basis, and it does not follow from first principles of biology.

Razeto-Barry and Vecchi's proposal (2016) that the emergence of mutations occurs independently of the realization of fitness, conditional upon common environmental factors, is attractive for two general reasons. The first reason is that it appears to provide objective clarity while avoiding obvious errors. For instance, we identified some clear violations of independence that are not violations of conditional independence (e.g., the lateral transfer argument of Koonin and Wolf, 2009). Likewise, hypothetical schemes that violate conditional independence (Cairns's nonspecific RT mechanism) are understood by biologists to represent heterodoxies contrary to traditional thinking about mutation.

The second factor that makes this proposal attractive is that it follows approximately from a consideration of the biology of mutation. One cannot insist that mutation is walled off from the rest of biology: this simply is not true. However, one could argue, with some justification, that the processes by which mutations emerge—proceeding from precursor states (e.g., damage) to heritable mutation-states—are influenced by prevailing conditions, but largely *uninfluenced* by the incipient fitness effects of mutational intermediates. This is particularly true for animals, to the extent that mutations take place instantaneously in an isolated and largely inert (i.e., unexpressed) germline genome.

Nevertheless, this conception of randomness does not offer a rationalization of the historic randomness doctrine. First, we have no basis to project this new interpretation backwards in time, as if previous generations of evolutionary biologists secretly maintained a clear conception of random mutation that can be stated concisely in a single sentence, but which they repeatedly bungled and conflated with other conceptions of randomness.

Second, though it may represent a widely (if not universally) applicable claim regarding the mutation process, conditional independence is not a useful simplification. That is, "conditional independence" may sound like a slight modification of "independence," but in practice, it means that no assumptions about mutation are safe, prior to analyzing the detailed mechanistic dependence of mutation and fitness on common factors. In particular, it does not preclude broad correspondences between rates of mutation and fitness effects. For instance, the mutation-fitness correlation that we expect for lateral transfer, described by Koonin and Wolf (2009) as "Lamarckian," appears to be fully consistent with conditional independence.

CHAPTER 5

Mutational mechanisms and evolvability

Whatever the internal plausibility of these theories [of directed variation], they are in fact wrong. Neither the inheritance of acquired characters, nor any other theory of directed hereditary change (or directed mutation), is the mechanism of evolution.

Mark Ridley, *The Problems of Evolution* (1985)

5.1 Introduction

As discussed in Chapter 4, most ideas about "evolutionary" randomness are either (1) inconsistent with what we understand to be true about mutation and evolution, or (2) inconsistent with the meaning of the randomness doctrine as established in evolutionary discourse. For instance, if we interpret "mutation is random" (see Section 4.2) to mean that the process of mutation does not inevitably generate beneficial outcomes in the context of an environmental challenge, then the doctrine is scientifically sound, but fails to distinguish the randomness doctrine from the position of avowed heretics such as Caporale (2003), Shapiro (2011), or Rosenberg and Hastings (2003), who *also* accept that the process of mutation does not inevitably generate beneficial outcomes in the context of an environmental challenge.

We found one promising idea, based on Razeto-Barry and Vecchi (2016), to the effect that the process of mutation is (locally) causally independent from fitness, conditioned on their common environment, i.e., mutation and fitness may reflect shared environmental influences, but the emergence of a mutation is not affected by its incipient fitness effects. Some hypothetical mechanisms of directed mutation would violate this criterion, indicating one respect in which this concept of randomness aligns with a historic dispute over the nature

of mutation. We also found that some familiar mutation processes that violate independence do not seem to violate conditional independence.

Two issues from our analysis of the randomness doctrine remain to be addressed. The first one, addressed in Chapter 6, is the possibility that the randomness doctrine, rather than being a low-level claim about the nature of mutation, is really a high-level explanatory claim about the irrelevance of mutation for answering important questions in evolutionary biology.

The other remaining issue is the existence of specialized mutation systems such as trypanosome antigen shuffling, or somatic recombination in adaptive immunity. In Chapter 2, we encountered the idea that microbes have many different types of specialized mutation systems, including combinatorial shuffling of cassettes, diversity-generating retroelements, CRISPR-Cas spacer acquisition, and so on (Box 2.2). These systems were introduced briefly, without discussion, and were subsequently ignored.

This chapter begins by describing each of these types of mutation-generating systems in greater detail, and goes on to consider them in relation to the randomness doctrine, and in relation to the topic of evolvability. In general, in this book, we are developing a toolbox of concepts useful for talking about mutation and its role in evolution. In this chapter,

Mutation, Randomness, and Evolution. Arlin Stoltzfus, Oxford University Press (2021). © Arlin Stoltzfus. DOI: 10.1093/oso/9780198844457.003.0005

we apply those concepts to specialized mutation systems, and extend the toolbox to cover issues concerning evolvability. How well does the specific idea of conditional independence (developed in the previous chapter) apply to these systems? How could such specialized systems of mutation evolve? How do they relate to the issue of evolvability?

5.2 What a specially evolved mutation system looks like

The adaptive immune system of jawed vertebrates targets previously unknown pathogens by generating antibodies with high affinity, through a process that combines cassette-shuffling and hypermutation with internal selection (Jung and Alt, 2004; Bassing et al., 2002). "Adaptive" immunity refers to a system that responds adaptively to new threats, as distinct from static mechanisms of "innate" immunity. The responsiveness of adaptive immunity relies on a mutation-selection process that is purely somatic: no germ-line genes are changed. The underlying system of mutation, called "somatic recombination," has the capacity to generate diverse functional variants of antibody genes, subject to subsequent somatic selection.

Somatic recombination occurs in the development of immunoglobulin genes in B cells, as shown in Fig. 5.1; a similar pathway occurs for T-cell-receptor genes in T cells. Immunoglobulin IgH (heavy chain) genes are assembled by combining V, D, and J segments out of hundreds of V regions, 13 D regions, and four J regions (Jung and Alt, 2004). This combination of *hundreds* $\times$ 13 $\times$ 4 allows for thousands of sequence combinations in the part of the antibody that confers specificity, the "complementarity-determining region." The system is sometimes called "V(D)J" recombination to encompass both the VDJ pattern of heavy-chain genes, and the VJ pattern in light-chain genes. This diversity is compounded further because the junctions are sloppy, subject to addition or removal of small numbers of nucleotides. If this "junctional diversification" results in an inactive allele due to frameshifting, then the other parental allele undergoes the somatic recombination process

(Bassing et al., 2002). Later, during proliferation of B cells, the immunoglobulin genes undergo hypermutation targeted at the complementarity-determining region, which allows affinity to be refined by point mutations (Di Noia and Neuberger, 2007).

Various sources refer to this and other mutation systems as "programmed DNA rearrangements" (Borst and Greaves, 1987; Plasterk, 1992; Zufall et al., 2005). Indeed, this is the topic of a special 1992 issue of *Trends in Genetics* (Plasterk, 1992). These sources address some of the systems of microbial germ-line mutation mentioned later in this chapter, such as phase and antigenic variation, and also various forms of somatic changes to DNA that are quasi-deterministic developmental changes. For instance, as discussed by Zufall et al. (2005), various multicellular organisms undergo genome remodeling by amplification (e.g., of rDNA) or large-scale loss of heterochromatic regions. Ciliates have a somatic "macronucleus" with chromosomes formed by extensive cleavage and rearrangement of the chromosomes from the germline "micronucleus." Somatic recombination is quasi-deterministic in the sense that a VDJ or VJ fusion occurs in one allele of each maturing B- or T-cell lineage.

Based on such examples, one may suggest a rubric for what researchers consider to be a specialized mutation system. Such systems tend to have most of the following properties:

- *Focused amplification*. For some highly specific subclass of possible mutational changes, the system raises the rate of occurrence above the background rate, by multiple orders of magnitude.
- *Recurrent use*. The type of mutation favored by the system is observed to be incorporated repeatedly in development or evolution.
- *Specific modulation*. The amplified mutational capacity is not merely specific in terms of genetic products, but is subject to modulation (relative to cell cycle, developmental stage, tissue type, etc.) in a manner consistent with its recurrent use.
- *Dedicated parts*. The mutational capacity depends on specific molecules or sequence patterns whose

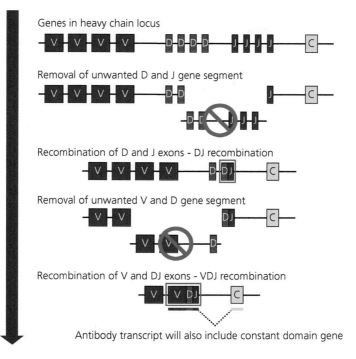

Genes in heavy chain locus

Removal of unwanted D and J gene segment

Recombination of D and J exons - DJ recombination

Removal of unwanted V and D gene segment

Recombination of V and DJ exons - VDJ recombination

Antibody transcript will also include constant domain gene

Figure 5.1 Scheme of VDJ somatic recombination of immunoglobulin genes. An internal segment is excised, and the ends are joined to form one out of many possible DJ fusions. The same process is repeated for joining one out of many possible V segments to the DJ segment. During gene expression, splicing joins this VDJ variable region to a constant C region. Image by gustavocarro reproduced under the Creative Commons CC0 1.0 Universal Public Domain Dedication.

only known function is to contribute to the mutational capacity.

- *Convergences*. Similar schemes have evolved independently.

For instance, in the case of antibody maturation, all of these criteria apply:

- *Focused amplification*. Somatic recombination is the joining of a D to a J, and a V to a DJ, specific mutations that would be much rarer without this system.
- *Recurrent use*. Diverse recombinants mature to become effective antibodies, that is, the somatic recombination pathway repeatedly yields products that are actually used in adaptive immunity.
- *Specific modulation*. The system is so tightly regulated that, even though B cells and T cells have similar VDJ somatic recombination patterns,

only the T-cell-receptor genes are rearranged in T cells, and only the immunoglobulin genes are rearranged in B cells (Bassing et al., 2002).

- *Dedicated parts*. Various components of the system seem to function only in ensuring the mutational capacity, including the recombination signal (RS) sequences used to guide recombination, the recombination-activating genes *RAG1* and *RAG2*, and the terminal deoxynucleotidyl transferase involved in generating junctional diversity (Bassing et al., 2002).
- *Convergences*. Though no other known system matches the complexity of *shuffling three different kinds of cassettes into a fusion and combining this with junctional diversity and hypermutation*, trypanosomal antigenic variation (Box 2.2) combines cassette shuffling with enhanced point mutation.

5.3 Specialized systems of germline mutation in microbes

In the nominal "evolvability" literature, authoritative reviews raise the question of whether evolvability has evolved, leaving this question unanswered (Payne and Wagner, 2019; Pigliucci, 2008). Meanwhile, in the microbial genetics literature, researchers have assumed for decades that systems of shuffling and hypermutation involved in immune evasion represent strategies or adaptations for immune evasion (Bankhead and Chaconas, 2007; McDowell et al., 2002; Dai et al., 2006; Norris, 2006; Barry et al., 2012; Horn, 2014; Bidmos and Bayliss, 2014; Foley, 2015; Vink et al., 2012; Palmer et al., 2016; Recker et al., 2011; Guizetti and Scherf, 2013). The most compelling examples of evolved mutation systems were introduced briefly in Box 2.2, and are addressed somewhat more thoroughly here:

- Multiple-inversion systems like the R64 shufflon (Section 5.3.1)
- Diversity-generating retroelements (Section 5.3.2)
- CRISPR-Cas and piRNAs (Section 5.3.3)
- Cassette donation systems (Section 5.3.4)
- Phase variation (Section 5.3.5)
- Mating-type switching (Section 5.3.6).

What do these systems do? In what sense do they suggest evolved mutation strategies? What is the evidence that they have evolutionary effects, and that they evolved due to those effects? In this section, we begin by first describing and distinguishing these systems. Later in the chapter, we consider how they are superficially consistent with adaptive stories used to describe them, and we consider the plausibility of the largely implicit claim that they are indeed evolved mutation strategies.

In regard to the rubric mentioned in Section 5.2, each of these types of systems amplifies a particular type of mutation that is used repeatedly in evolution. Each of the types has evolved more than once. In some cases, the mutational capacity is known to be modulated, and to rely on dedicated parts, whereas in other cases, these issues have not been resolved.

5.3.1 Multiple-inversion systems (shufflons)

The R64 "shufflon" in the Incl1 plasmid consists of four unique DNA segments interspersed with seven nearly identical repeats of a 19-bp sequence in the canonical order → A ← D ←→ C ←→ B ←, with arrows indicating forward and reverse repeats (Komano, 1999). Site-specific recombination between the first right and left arrows, for instance, results in inversion of segment A, and various other inversions are possible. The shufflon structure is located at the downstream end of the *pilV* gene required for mating, and thus for propagation of the Incl1 plasmid: shuffling has been shown experimentally to modulate host-specificity (Komano, 1999).

Homologous shufflons have been observed in related plasmids (Sekizuka et al., 2017; Komano, 1999). More interestingly, nonhomologous inversion systems appear to have evolved separately in other cases, including five cases reviewed by Komano (1999), and in cases identified by genome analysis in *Bacteroides* species (Cerdeño-Tarrága et al., 2005; Kuwahara et al., 2004).

On this basis, one may generalize the concept of a multiple-inversion system or shufflon. In a shufflon, diversity arises from a large number of possible inversions. The hypothetical system shown in Fig. 5.2 includes four segments, with six possible inversions shown by dashed lines. The number of possible inversion steps from a given starting sequence depends on the number of possible pairings of inverted repeats, which is simply the product of the numbers of forward and reverse

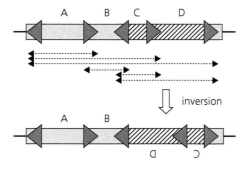

Figure 5.2 Scheme of an inversion shufflon, consisting of repeat units (triangles) in forward and reverse orientation.

repeats, i.e, $n_f \times n_r = 3 \times 2 = 6$. The total number of configurations will be larger than this number, due to the potential for additional inversion events.

5.3.2 Diversity-generating retroelements

Diversity-generating retroelements (DGRs) are widely distributed among prokaryotes and their viruses (Paul et al., 2017; Wu et al., 2018). In the DGR scheme of targeted hypermutation (Fig. 5.3), a specific section of a protein-coding gene called the VR (variable region) undergoes a high rate of mutation at sites of A nucleotides (Guo et al., 2014). The mutations are introduced by transcription, error-prone reverse-transcription, and gene conversion from a homologous template region (TR) maintained at a separate location in the genome. Typically the VR is located at the downstream end of a C-type lectin domain, a kind of domain that is used in diverse types of binding interactions.

The specificity of the system depends partly on donor-recipient sequence homology, a generic relationship that could be generated easily for any potential recipient locus by simply creating a separate donor copy. However, the specificity also depends on cis-acting signals at the donor locus, including sequence signals and tight linkage with a reverse-transcriptase gene. In fact, the system is nearly always found in association with C-type lectin domains rather than being used with a great variety of different domains.

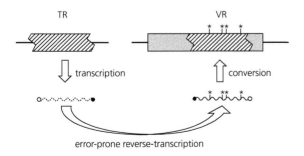

Figure 5.3 Scheme of diversity-generating retroelemens (DGRs), which generate diversity via error-prone reverse transcription (see text for explanation).

5.3.3 CRISPR-Cas and piRNAs

Clustered regularly interspersed short palindromic repeats—CRISPRs—were found widely in the prokaryotic world before their significance in a system of defense against phages and plasmids was established (Barrangou et al., 2007). In this system (Fig. 5.4), small segments of invading foreign DNA (e.g., from bacteriophages) are incorporated into the genome, then expressed as RNA transcripts, which are then used to target the invading genome for destruction.

The piRNA (piwi-interacting RNA) system active in some animal gonads (see Czech and Hannon, 2016) is similar, in that sequences whose activity threatens the host are added to repeat arrays, which are expressed and cleaved into short RNAs used to target the threatening sequences by complementarity. In CRISPR-Cas, the threat is an invading genome, and the short CRISPR RNA associates with Cas proteins to cleave the target, whereas in the piRNA system, the threat is a native transposable element, and the short piRNA associates with piwi proteins, either to suppress transcription or degrade transcripts.

CRISPR-based defense is usually described in terms of three stages defined by specific molecular outcomes (see Barrangou, 2015; Fineran and Charpentier, 2012). In the "adaptation" stage, pieces of phage DNA are incorporated into the genome. To avoid confusion, I will call this stage "CC adaptation." The pieces excised from the phage genome are typically several dozen bp in length, and they are integrated into repetitive CRISPR loci consisting of a repeat unit separated by phage-derived (or plasmid-derived) "spacers." In the "expression" stage, CRISPR loci are transcribed and then processed by Cas (CRISPR-associated) proteins into CRISPR RNAs or crRNAs, each consisting of (metaphorically) a generic handle and a specific blade, the handle being the palindromic repeat recognized by Cas proteins, and the target-specific blade being the phage-derived part. In the third "interference" stage, a complex made of Cas proteins and crRNA recognizes and degrades the phage DNA.

However, this canonical sequence of *molecular* operations does not accurately convey emergent

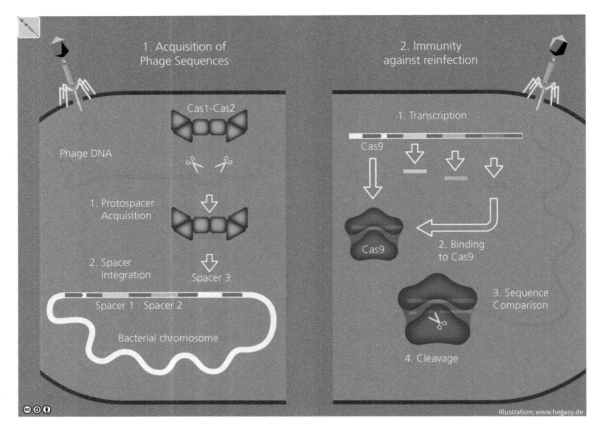

Figure 5.4 How CRISPR-Cas9 confers resistance. The left panel shows the process by which a piece termed a "protospacer" is cut out of foreign DNA, and inserted into a CRISPR array. This array contains spacers from similar events in the past. The right panel shows what happens when the spacer array is expressed. When individual RNA spacers are cut out, they combine with Cas9 protein to form site-specific DNA shears, i.e., an enzyme that will cut DNA with a matching sequence. Image by Guido Hegasy (Guido4) reproduced under Creative Commons Attribution Share-alike (CC-BY-SA-4.0) license.

features of the CRISPR defense system, for several reasons. First, the survival of cells acquiring a new spacer is, to an unknown but apparently large degree, *not due to expression and interference using the new spacer*, a process believed to be too slow to stop the incoming phage (Hynes et al., 2014). Instead, cells that acquire a defensive spacer survive for other reasons. For instance, lucky cells infected by a defective phage—a phage particle that can inject DNA but not complete its lifecycle—face no threat, but can acquire a spacer and pass it on to their descendants (Hynes et al., 2014). Cells may also survive due to innate immunity based on restriction endonucleases (restriction enzymes) that cleave nonself DNA preferentially. That is,

an incoming phage genome could be disabled by restriction enzymes, allowing the cell that acquires a spacer to survive the initial infection (e.g., see Fig. 1 of Koonin, 2019). Note that cleavage of an invading phage genome by the process of Cas-mediated spacer acquisition *is itself a mechanism of innate immunity*, depending on the degree of nonself specificity.

Second, the nonself specificity of spacers that provide defense does not arise entirely from molecular mechanisms that underly spacer acquisition mutations, but depends—to a degree that varies from system to system—on population-level selection. Specifically, in systems that are relatively indiscriminate, most spacers inserted into CRISPR arrays are from the host genome, which is typically several orders of magnitude larger than an invading phage or plasmid genome, representing a much larger mass fraction of potential donor DNA. The self-derived spacers are then lost over time at the population level, by auto-immune suicide, leaving lineages that carry nonself spacers (see Koonin, 2019). That is, when mutational spacer acquisition is indiscriminate, negative selection plays a large role in the specificity of CC adaptation.

However, other systems are much more discriminate, such that nonself spacers acquired by mutation out-number self-derived spacers by two or three orders of magnitude (Koonin, 2019). A number of systems show preferences for features such as free DNA ends and high replication rates, which tend to distinguish plasmids and phage genomes from the host chromosome (Levy et al., 2015).

Therefore, the CC adaptation stage described in the literature must not be equated with mutation, because it sometimes depends on differential survival among members of a population. The individual-level process of mutation is the process of initial spacer acquisition.

Third, CRISPR-Cas mediates anti-phage defense partly, and perhaps largely, by abortive infection, though there may be some systems in which it provides genuine immunity (Strotskaya et al., 2017). In an abortive infection strategy, the infected host is not saved. Instead, the lineage is saved by delaying or stalling completion of the phage life-cycle in infected individuals, preventing a burst that would infect their siblings and extinguish the lineage. Thus, the CRISPR-Cas system may be said to provide defense, but does not provide genuine immunity, unless we mean to apply the concept of immunity at the level of populations, i.e., herd immunity rather than individual immunity. More generally, as stressed by Koonin (2019), CRISPR-Cas systems are mechanistically and (presumably) evolutionarily tangled up with systems for programmed cell death that destroy cellular components in response to the stresses of infection.

Now, with these complications in mind, let us reconsider the specialized system of mutation that underlies CRISPR-based defenses. The mutational part of the CRISPR system is the process of spacer

acquisition from the activation of a Cas cleavage pathway to insertion of a spacer into a CRISPR array. This process, even when indiscriminate, results in correlations between useful mutations and environmental threats, for the reasons explained previously, in Section 4.7. Imagine three environments A, B, and C, with resident phages ϕ_A, ϕ_B, and ϕ_C. The rate of mutations acquiring a defensive spacer against ϕ_A is presumably near zero when ϕ_A is not present (or when there is no active Cas-mediated spacer acquisition pathway). The rate of mutational spacer acquisition (prior to selection) for ϕ_A reaches its peak in environment A, and likewise for ϕ_B in environment B, and for ϕ_C in environment C. Correspondingly, the threat from ϕ_A—and thus the fitness benefit of an anti-ϕ_A spacer—is maximal in environment A, and likewise for ϕ_B in environment B, and for ϕ_C in environment C.

Therefore, in this scenario with phages ϕ_A, ϕ_B, and ϕ_C in environments A, B, and C, respectively, there is a strong correlation between the rate of a specific mutation in an environment, and its utility in that environment, and this correlation is caused by the same condition, which is the presence of a specific phage in that environment. That is, if we were to write down an equation for the rate of anti-ϕ_A spacer acquisition as a function of environmental conditions, and another equation for the fitness benefit of anti-ϕ_A spacer acquisition as a function of environmental conditions, both equations would depend positively on the same term $[\phi_A]$, representing the concentration of phage ϕ_A in the environment.

This correlation will occur even if the mechanism of spacer acquisition fails to discriminate self and nonself. However, various mechanisms or conditions enhance the chance of nonself spacer acquisition. Chief among these is that the Cas-mediated pathway of spacer acquisition tends to be up-regulated upon infection (Patterson et al., 2017). Thus, the system, being activated by conditions correlated with the presence of nonself DNA in the cell, is more likely to acquire nonself spacers. In addition, various other factors reviewed by Weissman et al. (2020) influence the preference for nonself DNA. For instance, spacer acquisition tends to favor replicons with free DNA ends, high replication rates, and high expression rates, all of which tend to discriminate phage from host sources.

These factors can make spacer acquisition sensitive to correlates of virulence, e.g., Goldberg et al. (2014) report a CRISPR-Cas system that is effective against the more destructive class of lytic phages, but not the more benign class of lysogenic phages (the ones that tend to integrate into the genome in a quiescent state).

In summary, the system of defense offered by CRISPR-Cas depends on a mutation-generating system whose outputs are highly correlated with relevant conditions. This is due to the role of external DNA in mutation. In the other types of mutation systems discussed in this section, the sequence information involved in mutation comes entirely from the same genome. By contrast, in the case of CRISPR-Cas, the content of the mutation is not pieced together or hyper-mutated from designated loci within the genome, but is (in a sense) captured or acquired based on *behavior* associated with the sequence. The environment is *instructive* in a very precise sense, because the mutation is actually templated by a nucleic acid threat (or potential threat), with the result that the rate of a particular defensive mutation is extraordinarily higher when the threat is present than when absent.

5.3.4 Multiple cassette donation

Systems of multiple cassette donation, described already in Box 2.2, are often associated with the term "antigenic variation" (van der Woude and Bäumler, 2004; Wisniewski-Dyé and Vial, 2008). This language seems too broad, because any ordinary mutation can cause an antigen to vary, but actually the term refers to a specific *clinical* phenomenon: a pathogen is said to show "antigenic variation" if, in the course of a typical infection, it undergoes multiple shifts in the dominant antigen. Thus, when mutation systems are described as systems of "antigenic variation," this means they are associated with a higher-level phenomenon of immune evasion.

Most mutational systems underlying antigenic variation have a cassette mechanism with gene conversion from multiple silent donor sequences, like the schematic version shown in Fig. 5.5. However, the atypical *Campylobacter fetus* system generates alternative versions of SLP (surface-layer protein) by an inversion (guided by homologous sequences), the extent of which determines which of the 8 different pre-existing homologous versions of the SLP protein-coding region will be linked to the single functional promoter (Dworkin and Blaser, 1996).

As noted previously, cassette shuffling systems have been found repeatedly to underlie relapsing infections that may last for long periods of time. Such systems typically are not simply an *n*-fold switch among types determined by *n* donor loci, but offer > *n*-fold diversity due to the potential for hypermutation or combinatorial mosaics from multiple donor loci. Palmer et al. (2016) describe three such systems in bacterial species of the genera

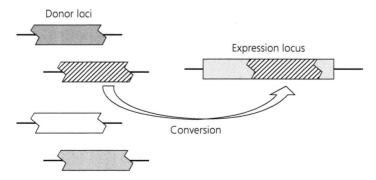

Figure 5.5 Cassette donation systems generate diversity in an expressed target locus via events of gene conversion from many homologous donor loci. Additionally, the donor and expression loci are sometimes subject to heightened levels of mutation.

Borrelia (causing relapsing fever, Lyme disease), *Neisseria* (gonorrhea, meningitis), and *Anaplasma* (anaplasmosis in livestock).

Box 2.2 presented the elaborate system of switching a variable surface glycoprotein in *T. brucei*, based on about 2,000 copies of variant surface glycoprotein (VSG) genes, most of which are silent (Horn, 2014). Switching occurs at a rate of about 10^{-3} per cell per generation, due to transcriptional switching between different expression sites, and ectopic recombination events that insert a new VSG into an expression site (ES) where the VSG is no longer silent. The VSGs are located in subtelomeric regions subject to higher rates of mutation than the rest of the genome (Barry et al., 2012).

5.3.5 Phase variation

The phenomenon of antigenic "phase variation" in *Salmonella enterica* (serovar Typhimurium) has been known for nearly a century (see the reference to Andrewes, 1922, in van der Woude and Bäumler, 2004). Stocker (1949) measured the rates of mutational switching between two flagellar antigen types, reporting values of 4.7×10^{-3} and 8.6×10^{-4} for one strain (strain 5713, p. 408), i.e., far higher than ordinary rates of mutation. Later work showed that switching of Salmonella flagellar antigens occurs by an inversion mutation (Zieg et al., 1977), catalyzed by a member of the *din* (DNA inversion) family of invertases (see Iida et al., 1984).

The varying "phases" originally referred to visibly distinct colony types, but now the term "phase variation" refers to binary switching between states, often by mutational changes—tandem repeat changes, inversions, and insertion-deletion—but sometimes an by an epigenetic mechanism (van der Woude and Bäumler, 2004). Phase variation has been identified in various prokaryotic species and in some single-celled eukaryotic pathogens. According to van der Woude and Bäumler (2004), who review a variety of systems, switching mutations occur at high rates, typically on the order of 10^{-3}, but sometimes as high as 1 in 10.

Phase variation can take place by different mutational mechanisms, as shown in Fig. 5.6. Some of the most well-studied systems of phase switching involve the inversion of a segment so as to reorient one or more genes relative to a promoter, catalyzed by a specific invertase encoded in the same genome. *Salmonella* flagellar antigen switching depends on the *hin* invertase. Other members of this protein family are involved in switching schemes (see Iida et al., 1984), including *cin* in P1 phage, *gin* in Mu phage, and the invertase used for switching in a phage-derived bacteriocin of *Erwinia carotovora* (Nguyen et al., 2001). In these cases, the inversion changes target-specificity. In the case of P1, inversion of the C segment determines the downstream end of a protein-coding region that encodes a tail fiber protein. Due to the shift in tail fibers, each type of phage particle can infect some hosts, but not others. The rate of this specific mutation is modulated in the sense that it occurs regularly during lysogeny, but not during lytic growth (Iida et al., 1984).

S. typhimurium has a second kind of inversion-mediated phase variation that determines whether or not *fimA*, encoding a fimbrial protein, is expressed (fimbriae are hairlike projections on the cell surface used in adhesion). Several related bacteria have a similar system for fimbrial phase variation, including *Escherichia coli*, *Proteus mirabilis*, and *Klebsiella pneumoniae* (Clegg et al., 2011). Reported rates of switching in *E. coli* for most conditions are on the order of 10^{-3} per cell per generation (Holden et al., 2007), but under highly favorable conditions, the "off" mutation may happen with a frequency of 0.3 per cell per generation (Gally et al., 1993).

Though DNA inversion is the most obvious mutational choice for a reversible mutation, a more common mechanism involves changes in the number of short tandem repeats (STRs; see Table 1 of van der Woude and Bäumler, 2004), typically in a manner that turns expression on or off by changing the spacing between transcriptional promoter elements, as illustrated in Fig. 5.6. In a minority of cases (e.g., Higgins et al., 2007), phase variation reflects the integration and precise excision of a mobile element (insertion sequence or transposon).

Bidmos and Bayliss (2014) report on genomic approaches to understanding the dynamics and biological significance of phase variation, focusing on systems based on STRs, whose states can be monitored easily using high-throughput sequencing.

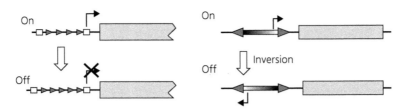

Figure 5.6 Two different mechanisms of phase variation. Left: addition of a repeated motif (right-facing triangles) changes the spacing between promoter elements (white boxes) so as to block transcription. Right, inversion of a segment deliminated by inverted repeats (triangles) reorients a promoter so that it no longer drives transcription of a target gene.

They cite various studies showing systematic differences (in on-off states of phase-variable loci) between disease isolates and carriage isolates (commensal infections, without pathological symptoms) for *Campylobacter jejuni*, *Neisseria meningitidis*, and *Haemophilus influenzae*. The biological significance of differences identified in these genome-wide screens is not firmly established.

5.3.6 Mating-type switching

Mating-type switching is a regular part of the life history of the yeast *S. cerevisiae*, known for over 60 years. When a diploid cell develops into spores through meiosis, it makes two **a** and two **α** spores. Mating to form a diploid requires **a** and **α**. However, germination of an isolated **a** or **α** spore will quickly result in diploid cells, due to mating-type switching—haploid cells propagate by budding, switch, then mate to form diploids.

The simplest kind of mutation allowing an alternation between two developmental pathways would be an inversion that switches the positions of two genes relative to a promoter, as found in some phase variation systems (see Section 5.3.5). However, the yeast system is more complex. The haploid mating type determined by the *MAT* locus switches by a cassette model proposed 40 years ago (Hicks and Herskowitz, 1977), and now a textbook staple. Each cell maintains silent (unexpressed) *MATα* and *MATa* alleles at the *HML* and *HMR* loci, respectively (for **H**idden *MAT* **L**eft or **R**ight). The molecular basis of switching is a biased gene conversion event, in which one of the silent donor loci converts the expressed *MAT* locus. The donor segments encode transcriptional regulators that, when expressed, determine the mating type.

Switching upon spore germination occurs at a sufficiently high rate that it can be considered an individual-level behavior, something that occurs reliably within one or two generations, whereas the rate of switching in diploid cells is $< 10^{-6}$ per generation (Klar, 2010). That is, the mutation is not simply targeted and directional: its rate is modulated in response to conditions. Also, cells of one type preferentially switch to the other, and mother cells alternate between bud-and-switch and bud-and-do-not-switch, ensuring that both mating types are present immediately in roughly equal numbers. The mechanisms underlying these behaviors are discussed by Haber (2012). The key modulated event in the genesis of a mutation is the introduction of a double-strand break in *MAT* by the HO endonuclease, which initiates the conversion process. Silencing of donor loci due to nucleosomes prevents the HO endonuclease from nicking its target sites in the donor loci, which is why *MAT* is always the recipient.

Mating-type switching occurs in some other ascomycete fungi. In *Ogataea (Hansenula) polymorpha*, switching between two different *MAT* loci is accomplished by an inversion (Maekawa and Kaneko, 2014). Hanson and Wolfe (2017) argue that mating-type switching was present in the common ancestor of *O. polymorpha* and *S. cerevisiae*, though the descendants show two different kinds of switching. Meanwhile, *Schizosaccharomyces pombe* has a separately evolved donor-locus system of mating-type switching (Yu et al., 2012; Hanson and Wolfe, 2017).

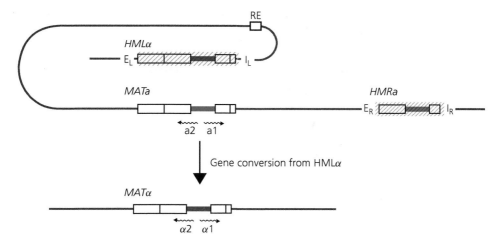

Figure 5.7 Yeast mating-type switching, showing gene conversion from type *MATa* to *MATα* (after Haber, 2012). The E and I sequences flanking *HML* and *HMR* induce heterochromatin formation (hatched lines), so that these donor loci are silent (unexpressed). When the *MAT* locus has the "a" allele, the recombination enhancer (RE) promotes the use of *HMLα* as a donor.

The genetic mechanisms underlying mating-type switching do not distinguish it from other specialized systems described above: *S. cerevisiae* switches mating types by a cassette system with gene conversion from two donors, and *Ogataea polymorpha* switches by inversion, like a phase variation system. Mating-type switching is distinguished here partly because it is treated separately in the literature, and partly because the evolutionary dynamics are presumably different. Regulated mating-type switching has the *prima facie* benefit of increased mating assurance at low densities (Yu et al., 2012; Nieuwenhuis et al., 2018). Hanson and Wolfe (2017) argue that the benefits of mating-type switching relate to the capacity to resporulate after germinating, allowing for resumption of a quiescent state under poor growth conditions. That is, mating-type switching is assumed to be beneficial in some way, but the nature of the benefit has not been firmly established.

5.4 Formulating plausible scenarios

Systems that enhance the generation of useful mutations pose a special challenge to evolutionary thinking. To address this challenge, we may begin by setting aside weak ideas and failed theories from the past, including everything from Lamarckism to directed mutation. The useful material from the past comes mainly from the past few decades, and consists of (1) a small body of relevant theoretical work on phase variation, phenotypic switching, and evolvability, and (2) the experimental work delineating the mechanisms and phenotypic consequences of specialized mutation systems in microbes.

A second helpful step is to formulate one or more plausible evolutionary scenarios for token cases, drawing on basic principles to imagine how a specialized mutation system could arise and be favored evolutionarily. This kind of exercise does not prove anything, but may serve as a framework to identify challenges and to formulate more precise questions for future research.

To imagine a system that facilitates mutating a complex living system to a beneficial alternative suited to some novel environment is difficult. Imagining a system that responds productively to surprises is difficult, because surprises are, by their nature, unexpected. Instead, scenarios of *mutation from* a known state and *mutation between* known states are much easier to contemplate than scenarios of *mutation to* an unknown state.

To understand the value of *mutation from*, consider an arbitrarily chosen case of adaptation-by-avoidance. A variety of small-molecule toxins and antibiotics achieve their specificity by their chemical complementarity with a specific binding pocket on a target protein. In such cases, e.g., the cases of tetrodotoxin and cardiac glycosides that we consider in Chapter 9, resistance may evolve via mutations that disrupt the binding pocket. That is, resistance is achieved by changing *away from* a specific pattern, disrupting chemical complementarity. The adaptation-by-avoidance pattern is indicated (for instance) when resistance is conferred by many different changes to the same site, e.g., cardiac glycosides bind to the H1-H2 extracellular domain of the α Na$^+$-K$^+$-ATPase, and five different mutations at site 111 are known to confer resistance, by changing the wild-type Gln to Leu, Arg, Val, Thr, or Glu (Ujvari et al., 2015). These variants represent five of the six possible ways to change Gln to another amino acid by a single-nucleotide mutation (the 6th type of mutation, to Pro, is not seen, perhaps because this change is too disruptive).

This kind of adaptation-by-avoidance provides a context to understand the value of *mutation from*. If the codon at position 111 (or the 12-codon region encoding the H1-H2 domain) is particularly mutable, regardless of what it mutates *to*, this would make the evolution of resistance more likely.

However, the preferential evolution of an enhanced mutation rate does not seem reasonable in this scenario, because it involves only a single episode of evolutionary change, i.e., the evolution of resistance to a specific toxin. A single instance does not provide a basis for generalized evolutionary learning. The only kind of scenario in which we might expect substantially enhanced mutation at site 111 from a single episode of evolution would be the hypothetical scenario in which the entire biosphere is suddenly saturated with cardiac glycosides, resulting in a global animal extinction event. In this fanciful scenario, we expect the survivors to be enriched for genomes that happened to have a higher mutation rate at positions conferring cardiac glycoside resistance. However, this would be merely a one-step episode of enrichment, presumably with limited impact.

Instead, the key to facilitating evolutionary learning is serial repetition. Suppose that some segment of the genome is repeatedly subject to conditions that benefit from the disruption of specificity, favored by enhanced mutation-from.

The most obvious ecological scenario would be the case of pathogens subject to specific immunity in hosts such as mammals, with adaptive immune systems (see Chs. 5 and 7 of Frank, 2002). The immune systems of jawed vertebrates generate antibodies that react specifically with pathogen-associated molecular structures, most often surface proteins, as explained previously (5.2). The specific antigenic region recognized by an antibody is called an "epitope." Binding of an antibody sets off a chain reaction that ends with the targeted destruction of pathogen cells that display the epitope.

Thus, surface proteins of microbes that infect jawed vertebrates are repeatedly subject to conditions that favor adaptation-by-avoidance, based on the disruption of specificity. The immune system tends to operate against an infective agent via a single dominant antibody. For the pathogen to evolve away from the currently dominant epitope represents an advantageous event of "immune escape" or "immune evasion." A pathogen lineage that has persisted through thousands of immunocompetent hosts must have evaded immunity thousands of times. Thus, as Frank (2002) writes, "specific immunity favors parasites that change their epitopes and escape recognition" (p. 93).

Now, having considered mutation-from, let us consider mutation-between. Immune evasion again provides a context to imagine a plausible scenario. Suppose that a microbial pathogen has sticky hair-like appendages called "pili" or "fimbriae" that are useful for attaching to mucosal surfaces and establishing an infection in a new host, but which are also highly antigenic. That is, the appendages are beneficial for establishment but not for persistence. Switching between on and off states for expression might be beneficial whether this occurs by gene regulation or mutation: a pathogen lineage that has survived through thousands of hosts has cycled thousands of times between establishment and persistence.

The scenario in which an evolutionary lineage is subject to unpredictable alternation of environments that favor one of two mutually exclusive phenotypes is addressed in the theoretical literature on "phenotypic switching" (Kussell and Leibler, 2005; Patra and Klumpp, 2015; Fudenberg and Imhof, 2012). The theory applies to cases in which stochastic switching improves the chances of long-term lineage survival at a slight cost to short-term fitness. The examples addressed in this body of literature are sometimes developmental switches, but the mechanism of the switch—genetic, epigenetic, or developmental—is irrelevant to the conditions for maintenance, which concern only the periodicity of switching.

To summarize, some familiar biological scenarios provide a context to understand the potential benefits of mutation-from and mutation-between, without having to consider the more problematic concept of mutation-to.

With this context, we may proceed to considering how to construct a more complete evolutionary scenario. In general, constructing a complete hypothesis for the evolution of switching will require (1) some scenario of evolutionary *innovation* that gives rise to the switch (mutational or developmental), and (2) some evolutionary dynamics that favor the *maintenance* or *persistence* of switching, but the theoretical literature on switching is about the latter conditions of persistence, not innovation. If mutational switching evolves like phenotypic switching, then (for instance) the theoretical literature tells us that switch rates will tend to match the frequency of switching environments. Thus, if the chance of a pathogen switching environments is 1 in 1,000 generations, the optimal switching rate is 1 in 1,000 generations.

That is, if a mutational switching system could emerge, it could persist under relatively broad conditions as a favored strategy, as argued by Palmer et al. (2013). Systems of phase variation (binary switching) in microbes, e.g., the systems listed in Table 1 of Wisniewski-Dyé and Vial (2008), are candidates for mutational switching systems that are maintained evolutionarily according to the theory for the maintenance of phenotypic switching. For instance, a reasonable hypothesis for classic fimbrial phase variation in *Salmonella typhimurium* is precisely that outlined above, in which fimbriae are advantageous during an initial establishment stage, but deleterious for longer-term persistence due to their antigenicity.

Now, let us consider the plausibility of the *evolutionary innovation* of a mechanism that enhances the production of useful mutations. That is, from the earlier considerations, we can imagine plausible scenarios for the *maintenance* of such systems once they arise (most obviously in the case of binary switching), but it remains to consider whether their initial emergence is also plausible. In general, if evolution may (under some conditions) favor X over not-X, then the same conditions will tend to favor X over an incomplete version of X, i.e., the same conditions will favor the transition from a crude and imperfect trait to a more refined and precise trait (this is the easiest way to think about the problem, although not the only way, because sometimes the path of evolution is indirect, and earlier stages of a trait may have different benefits than later versions). That is, the challenge in imagining the origin of a novel trait is to imagine intermediates that offer some of the advantages of the trait, and which do not rely on unlikely mutations.

Thus, one must begin with an awareness of the background conditions in the absence of any specialization. In any given prokaryotic genome with thousands of loci, some loci must exist such that an expression state A is favored in one environment, and an alternative expression state B is favored in a different environment. For simplicity, suppose that the relevant states are "on" and "off." Depending on conditions, these loci are candidates to evolve some scheme of regulation or mutational switching, so that the gene is typically turned on in cells inhabiting one environment, and turned off in cells inhabiting the other.

All such genes are already subject to non-specialized mutational switching by ordinary reversible mutations. For instance, the genes in most prokaryotic genomes are subject to inactivation by the insertion of mobile elements (insertion sequences and transposons) at rates on the order of 10^{-6} per locus per generation; and such insertions can be reversed by precise deletions that restore a disrupted gene (e.g., Sousa et al., 2013; Lu et al., 1989). Many, many cases of short-term adaptation in prokaryotes have been linked to insertion elements (e.g., see Vandecraen et al., 2017).

However, phase-variable loci typically have some specialized scheme of mutation that operates at higher rates. The most common mutational mechanism involves the mutational expansion and contraction of short tandem repeats or STRs (sometimes called SSRs or "simple sequence repeats"). Whether in prokaryotes or eukaryotes, STRs mutate at a high rate, and the most common mutations are the addition or subtraction of a single motif (repeated unit), e.g., going from six repeats of a CAG sequence to five or seven repeats. In general, the chance of such a length-changing mutation increases with the number of repeats, and decreases with the motif length (Willems et al., 2014).

Often the STRs involved in phase variation are located between transcriptional promoter elements upstream of a gene (Deitsch et al., 2009). This results in on–off phase variation because, in eubacterial genomes, a functional transcription promoter consists of two 6-bp elements with about 25 bp between them, where the activity of the promoter is sensitive to the spacing, as illustrated in Fig. 5.6. An STR mutation that increases or decreases the spacing disrupts the promoter, and the reverse mutation restores the spacing exactly.

How could such an arrangement arise by a series of steps, starting with the background conditions identified earlier?

One expects short sequence repeats to occur frequently by chance. For instance, for a sequence drawn uniformly from all possible sequences, the chance that a given 2-bp sequence is repeated in the next 2 bp is $(1/4)^2 = 1/16$, and the corresponding chance for a 3-bp sequence is $(1/4)^3 = 1/64$. Thus, most 25-bp spacers between promoter elements will have a chance tandem repeat of a 2-bp motif, and many will have a chance tandem repeat of a 3-bp motif.

Therefore, let us imagine a scenario of innovation that begins with a spacer region with a tandem repeat, e.g., CAGCAG. This short STR would have an enhanced mutation rate, though not as high as that of a typical phase-variable locus, which has a longer STR. An STR with a triplet motif could grow by accretion when point mutations change flanking triplets to match the motif. For an STR with a triplet pattern such as CAG, the chance that a flanking 3-bp sequence is one mutation away is 9/64, and if we consider both upstream and downstream flanking triplets, the chance of a mutationally accessible flanking repeat is $1 - (1 - 9/64)^2 = 0.26$.

Another possible mode of STR growth would occur when a lineage subject to a variable environ-

ment switches to "off" by expanding from two to three repeats, and then switches back to "on," not by loss of a repeat, but by a 3-bp deletion elsewhere in the spacer. Such microindels typically occur at rates (per site) about 3 to 20 times less than nucleotide substitutions (Katju and Bergthorsson, 2019). The result of this hypothetical series of events is that the STR has grown but the functional promoter spacing is preserved. Further STR growth, e.g., from three to four repeats, could occur by the same mechanism.

Similar reasoning could be applied to the emergence of phase variation by inversion (right panel of Fig. 5.6). Switching by inversion typically reorients a promoter relative to a protein-coding region, so that gene expression occurs in one orientation but not the inverted orientation. Just as STR mutations are catalyzed by short sequence matches, known as "microhomology," inversions are catalyzed by microhomology between a pair of inverted repeat (IR) sequences. The initial steps in the emergence of an inversion system thus depend on chance matches between (1) an inversion endpoint in the segment separating a protein-coding region from its promoter, and (2) a second endpoint, which may have a variety of different locations in inter-genic spacers one or more genes upstream or downstream. Because of this variety, the chances for finding suitable IR matches are even better than for finding an initial tandem repeat.

Thus, the stepwise emergence of a mutational system supporting precise phase variation, beginning with background conditions and proceeding via a series of ordinary mutations, is plausible. As noted previously, this kind of imaginative exercise does not prove anything, but simply helps us to formulate questions more clearly. For instance, the earlier scenario could be treated as a specific model for the kinds of mutations involved in stepwise evolution of precise switching from background conditions. Then, the more precise theoretical issue of plausibility is to determine if this model suggests a reasonable probability for the emergence of switching, given realistic mutation rates and demographics applicable to microbial pathogens. If the model is plausible in this more rigorous sense, then the final question is whether the model actually accounts for the emergence of known systems of phase variation in microbial pathogens.

Box 5.1 Understanding evolvability claims

The topic of evolvability has been characterized as a "research front" (Nuño de la Rosa, 2017), or an emphasis on certain "themes" (Payne and Wagner, 2019), rather than as a theory, a set of principles, or a set of conjectures. However, as suggested by Brown (2014), the distinctiveness of evolvability research, in historical context, reflects the novel focus on previously neglected internal factors—the internal details that, in the neo-Darwinian theory (Box 6.1), do not matter. Given this focus, the evolvability research front can be understood in regard to a series of increasingly strong (i.e., high-risk, high reward) claims regarding the role of internal factors.

E1: The efficacy of intrinsic factors in evolutionary change (evolvability as fact) For an evolving lineage with state i, the chance of evolutionary emergence of an alternative j is not merely a function of environment and demographics (e.g., population size), but depends on the internal developmental-genetic organization (where "emergence" means invasion or establishment, not just an isolated event of introduction; and where we could define things in terms of rates instead of probabilities). This definition is stated in a simplistic way, ignoring population diversity, but we could also conceive of i as a distribution of states in a population, so long as j is a state (or a distribution) representing novel types not in i. Evolvability has a "from" and a "to" state, here i and j; evolvability **from** i (to all possible j), or evolvability **to** j (from all possible i), are special cases; evolvability to all states j of higher fitness is another special case.

E2: The explanatory power of internalism (evolvability as *explanans*) The features of an internal organization, including the purely formal feature of its location in an abstract space of possibilities, can be understood productively by means of certain bridging concepts (e.g., modularity, weak linkage) that make evolution more explainable (retrospectively) and more predictable (prospectively), i.e., these features are the foundation of an *explanans*, an explanation for an evolutionary outcome. Bridging concepts are a necessity, given how evolvability is defined: to depict evolvability as a cause is to link aspects of internal organization to causal accounts of evolutionary change. Bridging concepts ideally have metrics that can be applied to predicting or explaining evolution, e.g., relating the form of organization to some aspect of the mutation spectrum or to patterns of epistasis or pleiotropy.

E3: The evolution of evolvability (evolvability as *explanandum*) Developmental-genetic organization evolves. If this influences evolution (i.e., if intrinsic factors are real), then evolvability evolves. Given a specific concept of evolvability, any mode of organizational change will change evolvability in some way. E3 claims focus on evolvability as the phenomenon to be explained (the *explanandum*). However, "the evolution of evolvability" traditionally has a much narrower connotation: the conjecture that organisms have evolved to be surprisingly adaptable (i.e., able to avoid extinction in spite of changing conditions), not by chance, but by *modes of evolutionary change that preferentially favor increased evolvability*. That is, over the long term, evolving systems will be influenced by modes of organizational evolution that are self-perpetuating, e.g., modularity-fueled mechanisms that increase modularity. Contrary to a myth popular among critics of evolvability research, the evolution of evolvability (*qua* adaptability) does not require some implausible form of group selection.

One might also define a class of E0 claims to the effect that organisms differ in intrinsic variability or in internal organizing factors, e.g., establishing that a particular organism exhibits developmental biases in phenotypic variation, without establishing an evolutionary impact (e.g., Kavanagh et al., 2013; Lange et al., 2014).

The distinction of E1, E2, and E3 claims is useful because they are often proposed separately, and are established differently. For instance, Alberch and Gale (1985) use phylogenetic comparisons to show that frogs and salamanders evolve the loss of digits in consistently different ways, where the difference lacks any obvious adaptive explanation, thus implicating internal factors in a pattern of evolutionary divergence—an E1 claim. The authors suggest an E2 mechanism involving developmental biases in variation, and provide evidence for it, in the form of different tendencies for developmental digit loss in salamanders and frogs. However, Alberch and Gale (1985) do not propose or imply any E3 claim, i.e., they do not suggest that salamanders (or frogs) evolved a *better* way to lose digits. Likewise, we recognize a

Continued

Box 5.1 *Continued*

focus on E2 claims when Pigliucci (2008) suggests, following Hendrikse et al. (2007), that

> what sets evo-devo apart from developmental biology on the one hand, and evolutionary biology on the other, is precisely the focus on which characteristics of the developmental system allow for an explanation of evolvability.

We find the same three kinds of claims outside of the traditional "evolvability" literature, e.g., the literatures of microbial and molecular evolution include many evolvability claims, some quite old. For instance, upon the elucidation of the genetic code, it was argued by Woese (1965) and others that the genetic code improves evolvability by placing similar amino acids in mutationally accessible positions. Because the possibility-space of genetic codes is well defined, the premise of this E3 claim—that the canonical genetic code has unusual evolution-enhancing properties—can be established, in principle (Freeland and Hurst, 1998). As a final

example, Krašovec et al. (2017) write, in a study of the surprising inverse correlation of mutation rates with cell density in microbial cultures (see Fig. 2.11):

> The evolutionary causes of plasticity in mutation rates need not be adaptive [*referring negatively to an E3 claim*]. Nonetheless, mutation is an evolutionary mechanism, so any plastic variation in mutation rates will have consequences for evolutionary trajectories [*an E1 assertion that internal causes must have evolutionary consequences*]. What the evolutionary consequences might be depend on how mutation rate associates with the environment [*referring to the need for E2 source laws*].

Thus, defining evolvability in terms of E1, E2 and E3 claims expands the topic to cover diverse phenomena of interest to molecular evolutionists, microbiologists, and evo-devo enthusiasts. This greatly expands the opportunities for empirical and experimental studies of evolvability, which often are more practical in regard to molecules and microbes.

5.5 Challenges and opportunities

Authors such as Sober (1984, e.g., p. 208) define the concept of adaptation in a historical-causal way, as the outcome of a process of adaptation, rather than as a thing-in-itself. A feature X is an adaptation for doing Y if (1) X has the consequence of Y and (2) X evolved due to selection for Y.

When "adaptation" is defined this way, applying the concept to specialized mutation systems (and to evolvability features generally) is dubious, because many models for the establishment of features that enhance evolvability combine selection with some kind of hitchhiking or self-organizing aspect (see Section 5.7). This would mean that the second condition "X evolved due to selection for Y" is not achieved, if we mean for "due to selection" to be exclusive of other necessary causes. Likewise, to speak of organismal adaptations as being "beneficial" or "advantageous" is common, drawing on a strong overlap between what is favored by natural selection and intuitive notions of what is beneficial for reproduction or survival. But this kind of intuitive correspondence is not very accessible when considering what might be favored by evolution in the long-term.

That is, ordinary adaptations are features that are beneficial to organisms, almost by definition, though the concept of beneficiality is intuitive. In my opinion, to use exactly the same language for higher-order effects on evolution that do not accrue to time-bounded individuals over their lifespans (but to lineages over evolutionary time), and that do not necessarily correspond to our intuitive understanding of what is beneficial, is to invite confusion.

However, once we understand the danger of conflating ordinary processes of evolutionary adaptation with modes of evolutionary change that favor evolvability, we have no reason to avoid considering hypotheses in which a particular mutation-generating feature has evolved because of its mutation-generating properties. Though evolvability-enhancing features are not conventional adaptations, we still face something like the two-fold challenge posed by the ordinary hypothesis that feature X is an adaptation for effect Y: one must show that (1) X entails Y, and (2) X evolved because of entailing Y.

For instance, Wright (2000) proposes that transcription-associated mutagenesis is an evolved mutation strategy in bacteria, because the pervasiveness of metabolite-linked schemes of gene regulation in bacteria, together with transcription-associated mutation, means that mutation is enhanced in metabolically relevant genes (see also Wright, 1997; Wright et al., 1999; Wright, 2000). That is, genes and operons are often regulated so that they are transcribed when their products are beneficial for processing inputs, building useful metabolites, and breaking down toxins and wastes. Transcription tends to increase the rate of mutation (Section 2.9). Thus the rate of mutation is enhanced when the products are useful.

The obvious experimental test for this hypothesis is to engineer a strain of *E. coli* that retains metabolism-associated regulation of transcription, but lacks transcription-associated mutagenesis, and then to compare the evolvability of the engineered strain with its unmodified parent. More generally, the obvious experimental intervention is to alter (quantitatively) the relationship between transcription and mutagenesis, then carry out repeated evolution experiments to determine the effect on evolvability. When using the comparative approach rather than experimental manipulation, the obvious strategy would be to identify differences among taxa in the quantitative relationship between transcription and mutagenesis, apply some measure of the effectiveness of adaptation, and then interrogate the relationship between the two.

If successful, this would tell us only whether X entails Y, i.e., whether transcriptional mutagenesis improves the chances of adaptation by targeting metabolically relevant genes. Establishing that X evolved because of Y would be even more difficult.

On this basis, Wright's hypothesis appears intractable, due to the difficulty of altering the relationship of transcription and mutation in a precise way, without altering other basic aspects of biology. The same might be said for a variety of other hypotheses that refer to fundamental processes that cannot be interrogated effectively either (1) due to being deeply tangled up with essential features of molecular biology, genetics, and development, or (2) due to having evolved only once, or (3) both. These hypotheses may provide alluring topics for discussion and contemplation, but they will not help us to build a solid foundation of scientific knowledge on evolvability, due to their lack of testability.

Fortunately, other hypotheses are much more amenable to experimental and comparative analysis. For instance, genetic experiments with *in vivo* mouse hosts by Bankhead and Chaconas (2007) have demonstrated that the cassette-based VlsE system of antigenic variation in *Borrelia burgdorferi* enhances long-term (cross-generational) survival, precisely because of the way that it shuffles antigens, evading the host's immune response. That is, Bankhead and Chaconas (2007) leveraged the tools of genetic dissection to show not only that the VlsE-bearing system is genetically required, but also specifically that the capacity to shuffle is required in immuno-competent mouse hosts, but not in immuno-compromised hosts. The VlsE system enhances lineage survival, and it does so because of the way that it shuffles antigens, evading the host's immune response.

What do we know about the other systems described in Section 5.3? In plasmid R64, shuffling of the *pilV* shufflon alters host-specificity (Komano, 1999). Similarly, diversity-generating retroelements alter host specificity in *Bordetella* phage BPP-1 (Doulatov et al., 2004). The CRISPR-Cas system has been shown experimentally to defend against novel phages after an initial exposure (Barrangou et al., 2007). Yeast mating-type switching obviously increases the chances of mating at low densities. Phase switching is required for persistence of *Mycoplasma agalactiae* in immunocompetent sheep (Chopra-Dewasthaly et al., 2017). Dozens of known systems of phase variation have been implicated in a wide variety of potentially useful behaviors involving switches in interactions with hosts, host tissues, phages, and environments (Bidmos and Bayliss, 2014; Hallet, 2001; Henderson et al., 1999; van der Woude and Bäumler, 2004; Wisniewski-Dyé and Vial, 2008).

That is, for each of the types of mutation-generating systems considered here, experimental evidence establishes a link (in specific cases) between a mutational capacity, and an outcome implicated in a scheme of long-term persistence reliant on that mutational capacity.

Now, let us return to the formula of showing that X entails Y, and then showing that X evolved because of Y. For the mutation-generating systems described previously, we can argue that X entails Y, where X is some set of features that determines a mutation-generating capacity (e.g., a configuration of repeats that makes a shufflon), and Y is an evolutionary outcome that depends on that capacity: long-term survival of a clan or lineage under challenging conditions via (1) mutational switching to allow exploitation of a new resource when an old resource such as a host population is depleted (shufflons, DGRs, phase variation), (2) mutational acquisition of defense against a phage threat (CRISPR), (3) mutation changing or turning off an antigen, facilitating immune evasion (cassette systems, phase variation), or (4) mutation switching mating type so as to allow mating and sporulation (mating-type switching).

The next step in establishing that these are evolvability features is to show that X evolved because of Y, that is, a set of features that determines a mutational capacity evolved toward this capacity, and maintained it, because of the way this mutational capacity influenced past evolution.

For theoreticians, this kind of demonstration calls for a model of evolutionary genetics including (1) variable genetic factors that determine a mutational capacity (e.g., the way that the length of an STR array determines its mutation rate), (2) other genetic factors that vary as consequences of that mutational capacity (e.g., a gene subject to STR mutations that turn it on or off), (3) some fitness function that assigns values to variants, and (4) various other genetic and demographic parameters that may be important, such as linkage, population size, a variable environment, and so on. As part of a hypothesis-testing strategy, such a model could be used to determine the likelihood that X evolved because of Y. The likelihood of this explanation could be compared with the likelihood of other explanations for X. The most obvious null models for X are that X is nothing more than noise, or that the mutational capacity determined by X is merely a side effect.

This discussion, which may seem elementary, is necessitated by the fact that knowledge of mutation-generating systems has advanced to a considerable degree of empirical sophistication (in the biomedical and microbial literature), without much atten-

tion from evolutionary biologists. As a result, in the literature describing these systems, certain issues remain unresolved or unaddressed that would have received attention a long time ago, if more evolutionary biologists had been paying attention to this interesting and important topic.

Or, to state this same situation differently, this topic provides numerous opportunities for evolutionary biologists to do important work. The current state of the field is that only a few researchers have begun to formulate models of the evolution of the specialized mutation systems described here (e.g., Palmer et al., 2013).

In addition to evolutionary modeling, several other kinds of investigation will be important to advancing our understanding, which would benefit from having, not just a catalog of different mutation systems and abstract theoretical models, but historical accounts of how such systems have diversified. Phylogenetic comparative analyses can identify changes in specialized mutation systems, including changes in proteins and sites involved in determining mutational capacities. An example would be the analysis of mating-type switching in yeasts by Hanson and Wolfe (2017), which reveals that the **a** and α mating types of yeast are ancestral to several different switching systems. Another example would be the comparative analysis of CRISPR-Cas systems by Makarova et al. (2015).

Though the literature on microbial mutation-generating systems is very rich, it frequently does not provide the precise kinds of systematic data relevant to making evolutionary models or verifying their expectations. For instance, it has been mentioned a number of times that the proteins affected by phase variation and antigenic variation frequently are virulence factors, and are mostly surface proteins (van der Woude and Bäumler, 2004). However, it would be better to have some actual numbers comparing (1) the total genomic frequency of genes for surface proteins and (2) the frequency with which surface proteins are implicated in schemes of phase variation.

Likewise, available evidence makes a strong intuitive case for CRISPR-Cas as an evolvability mechanism, but the case would be stronger, and the capacity to make specific models would be expanded, with more systematic data. For instance, it has been suggested repeatedly in the literature that CRISPR arrays evolve rapidly and frequently contain spacers that match phages and plasmids: this is *prima facie* evidence that CRISPR systems are actively evolving as defense systems. Likewise, the fact that *anti-CRISPR-Cas proteins* are widespread in virulent phages (Hynes et al., 2018) is a striking (albeit indirect) confirmation of the hypothesis that CRISPR-Cas systems are phage defense systems, because they indicate a competitive arms race with specific weapons and counter-weapons. Yet, the CRISPR system is quiescent in some cases, e.g., *E. coli* (Westra et al., 2010). Modeling alone is unlikely to tell us why this is the case, without more systematic ecological and comparative data.

Finally, consider the issue of whether a mutation-generating system has dedicated parts and is subject to modulation, two criteria in the rubric suggested (see Section 5.2). In general, the maintenance of proteins or sequence patterns that function only in a specific mutational capacity would be strong evidence of evolutionary dynamics that favor the persistence of that capacity; therefore, it would be helpful to have more genetic and comparative analyses of whether mutation-associated proteins have other functions. For instance, it appears that the inversion-catalyzing enzymes implicated in phase shifting (above) do not have other functions, and if so, this would be striking evidence for the evolutionary maintenance of a mutational capacity, but this is not known for certain.

Modulation of mutation also may provide evidence of specially evolved capacities. Antigen switching in the VlsE system in *B. burgdorferi* has been observed only in mammalian hosts, and attempts to induce switching *ex vivo* have been unsuccessful (Palmer et al., 2016; Coutte et al., 2009), for unknown reasons. Inversions that alter host specificity in phages P1, Mu and P7 occur at high rates in induced lysogens, but not during lytic growth (Iida et al., 1984). Fimbrial phase switching in *Klebsiella* and *E. coli* is responsive to oxygen concentration (Lane et al., 2009), as well as temperature and growth medium (Gally et al., 1993). van der Woude and Bäumler (2004) provide some additional examples.

All of these facts suggest that the mutational capacities are modulated, but in these cases, available knowledge is incomplete or difficult to interpret. The exception would be yeast mating-type switching, for which the mechanism of precise modulation is known in detail. Note that a mutation system subject to modulation (like one with dedicated parts) is more amenable to experimental interrogation to determine if the mutation system conveys an advantage.

5.6 Conditional independence and specialized mutation systems

We ended Chapter 4 with the sense that mutations typically emerge independently of their incipient fitness effects, conditional upon a common environment that influences both mutation and fitness. This kind of independence will always hold when mutations happen instantaneously, without intermediates, because this prevents any opportunity for their incipient fitness consequences to influence the realization of mutations.

When mutations do not happen instantaneously, but emerge through a series of stages, there is a chance for the fitness consequences of intermediates to influence the chances for realization. We articulated a hypothetical mechanism that violates conditional independence, namely Cairns's nonspecific reverse-transcription model, in which an RNA transcript with a beneficial error enhances the health of the cell and thereby increases the chance for the RNA to serve as the template for a DNA mutation.

Because exceptions like this are biologically possible, conditional independence is not a law of biology. At best, it may be a *contingent law*, in the same sense that the "central dogma" of information transfer is a contingent law: nothing prevents us from engineering a biological machine to reverse-translate from proteins to DNA, but no such machine exists naturally in the biological universe (so far as anyone knows). Likewise, it might be the case that fitness-dependent mutation is possible but is not part of the known biological universe.

What do specialized mutation systems tell us about conditional independence? Do they represent violations?

These questions deserve further attention, but a brief consideration of the cases we have discussed suggests an initial answer. For instance, yeast mating-type switching may be tightly regulated in response to conditions, but this mutation begins and ends before any fitness consequences of mating assurance are felt. Likewise, it may be that cassette-based VlsE switching in *Borrelia hermsii* is triggered by entering the host, but the switch mutation happens before an altered protein is expressed, thus before any interaction with the host's immune system. Many more examples could be given that follow this pattern. Even if the mutation does not happen instantaneously, if we can imagine it happening instantaneously (i.e., using a thought experiment) without affecting any fundamental properties of the system, then the incipient stages of mutation must not be involved in feedback loops that probabilize favored mutational outcomes.

That is, the typical form of specialized mutation systems is that they violate independence, but not conditional independence.

These same examples also reveal why the conditional independence claim is not a suitable replacement for the classic randomness doctrine, due to its general lack of informativeness. Let us say that we are modeling a complex bacterial community subject to phage-based mortality, and we want to know what is the rate of occurrence of phage-defensive mutations in the various host species. Naively, we might assume that the rates will be the same for each host species, and they will be the same for each type of phage regardless of how virulent. But the potential presence of a CRISPR system changes all of that. The rates of mutation for a host will depend utterly on whether it has a CRISPR-Cas system, and the rates for specific phages will depend on the details of that system. Once we know the details, we can calculate the expected rates of mutation for various phages based on their genome sequences, replication rates, and so on, but if we do not know all of the internal details, the assumption of conditional independence does not give us any practically useful information.

5.7 Evolvability

The contemporary evolvability research front (Nuño de la Rosa, 2017) blossomed in the 2000s, following key papers from Houle (1992), Wagner and Altenberg (1996), and Kirschner and Gerhart (1998). An aspect of evolvability research that makes it difficult to comprehend is the lack of a central unifying theory, or even a shared definition of evolvability (see Pigliucci, 2008). Yet, considered within the long-term development of evolutionary thinking, the evolvability research front is clearly a reaction to neo-Darwinism (Box 6.1) that stresses the role of internal organizational factors. This focus on internal factors naturally provokes an interest in the kinds of claims described in Box 5.1.

Over time, the evolvability research literature has grown to include specific models for the evolutionary causes and consequences of internal organizational factors. The topic has grown in respectability, yet the reception of evolvability claims is marred by persistent misapprehensions and mischaracterizations, and even direct attempts to undercut the novelty and importance of this topic.

Advocates of E3 claims (see Box 5.1) to the effect that some systems have evolved to be more adaptable (e.g., Caporale, 2003; Conrad and Volkenstein, 1981; Jablonka and Lamb, 2005) often appeal to adaptationist sympathies, arguing that, given the immense power of evolution by natural selection, we expect it to bring forth, not only ordinary adaptations to conditions, but adaptations in how organisms evolve, i.e., in the ability to produce useful variations. Jablonka and Lamb (2005, p. 101) write that

it would be very strange indeed to believe that everything in the living world is the product of evolution except one thing—the process of generating new variation! . . . In fact, it is not difficult to imagine how a mutation-generating system that makes informed guesses about what will be useful would be favoured by natural selection.

However, this argument pits two aspects of neo-Darwinism against each other: (1) belief in the power of selection to adapt organisms pervasively, down to the finest details, and (2) commitment to the doctrine that variation merely supplies raw materials, with no dispositional influence on evolution, so that selection may be awarded credit for the important work of evolution. The fact that E3 claims have faced so much resistance is a testament to the ongoing commitment of evolutionary biologists to neo-Darwinism, that is, to viewing selection as the potter and variation as the clay.

Nevertheless, if we ignore traditional allegiances and look at things abstractly, serious research on E2 factors (internal factors that influence the chances of evolution) began long ago with models concerning genetic recombination and other phenomena associated with sexual reproduction (as argued previously by Barton and Partridge, 2000, among others), and this research has led to E2 and even E3 claims that are well regarded. Theories about sex and recombination refer to measurable aspects of internal organization that can be related to mechanisms of evolutionary change. Thus, Colegrave and Collins (2008) include sexual reproduction in their review of evolvability, writing that

Indeed the machinery involved in meiosis and syngamy appears so obviously designed to increase the variation of offspring, that for many years, it was accepted without question that the function of sex was to increase the ability of a species to evolve.

That is, the most popular mainstream view of the evolution of sex is itself an E3 claim, whereas more recent E3 claims from evo-devo and other fields are sometimes dismissed as dubious appeals to group selection or higher-level selection, e.g., as by Dickinson and Seger (1999), Lynch (2007a), numerous examples cited in Zhong and Priest (2011, p. 500), or

Because populations, not individuals, evolve and adapt, it follows that evolvability-as-adaptation must be the consequence of selection among populations rather than selection among individuals (Sniegowski and Murphy, 2006).

In fact, a variety of mechanisms have been suggested to promote evolvability in the sense of adaptability, from "constructional selection" (Altenberg, 1995), to the theory of capacitors (Masel, 2005), to "contingency locus" models (Moxon et al., 2006), to the argument that incremental mechanisms of genetic code evolution explain the apparent evolution-enhancing properties of the genetic code (Stoltzfus and Yampolsky, 2007). None of these mechanisms involve group selection.

Likewise, the evolution of enhanced recombination does not require group selection (Charlesworth and Barton, 1996). What happens is simply that modifier loci that increase recombination hitchhike

along with favorable events that they probabilize. The same thing happens with mutator strains in bacteria: their emergence is favored under conditions of maladaptation because they hitchhike with the favorable mutations that they probabilize. Michael Conrad, an early thought-leader on the topic of evolvability, long ago dismissed the objection about group selection on exactly this basis, asserting that evolvability traits "will hitchhike along with the advantageous traits whose appearance they facilitate" (Conrad, 1990).

A third misconception, widespread in the evolvability literature itself, is to frame discussions of evolvability as though the central issue is whether there exist evolvability traits that are adaptations built by natural selection for enhancing evolution, analogous to ordinary adaptive traits that enhance life in the historically prevailing environment (e.g., Colegrave and Collins, 2008; Metzgar and Wills, 2000). That is, the topic of evolvability is often equated with E3 claims, and is sometimes identified even more narrowly with E3 claims that refer to adaptation by natural selection.

This framing is, first of all, too narrow, because other kinds of claims appear prominently in the literature of evolvability. As noted (Box 5.1), the famous study by Alberch and Gale (1985) involves an E1 claim (an empirical pattern of differential evolvability) and an E2 hypothesis (about the possible developmental and evolutionary basis for this difference), but lacks any E3 claim. Schaerli et al. (2018) engineer two *E. coli* strains in which the same phenotype is encoded using two different kinds of regulatory circuits, and show that mutations to these systems produce different spectra of phenotypic variants. This work, which strikes directly at the heart of evo-devo and evolvability— the importance of, not the phenotype, but *how the phenotype emerges*—does not address any E3 claim, but merely establishes the premise for E1 and E2 claims that internal organization is consequential for the mutation spectrum, and thus for evolutionary potentialities.

Furthermore, the assumption that E3 features must be adaptations by natural selection is a drastic and arbitrary restriction, unless we mean to use "adaptation" and "natural selection" very loosely (see Section 5.5). For instance, in "constructional selection" per Altenberg (1995), the composition of a genome shifts to include a greater fraction of evolvable gene modules because the evolvability of a module probabilizes its establishment in new genes, simultaneously and necessarily increasing its frequency in the genome. That is, constructional selection includes ordinary Mendelian selection, but also implicates a second intragenomic amplifying effect mediated by mutational rearrangements that mobilize modules into new locations. For instance, an evolutionary change that increases the copy number of a successful module from one to two, doubles the chance that this module will be the substrate for a subsequent mutation that mobilizes the module into a third location. Thus, to the extent that such mutations can be understood as events of sub-genomic reproduction, constructional selection involves two kinds of reproductive amplification, one of which amplifies the chances of future mutations in the same direction, i.e., generating a mutational bias. This is why the whole process deserves a different name.

Some other hypotheses for E3 claims are even less like adaptation by natural selection. For instance, Stoltzfus and Yampolsky (2007), expanding on an idea from Crick (1968), argue that the nonrandomness of the canonical genetic code can be explained almost entirely as a side effect of incremental mechanisms of code evolution. That is, the possible evolutionary mechanisms implicated in the emergence of a specific code, and in subsequent changes to the code, tend to assign or reassign similar amino acids to similar decoding mechanisms, e.g., changing the specificity of a tRNA-charging enzyme from one amino acid to another is more likely if the amino acids are structurally similar. When applied repeatedly, these incremental mechanisms will lead to a positive correlation between genetic closeness of codons and structural closeness of amino acids. This explanation for the nonrandomness of the genetic code does not require selection for evolvability. Indeed, correlations would be expected even in the hypothetical case in which all changes to a code are due to the neutral "codon capture" model (Osawa and Jukes, 1989), bearing in mind that neutral evolution is subject to differential filtering by selection.

Again, a key lesson of theoretical modeling is that features that enhance evolution can evolve without magic, and without group selection. The kinds of mechanisms that have been proposed to account for enhanced evolvability (E3) always involve selection, but they also involve hitchhiking, side effects, and self-organizing effects.

Finally, let us briefly consider how to rebut attempts to undercut the novelty or importance of evolvability research. For instance, Lynch (2007a) draws on the "we have long known" trope in an attempt to appropriate "evolvability" for classical quantitative genetics, citing Houle (1992) and treating everything else as speculation and loose talk. Likewise, Barton and Partridge (2000) attempt to appropriate "evolvability" for "evolutionary biology" as distinct from what the "molecular and developmental biologists" are thinking:

The term has a technical meaning in evolutionary biology, as a dimensionless measure of quantitative genetic variance [citing Houle (1992)]. However, it is now used in a broader sense by molecular and developmental biologists.

In fact, Dawkins (1988), Conrad (1990), Kauffman (1990), and Alberch (1991) all preceded Houle (1992) in their use of the term "evolvability" to refer to effects of internal organization or genetic encoding on evolutionary potential. Conrad's thinking goes back much further, to the 1970s, with references to a "self-facilitating" aspect of evolution that he called a "bootstrap principle" (e.g., see references in Conrad and Volkenstein, 1981). Conrad, a computer scientist, suspected that the disappointing performance of evolution-inspired computing methods was due to the failure of these methods to match the evolvability of naturally evolved systems (Conrad, 1978).

The study by Houle (1992) is indeed foundational for the evolvability research front (Nuño de la Rosa, 2017), and this particular study exemplifies a purely instrumental approach, based on defining evolvability as the capacity to respond to selection, measured by **G** (the matrix of variances and covariances observed in a population for a set of quantitatively variable traits—see Section 7.5.1), without any explicit or implicit theory about what causes **G**. That is, this classic study lacks E1, E2, and E3 claims, though one could construct an E2 claim from the implications of the passing comment that "the genetic variance in a trait will tend to be correlated with the number of loci which affect it," as Houle himself does, in his later treatment of mutational target size (Houle, 1998).

This purely instrumental approach does not betray any particular position on internal factors, i.e., on evolvability as defined here. For instance, one can admit that **G** is a measurable quantity that is predictive for evolution, and thus interesting to study, while adopting the neo-Darwinian theory that, because selection is the governing principle and the ultimate source of explanation in biology, any influence of **G** on evolution must be due to the influence of past selection on **G**, rather than reflecting such things as biases in variation or patterns of epistasis induced by developmental-genetic organization. Or, one could suppose that internal effects are important only on short timescales, whereas selection rules over long timescales. This neo-Darwinian position, by the way, is not a straw man. In the contemporary literature of quantitative genetics, serious consideration is given to the possibility that differential effects of mutation and epistasis, though real in the short term, are not important in long-term evolution (Barton and Partridge, 2000; Crow, 2010; Houle et al., 2017; for discussion, see Hansen, 2013).

Yet, the idea that evolvability is really about the intrinsic variability of a biological system has been stated repeatedly in the evolvability literature, going back to Wagner and Altenberg (1996). Within the context of quantitative genetics, if one identifies (or postulates) internal factors that shape **G** and, via this influence, shape evolution, then one is addressing evolvability in the sense intended here and in Box 5.1. In fact, several distinct E2 themes have emerged prominently in the literature of quantitative genetics. One of them involves a focus on **M**, the variance-covariance matrix for new variation introduced by mutation, represented by studies such as Jones et al. (2007), or "Mutation predicts 40 million years of fly wing evolution" (Houle et al., 2017), though the authors of the latter study argue that this correspondence arises due to the way that a history of correlated selection shapes **M**. Another theme focuses on epistasis (e.g., Carter et al., 2005; Le Rouzic and Álvarez-Castro, 2016).

A third theme focuses on pleiotropy (e.g., Hansen, 2003; Wagner and Zhang, 2011).

5.8 Synopsis

The primary purpose of this chapter has been to explore a narrow but important topic at the intersection of mutation, randomness, and evolution: specialized mutation-generating systems. Our aims were to understand these systems genetically and biologically, and then to relate them to the randomness doctrine and to research on evolvability.

Somatic recombination and other cases of so-called "programmed DNA rearrangements" (Borst and Greaves, 1987; Plasterk, 1992; Zufall et al., 2005) exemplify what researchers consider to be specialized mutation systems. Such systems tend to have most or all of the following characteristics: (1) they vastly amplify the rate of a highly specific class of mutations, (2) the amplified mutations are used repeatedly in development or evolution, (3) the mutational capacity is highly modulated in a manner consistent with its repeated use, (4) the mutational capacity depends on dedicated parts, and (5) similar schemes have evolved independently.

We considered six types of mutation-generating systems in microbes that have most of these properties, and which are understood by researchers as specially evolved systems, e.g., mutational systems involved in immune evasion are often described as a "strategy" for immune evasion (Recker et al., 2011; Guizetti and Scherf, 2013; Foley, 2015; Vink et al., 2012; Palmer et al., 2016). The identified systems include diversity-generating retroelements, multiple-inversion systems, CRISPR-Cas and piRNAs, cassette donation systems, phase variation, and mating-type switching.

How do these systems evolve? How do they emerge by mutational innovations? Under what population-genetic conditions are they established and maintained? Solid answers to these questions are not available. With some notable exceptions (Moxon et al., 1994; Metzgar and Wills, 2000; Frank and Barbour, 2006; Palmer et al., 2013; Graves et al., 2013), evolutionary biologists have ignored these systems, so they have not been subjected to the kind of scrutiny that would focus attention on resolving the questions of greatest interest in evolutionary analysis. In this context, we took a step in the direction of greater rigor by formulating a plausible scenario for the origin and maintenance of a system of binary mutational switching, relating this to an existing body of theory on the evolution of phenotypic switching.

With respect to the randomness doctrine, such systems present us with two different conclusions. On the one hand, these specialized mutation systems do not challenge the mechanistic understanding of mutation built up in the previous chapters. We already dismissed the claim that mutation and fitness are independent, though they might be conditionally independent, conditioned on a common environment. Specialized mutation-generating systems exhibit a systematic nonindependence that is, by presumption, an evolved trait. However, these systems typically do not seem to violate conditional independence. To state this differently, yeast can learn (evolutionarily) to switch its mating type, and bacteria can learn (evolutionarily) to put pieces of phage DNA into their genomes, without requiring magic or consciousness, using only ordinary aspects of molecular biology, and typically without relying on feedback loops that probabilize mutations based on their incipient fitness effects.

On the other hand, cases such as CRISPR-Cas spacer acquisition provide "a precise example of the kind of dedicated, nonrandom, beneficial change specifically excluded by generations of evolutionary theorists" in the words of Shapiro (2011, p. 78). An example would be the claim of Mayr (1959a, p. 6–7) (item C.26) that

It would be exceedingly difficult to visualize a mechanism by which the environment could induce directly a structural change in the DNA molecules that would result in the production of a superiorly adapted phenotype and more specifically in the appropriate response to a temporary need. Nor is there any evidence that this occurs. Indeed, there is no need for such an induction within the framework of the synthetic theory of evolution. Infinitely variable natural populations are of such evolutionary plasticity that natural selection can mold them into almost any shape

The CRISPR spacer acquisition pathway contradicts both Mayr's limited vision of mutation processes, and his neo-Darwinian theory of the role of variation (Box 6.1). Natural selection does not mold anti-phage spacers out of the raw materials of infinite natural variability. Instead, the spacers come directly from the phage DNA itself, by mutation.

Reconsidering evolvability research in terms of E1, E2, and E3 claims helps us to understand the overlap in interests and ideas between (on one hand) arguments emanating from microbiologists and molecular biologists and (on the other) the evolvability research front of Nuño de la Rosa (2017). For instance, the classic "introns speed evolution" conjecture (Gilbert, 1978) argues that, via effects of linkage and modularity, split genes are better for evolution (an idea largely subverted by subsequent analyses of Stoltzfus et al., 1994 and Conant and Wagner, 2005). The "selfish operons" hypothesis of Lawrence and Roth (1996) is about the emergence of beneficial modularity through lateral gene transfer. Claims for effects of mutation bias on genome composition or codon usage (see Section 7.5.5) are E2 claims. Caporale (2003) and Shapiro (2011) present a great variety of mostly implicit E3 claims to the effect that mutation systems have evolved to enhance evolution.

This overlap presents a strategic opportunity for research, because molecular and microbial systems are simply better model systems for establishing E2 and E3 claims. Evo-devo studies of mammalian teeth or vertebrate autopods, for instance, are highly constrained by the fact that these systems (1) evolved only once (which greatly limits the opportunity to use retrospective comparative analyses), (2) are entangled with fundamental aspects of biology in a way that makes them difficult to manipulate, and (3) are embedded in large slow-growing organisms that cannot be mass-cultured in evolution experiments. By contrast, microbes present us with evolvability features that are not only of immense importance for human health, but also, they have emerged repeatedly in evolution, they often have specialized or dedicated parts (which means that they can be manipulated without affecting basic biology), and they are embedded in organisms that can be evolved in the laboratory. Indeed, whereas authoritative reviews in the evo-devo literature hesitate to answer the question "does evolvability evolve?" (Payne and Wagner, 2019; Pigliucci, 2008), our main question about specialized mutation-generating systems is not whether they evolved, but how they evolved in so many different ways.

How important are these systems in evolution? Surely the co-evolutionary contest between phages and their prokaryotic hosts is quantitatively the most important arena of evolution on earth, given that phages are a primary cause of mortality for a global population of $\sim 10^{31}$ prokaryotic organisms (Whitman et al., 1998) with short generation times. CRISPR-Cas systems, shufflons, and phase variation all have been implicated in phage–host co-evolution. A mutation system that is involved importantly in the greatest co-evolutionary battle on the planet is *an objectively important mutation system for evolutionary biologists to study*. Conversely, the enormous amount of fuel for adaptation provided by perhaps 10^{31} deaths per hour is presumably the reason that we find such amazing examples of evolvability in this arena. One could make a similar (less quantitatively compelling) argument for importance based on (1) the scale of the battle between jawed vertebrates and their microbial pathogens, and (2) the role of immune evasion mediated by cassette-shuffling, hypermutation, and phase variation.

CHAPTER 6

Randomness as Irrelevance

> Modern zoologists who claim that the Darwinian theory is sufficiently broad to include the idea of the survival [i.e., selection] of definite variations seem inclined to forget that Darwin examined this possibility and rejected it.
> **T.H. Morgan (1904)**

6.1 Introduction

What we know about mutation reveals enormous complexity. Mutations do not emerge by a single pathway, but by a great variety of pathways subject to diverse effects, different in every species. The process of mutation defies simple descriptions. It is not entirely spontaneous, uniform, indeterminate, or unpredictable. If we examine claims that mutation is independent of fitness or environment, we find that these claims are not justified in any rigorous way by biological principles or by systematic data.

All of this raises the question of why the randomness doctrine persists. The question becomes more acute when we consider that, as argued in Chapter 3, science does not require randomness claims. The randomness doctrine is not required for conducting evolution experiments, nor for analyzing data. When we undertake statistical hypothesis-testing, we may use null models that assume uniformity or independence, but this does not require us to commit to any ontological claim of randomness, because (generally) reasoning under conditions of uncertainty does not require us to posit a property of "randomness" in the world.

Why has the randomness doctrine persisted for decades, being repeated in textbook after textbook, and article after article? Why would the literature feature so many different *versions* of the randomness doctrine? There is obviously some reason for this, but the reason is not that the randomness of mutation is a well-founded claim. What has compelled so many evolutionary biologists to make statements combining the concept of mutation with the word "random"?

This chapter presents an extended argument to the effect that randomness claims are often best understood, not as assertions about the nature of mutation, but as assertions of explanatory irrelevance with respect to evolution. del Re (1988) argues generally that "irrelevance is the source of randomness" in science, explaining that "randomness must be assumed whenever there are in (our description of) reality details that are irrelevant for the application of physical laws holding at a given level of reality." Suppose that variation is merely a raw material like the sand used to build a sandcastle. Surely each individual grain of sand is unique, and each has a distinctive history that reflects physical laws, but if our task is to explain the shape of the sandcastle, these are irrelevant details: the shape of the sandcastle is explained by the builder or architect. Darwin expressed this position clearly with his "architect" analogy (see Section 6.6.1).

This interpretation explains several distinctive features of randomness claims, one of which is already firmly established: because irrelevance can have many different justifications, the claim that mutation is random is *polymorphic*, taking on multiple forms. A second distinctive feature is that, in the literature of evolutionary biology, the claim that mutation is random typically travels together with a variety of other claims, e.g., the "raw materials" claim, to the effect that mutation

Mutation, Randomness, and Evolution. Arlin Stoltzfus, Oxford University Press (2021). © Arlin Stoltzfus. DOI: 10.1093/oso/9780198844457.003.0006

is unimportant or inconsequential. In general, the claim that mutation is random is part of a larger nexus of claims that aims to minimize the perceived role of mutation in evolution. A third key feature is that this nexus of claims depends on invoking selection as a *contrast case*.

That is, the "mutation is random" claim is part of a set of interrelated claims, all to the effect that, whereas selection is influential, creative, powerful, and so on, mutation—though admittedly necessary for evolution in the long term—is not influential, creative, powerful, and so on, and thus its involvement in evolution is a nuisance that deserves as little of our attention as possible.

The presentation of this nexus of interrelated claims is complex, because the various threads all tend to run together. Here I divide arguments into five types, some of which come in multiple flavors: (1) arguments from analogy and metaphysics (e.g., mutation as material cause), (2) direct empirical arguments, (3) mechanistic arguments (e.g., the "gene pool" arguments), (4) the methodological argument (the influence of mutation can not be studied productively), and (5) the explanatory argument (mutation does not explain what is important for evolutionary biologists to explain). Various quotations on the theme of unimportance are presented in Appendix D.

6.2 Arguments from analogy and metaphysics

6.2.1 The "raw materials" metaphor

To identify variation or mutation metaphorically as "raw material" for evolution or selection is a common claim in textbooks as well as the research literature and authoritative monographs (e.g., Dobzhansky, 1970, pp. 65, 92, or 200; Gould, 2002). In an early version of this metaphor, Dobzhansky (1955a, p. 131) writes

We should clearly distinguish the two basic evolutionary processes: that of the origin of the raw materials from which evolutionary changes can be constructed, and that of building and perfecting the organic form and function. Evolution can be compared to a factory: any factory needs a supply of raw materials to work with, but when the materials are available they must be transformed into

a finished product by means of some manufacturing process.

What is a *raw material*? What is the relation of a raw material to a finished product? Wood pulp, sand, aluminum ore, and crude oil are examples of raw materials. For a material to be raw means that the material is unprocessed or unrefined. Wool is a raw material; woolen thread and woolen cloth are materials, but not raw materials; and a woolen blanket is a finished product. Paper is a material that is made from wood pulp. Raw materials are used in bulk and they are often "transformed" in the making of a "finished product" (Dobzhansky's language), e.g., in Fig. 6.1, iron ore is being loaded onto a truck made largely of iron, but the two look nothing alike. Thus, "raw material" is not merely a synonym of "input:" not all inputs are raw materials. Whether something is a raw material depends both on its origin and its subsequent use.

Somewhat similar to the "raw materials" metaphor is the simile offered by Stebbins, in which evolution is likened to a car being driven down a highway: mutation is like the gas in the tank, and selection is like the driver (Appendix D.22, or Stebbins, 1966, p. 12). The gasoline provides explosive energy that is converted into rotary motion, but it does not determine the direction or rate of the vehicle.

The architects of the Modern Synthesis promulgated the raw materials metaphor so thoroughly

Figure 6.1 An illustration of raw materials. Iron ore being loaded onto a truck made of steel. Image by Peter Craven reproduced under Creative Commons Attribution (CC-BY-2.0) license.

that it seems inescapable today. In regard to the role of mutation in evolution, the metaphor suggests that (1) variations are manifested in raw form, without being refined or processed in any way, (2) each evolved feature is built from an abundance of variations, and (3) the variations provide substance and not form, even to the extent that the manufactured product may look nothing like the inputs. Some of the same implications emerge from arguments for the creativity of selection, addressed in the next section. That is, the "fuel" and "raw materials" tropes used by Stebbins and Dobzhansky were chosen to represent a neo-Darwinian theory (Box 6.1) in which the role of variation is merely to supply substance, not shape or direction, which is supplied by selection.

6.2.2 Creativity

The argument for the creativity of selection is so intertwined with the concept of variation as raw materials that the distinction between the two is somewhat arbitrary: they are two sides of the same coin. As Darwin explains,

If selection consisted merely in separating some very distinct variety, and breeding from it, the principle would be so obvious as hardly to be worth notice; but its importance consists in the great effect produced by the accumulation in one direction, during successive generations, of differences absolutely inappreciable by an uneducated eye (Darwin, 1872, Ch. 1).

Or, as stated in the critique by Løvtrup (1987, p. 166–167),

The reason why Darwin rejected beneficial macromutations was that he had greater ambitions on behalf of his mechanism of natural selection; it was not enough that it ensures the survival of variations created by 'internal forces' residing within the various organisms, it must itself *create* the evolutionary innovations. Under these circumstances the variations themselves must be slight and insignificant as regards their effect; only through the accumulation by natural selection of such micromutations are large-scale changes possible in evolution.

In their seminal analysis of the creativity claim, Razeto-Barry and Frick (2011) present a table of authors who defend either the creative view of selection, or the noncreative view. On one side

they find Darwin, the architects of the Modern Synthesis, and various contemporary authorities. On the other side, they find early geneticists (de Vries, Morgan, Punnett, and Hogben), various philosophers (who typically misinterpret Darwin's view as a mutationist theory of the preferential survival of beneficial mutations), and contemporary reformers (e.g., Saunders, Fontana, Gilbert, Arthur, Müller, Reid and Newman). That is, the creativity argument and the neo-Darwinian theory co-occur.

Indeed, the architects of the Modern Synthesis all compared natural selection to an artistic creator. Reviewing these arguments, Gould (1977, p. 44) writes

But why was natural selection compared to a composer by Dobzhansky; to a poet by Simpson; to a sculptor by Mayr; and to, of all people, Mr. Shakespeare by Julian Huxley? I won't defend the choice of metaphors, but I will uphold the intent, namely, to illustrate the essence of Darwinism—the creativity of natural selection. Natural selection has a place in all anti-Darwinian theories that I know. It is cast in a negative role as an executioner, a headsman for the unfit (while the fit arise by such non-Darwinian mechanisms as the inheritance of acquired characters or direct induction of favorable variation by the environment). The essence of Darwinism lies in its claim that natural selection creates the fit. Variation is ubiquitous and random in direction. It supplies the raw material only. Natural selection directs the course of evolutionary change. It preserves favorable variants and builds fitness gradually. In fact, since artists fashion their creations from the raw material of notes, words, and stone, the metaphors do not strike me as inappropriate.

Thus, Gould defends the creativity of selection, and the comparison with a creative artist, ultimately invoking the neo-Darwinian dichotomy (see Box 6.1) of selection as the governing force, with variation in a supporting role as the source of ubiquitous random "raw materials." The alternative conception of selection, depicted by Gould as a medieval executioner, is rejected. The architects of the Modern Synthesis also rejected the Mendelian metaphor of selection as a "sieve," e.g., Dobzhansky (1974, p. 319) condemns this as "misleading," "exaggeration," and "oversimplification." That is, for the architects of the Modern Synthesis, anthropomorphizing natural selection as a human artist—a writer, painter, sculptor, or composer who creates complex products out of the raw material supplied by words, pigments, stones, or notes—is helpful to convey an understanding of how the process of evolution works, whereas comparing selection to a mechanical sorting device is unrealistic and unhelpful.

This kind of creativity argument continues to appear, particularly when reformers challenge neo-Darwinian orthodoxy, e.g., in the review of Kirschner and Gerhart (2005) by Charlesworth (2005) (see 6.4.1). Likewise, in response to work drawing attention to the alleged role of developmental constraints in evolution, Leigh (1987) writes that "The reason why I am disinclined to describe developmental constraints as directing evolution is the reason why I would not describe an artist's material as directing his handiwork" (p. 229). In a review of *Mutation-Driven Evolution* by Masatoshi Nei (2013), a book that directly challenges the primacy of selective explanations over mutational ones, Wright (2014) expresses the neo-Darwinian position, arguing that

Nei's viewpoint is essentially a philosophical position, which puts the source of the raw materials (mutation) as primary over the agent that drives the evolution of a highly functioning organism (selection). But by analogy, if we were to ask why houses are found on a street, the availability of bricks cannot but be a secondary explanation compared with the need for shelter.

This argument is not purely metaphorical, but draws on the more formal philosophical concept of a "material cause," distinguished from other causes in Aristotle's 4-fold scheme of causation (see Box 4.2). Material causes provide substance only, not form, initiative, or direction. Materials are inert: some *agent* must take the materials in hand, and form them into products, and this agent takes all the credit for his creative work. When Darwin says (see Beatty, 2010) that, in admiring a well-contrived building, "one speaks of the architect alone and not of the brick-maker," this is a rather direct denial of the importance of material causes in determining the shape of the final product. Wright's objection above literally contrasts "materials" with an "agent."

Note that Wright (2014) makes opposition to neo-Darwinism appear perverse by stating it in neo-Darwinian terms, a familiar rhetorical tactic, e.g., as when Poulton (1908) refers to the "revolt of the clay against the power of the potter." To assert that the raw materials rule over the agent, that the clay dominates the potter, or that the bricks are primary over the architect, would be perverse. However, this is not a fair or helpful description. Instead, opponents of neo-Darwinism reject the theory that the role of variation is correctly described in terms of clay or bricks, with selection in the role of the artist or craftsman. For instance, the criticism offered by Nei (2013) is partly (1) a rejection of the shifting-gene-frequencies theory of evolution, with a contrary emphasis on mutation-driven dynamics (the "sushi conveyor" dynamics of Section 8.7), and partly (2) an explanatory position akin to Bateson's explanatory mutationism, in which the reality of selection is accepted, but the ability to demonstrate its role convincingly (e.g., via experiments and retrospective analyses) is stubbornly disputed. Nei believes that historical explanations implicating the role of specific mutations with distinctive effects provide more explanatory power than generalized population-genetic laws (see Stoltzfus, 2014).

6.2.3 Levels and types of causes

Several arguments from the architects of the Modern Synthesis depict mutation as a lesser *kind* of cause than other causes such as selection. We have already seen one argument of this type—the argument that mutation is a material cause, whereas selection is an agent or sometimes a final (ultimate) cause. In this section, we briefly consider a few other arguments that distinguish mutation as a different and less important kind of cause.

Perhaps the most familiar of these arguments is that evolution results from processes that take place on different "levels," with selection acting at a higher "population" level that is alleged to be determinative, e.g., Dobzhansky writes that "It is not on the level of mutation, but rather on that of natural selection and other genetic processes that take place in living populations, that the creativity of evolution manifests itself" (Dobzhansky, 1974, p. 330), and also that "The process of mutation supplies the raw materials of evolution, but the tempo of evolution is determined at the populational level, by natural selection in conjunction with the ecology and the reproductive biology of the group of organisms" (Dobzhansky, 1955b, p. 282).

This relates to a trope common in didactic resources (e.g., Colby, 1996) to the effect that "Genes mutate, individuals are selected, populations evolve." This formula is sometimes used to explain why mutation and development are not evolutionary causes, e.g., in rebutting claims of novelty and significance for evo-devo, Maynard Smith (1983, p. 45) writes

If we are to understand evolution, we must remember that it is a process which occurs in populations, not in individuals. Individual animals may dig, swim, climb or gallop, and they also develop, but they do not evolve. To attempt an explanation of evolution in terms of the development of individuals is to commit precisely that error of misplaced reductionism of which geneticists are sometimes accused.

For an equally patronizing dismissal of evo-devo, see Wallace (1986). Eldredge (2001) rejects the importance of early conflicts between geneticists and followers of Darwin, explaining that inheritance and selection are now known to "operate in different domains" (p. 67).

Dobzhansky et al. (1977, p. 6) argue that

Each unitary random variation is therefore of little consequence, and may be compared to random movements of molecules within a gas or liquid. Directional movements of air or water can be produced only by forces that act at a much broader level than the movements of individual molecules, e.g., differences in air pressure, which produce wind, or differences in slope, which produce stream currents. In an analogous fashion, the directional force of evolution, natural selection, acts on the basis of conditions existing at the broad level of the environment as it affects populations.

Here Dobzhansky et al. (1977) are making an analogy with statistical physics, in which the motions of huge numbers of anonymous particles are described by high-level laws, e.g., the ideal gas law, $PV = nRT$ (pressure times volume equals the product of amount in moles, temperature, and a constant).

A closely related argument refers to the "forces" theory described in more detail below (Section

6.4.2). Mutation may be conceived both as a mass-action force ("pressure") that shifts the frequencies of alleles already present, or as a process that introduces new alleles at rare intervals. The former conception is what is meant by a force, e.g., the role of mutation in the mutation-selection balance equation is to act as a continuous force, with a generic ability to change a frequency f to $f + \delta$, where δ is an infinitesimal amount. An example of this distinction in a contemporary textbook would be the statement that

In summary, mutation creates the genetic diversity that is the raw material for evolution. In principle, mutation can also produce changes in allele frequencies across generations. In practice, mutation's role as the source of genetic variation is usually more important than its role as a mechanism of evolution (Freeman and Herron, 1998, p. 145).

Here "mechanism of evolution" means "force" capable of causing fixation.

In one sense, this argument seems to elevate mutation and selection to the same level, contrary to what Dobzhansky et al. argue. However, the arguments are not really contradictory when we divide them according to the dual role of mutation. As a process that generates rare variant individuals, mutation is denied importance on the grounds that it acts at the wrong level, as explained by Dobzhansky et al. (1977) above; considered as a mass-action force, mutation acts at the right level, but is a weak pressure unable to overcome the opposing force of selection (see 6.4.2).

Finally, we have the familiar distinction of "proximate" and "ultimate" causes by Mayr (1961), where the latter are "causes that have a history and that have been incorporated into the system through many thousands of generations of natural selection." According to Ayala (1970), "the ultimate source of explanation in biology is the principle of natural selection." This distinction has long been used to dismiss suggestions of a role for development in evolutionary causation, as pointed out by West-Eberhard (2003) among others. The claim of Mayr (1994) that proximate and ultimate are "hopelessly mixed up" in the evo-devo literature indicates that, for him, developmental biases in the production of variation are strictly a matter of proximate causation, not part of evolutionary causation. In Chapter 8, the precise reasons for rejecting such arguments will become clear. However, our purpose here is simply to document these arguments, rather than rebut them.

6.3 Direct empirical arguments

Are there any purely empirical arguments for the unimportance of mutation or variation? That is, are there cases in which an attempt has been made to (1) measure or otherwise assess the character of mutational changes, (2) measure or otherwise assess the character of evolutionary changes, and then (3) conclude from a comparison of the two that the character of mutation is not important in determining the character of evolutionary change?

One early form of such an argument, from the nineteenth century, was in regard to discrete variants that arise suddenly, which Darwin called "sports." The question of whether sports look like the kinds of changes that happen in evolution was answered negatively by Darwin, who dismissed sports as "monstrosities" and reported that he "failed to find, after diligent search, cases of monstrosities resembling normal structures in nearly allied forms, and these alone bear on the question" (Darwin, 1872, Ch. 2). Here "allied forms" means something like "closely related species or subspecies," i.e., Darwin examines close differences and asks if they ever look like the effect of sports, an idea he rejects. Meanwhile, other scientists drew the opposite conclusion from exactly the same kinds of evidence (e.g., Bateson, 1909b, p. 286).

A second comparison that one might make is between the distribution of effect-sizes for mutations, and that of evolutionary differences between species. When species are closely related enough to be cross-bred, it becomes possible to use the techniques of quantitative genetics to identify implicated loci and the sizes of their effects. The result of such an analysis then may be compared with knowledge of the effect-sizes of mutations. In fact, genetic dissection of species differences was used by Dobzhansky and others to argue that these differences are overwhelmingly made up of small effects, though the retrospective analysis by Orr and Coyne (1992) indicates that they overstated their case.

Thus, direct kinds of evidence have been used to argue for the unimportance of mutation. However, when facts are used to argue for the unimportance of mutation, they tend to be embedded in a theory-based argument about mechanisms. Such arguments are covered in the next section.

Box 6.1 The essence of neo-Darwinism is a dichotomy of roles

If adaptation is detailed and pervasive, that is, if we accept the premise of empirical adaptationism (Box 3.1), then (1) a generalized process of adaptation must exist, allowing detailed adjustment to conditions, and (2) the disruption of detailed adaptedness by nonadaptive processes must be prevented somehow.

In the *Origin of Species*, Darwin responds to the challenge of empirical adaptationism with a theory of "natural selection" combining the struggle for life with blending inheritance and indefinite variability (fluctuation)—the "endless slight peculiarities which distinguish the individuals of the same species, and which cannot be accounted for by inheritance from either parent or from some more remote ancestor" (Box 4.1). For every organismal feature such as body shape, individuals show endless slight peculiarities stimulated by the effect of "altered conditions" on the "sexual organs." The struggle for life subtly shifts the distribution of these fluctuations, which are blended together during reproduction. In the offspring, we see a new distribution of endless slight peculiarities superimposed on the previously blended distribution, continuing the process of adapting to conditions. Over long periods of time, large changes may accumulate. Adaptedness is maintained because the struggle for life counteracts the dispersive force of heredity.

This theory depicts a mode of change that is inherently gradualistic: *it cannot possibly produce discontinuities*, which is why Darwin (1872, Ch. 6) writes that *natura non facit saltum* "must on this theory be strictly true." Darwin knew that sudden nonblending variations occur, but held them to be monstrous and unworkable.

Likewise, Fisher (1930b) responded to the challenge of empirical adaptationism by invoking continuously variable traits, the multiple-factor theory, and phenotypic mass selection, concluding that the resulting mode of gradual adaptive modification is inevitable in a Mendelian world. He used a geometric model (see Section 7.5.2) to argue that the smallest changes are the most likely. These ideas contributed to the development of the shifting-gene-frequencies theory of the Modern Synthesis (see Provine, 1971; Box 6.2).

The relationship between these two theories, which generate the same high-level behavior by different means, is like the relationship of two versions of a software application that maintain the same user interface for two different computer operating systems. Darwin designed his theory for a non-Mendelian operating system that does not exist, as shown by the famous experiments of Johannsen (see Roll-Hansen, 1989; Gayon, 1998).

What, precisely, is the high-level behavior shared between Darwin's non-Mendelian theory, and the later Mendelian version? Sewall Wright (1967) says that "the idea that evolution comes about from the interaction of a stochastic and a directed process was the essence of Darwin's theory" (p. 117). Gould (1977) reviews how the architects of the Modern Synthesis compared natural selection to an artistic creator (see Section 6.2.2) and concludes that

> The essence of Darwinism lies in its claim that natural selection creates the fit. Variation is ubiquitous and random in direction. It supplies the raw material only. Natural selection directs the course of evolutionary change. (p. 44)

Indeed, Darwin's theory was offered originally as an alternative to divine creation, thus it was imperative for "natural selection" to be creative, not merely a sieve. The theory that natural selection is creative, whereas variation merely supplies raw material, applies both to Darwin's "natural selection" and to the view promoted by the architects of the Modern Synthesis (see Sections 6.2.2 and 6.4.1).

In this book, we use "neo-Darwinism" as the label for a high-level theory defined, not by a mechanism, but by a dichotomy of roles. This dichotomy may be expressed in various ways, e.g., by the metaphor that selection is the potter and variation is the clay, or by the more technical claim that selection alone provides shape and direction in evolution, whereas variation is merely a source of random raw materials.

The power of this theory resides in its ability to generate an account of *any evolutionary outcome that is plausibly adaptive*, by invoking a history of abundant infinitesimal hereditary variation together with the struggle for life. Such an account can be formulated without knowing the internal details of genetics, development, and so on, e.g., we do not need a detailed theory of the timing, character, and size distribution of variations, only the assumption that every trait exhibits abundant infinitesimal variation. That is, under the theory, the internal details do not matter. Thus, neo-Darwinism was a particularly useful theory when biologists understood little of the internal details of life, and had not yet begun using powerful algorithms to reconstruct detailed historical sequences of changes. Considered as a substantive theory subject to refutation, neo-Darwinism is valid to the extent this dichotomy of roles is justified, i.e., the theory is true to the extent that evolutionary change actually results from an unequal marriage in which selection makes all the important decisions, and deserves all the credit.

Note that, in the evolutionary literature, "neo-Darwinism" is also used as a "a sociological label one applies to oneself in order to be considered acceptably mainstream, regardless of the content of one's work" (Brooks, 2011). That is, the same label that we will use for a theory is also commonly used for a cultural brand whose associations are adjusted periodically, so as to maintain brand value in the face of scientific developments.

What is important to bear in mind is not the label, but how the neo-Darwinian dichotomy of roles has defined major axes of scientific disputes. Today, scientists tend to see a conflict between neutrality and selection as paramount, but prior to the molecular revolution, neutralism was *never* the primary antithesis of neo-Darwinism. "Non-Darwinian evolutionists did not deny the reality, or the operationality, of natural selection as a genuine cause" (Gould, 2002, p. 139). Instead, scientific opposition to Darwin's thinking, from Mivart (1871) to the Mendelians (Box 7.1), to evo-devo, typically focused on objections to *natura non facit saltum*, and to the idea that variation merely supplies raw material, with no dispositional role. Defenders of Darwin's thinking argue that selection is creative, an argument that critics reject (see Section 6.2.2; Razeto-Barry and Frick, 2011). Historically, to reject neo-Darwinism is to deny that selection is the sole source of creativity, initiative, discontinuity, or direction in evolution, and instead, to bring attention to internal factors. Thus, a century ago, Poulton (1908) characterized non-Darwinian views of the role of variation as "the revolt of the clay against the power of the potter."

6.4 Mechanistic arguments

6.4.1 Creativity arguments

The claim of creativity introduced previously often takes the form of a metaphor or a simile, to the effect that selection is, or is like, an artist or craftsman. The corresponding mechanistic argument is that specific mechanistic features of the evolutionary process ensure that selection is creative in the manner of an artist or craftsman. Gould (1977) identified the creativity of selection as the "essence" of neo-Darwinism in his analysis of the creativity metaphor; later, Gould (2002, p. 140) presents a version that is not merely an analogy, but refers more specifically to properties of variation in a mechanistic context:

As the epitome of his own solution [to the problem of creativity], Darwin admitted that his favored mechanism 'made' nothing, but held that natural selection must be deemed 'creative' (in any acceptable vernacular sense of the term) if its focal action of differential preservation and death could be construed as the primary cause for imparting direction to the process of evolutionary change. Darwin reasoned that natural selection can only play such a role if evolution obeys two crucial conditions: (1) if nothing about the provision of raw materials—that is, the sources of variation—imparts direction to evolutionary change; and (2) if change occurs by a long and insensible series of intermediary steps, each superintended by natural selection—so that "creativity" or "direction" can arise by the summation of increments. Under these provisos, variation becomes raw material only—an isotropic sphere of potential about the modal form of a species. Natural selection, by superintending the differential preservation of a biased region from this sphere in each generation, and by summing up (over countless repetitions) the tiny changes thus produced in each episode, can manufacture substantial, directional change. What else but natural selection could be called 'creative,' or direction-giving, in such a process? As long as variation only supplies raw material; as long as change accretes in an insensibly gradual manner; and as long as the reproductive advantages of certain individuals provide the statistical source of change; then natural selection must be construed as the directional cause of evolutionary modification.

These conditions are stringent; and they cannot be construed as vague, unconstraining, or too far in the distance to matter. In fact, I would argue that the single most brilliant (and daring) stroke in Darwin's entire theory lay in his willingness to assert a set of precise and stringent requirements for variation—all in complete ignorance of the mechanics of heredity. Darwin understood that if any of these claims failed, natural selection could not be a creative force, and the theory of natural selection would collapse.

In Gould's argument, the creativity of selection is only guaranteed to the extent that certain "requirements for variation" are met, namely that variation is "copious," "undirected" and "small in extent" meaning small in size of effect (Gould, 2002, p. 141). Note that Gould pushes things a bit too far, resulting in an error in his description of Darwinian logic: the requirement is not that variation is perfectly uniform ("isotropic sphere of potential"), but that any tendencies are inconsequential ("nothing about the provision of raw materials . . . imparts direction").

Whereas some arguments identify creativity with imparting direction to evolution, a second mechanistic argument treats creativity as a matter of making improbable things happen. Following an argument made earlier by Fisher (1930b, p. 96) and Haldane (1932, p. 94 to 95), Stebbins argues that "Natural selection has the positive, creative function of sorting out a few adaptive gene combinations from the infinite number of possibilities inherent in the gene pool" (Stebbins, 1966, p. 12). The ability of selection to "produce new combinations of characters" is emphasized by Charlesworth (2005) in his response to the critique of neo-Darwinism by Kirschner and Gerhart (2005). The idea of choosing from an infinitude of possibilities appears in other sources, e.g., the quotation from Mayr (1959a) (item D.27), or in comments of Darwin comparing selection to a creative builder working from an "infinitude of shapes" of stones (see 6.6.1 or Beatty, 2014, p. 177).

6.4.2 Directionality: the "opposing forces" argument

The argument that mutation is a "weak force" unable to overcome selection or influence the direction of evolution emerged from Fisher (1930a, b) and Haldane (1932, 1933) (see also Wright, 1931). Fisher and Haldane showed that, starting with a population of $\underline{A1}$ individuals, an inequality of forward and backward mutation rates favoring allele A2 will not result in fixation of A2 if this is opposed by selection. Instead, given that mutation rates are small in comparison to selection coefficients, a deleterious allele will persist at a low level reflecting a balance between the opposing pressures of mutation and selection. In the simplest haploid case, the "mutation-selection balance" is approximated by

$$f \approx \mu/s \qquad (6.1)$$

where μ is the forward rate of mutation to A2 (the backward rate hardly matters) and s is the selective disadvantage of A2. This kind of equation makes it easy to depict mutation and selection as opposing forces, with selection pushing *down* the frequency of A2 with magnitude s, and mutation pushing the same frequency *up* with magnitude μ.

Given that selection coefficients on the order of 10^{-2} or 10^{-3} were thought to be common, whereas mutation rates were much smaller, a broad conclusion seemed warranted:

For mutations to dominate the trend of evolution it is thus necessary to postulate mutation rates immensely greater than those which are known to occur, and of an order of magnitude which, in general, would be incompatible with particulate inheritance . . . The whole group of theories which ascribe to hypothetical physiological mechanisms, controlling the occurrence of mutations, a power of directing the course of evolution, must be set aside (Fisher, 1930b, Ch. 1).

Appendix D gives this (item D.36) as well as two related quotations from Haldane (item D.35, D.34).

That is, Fisher and Haldane confronted the classic idea of internal orienting factors in evolution, and argued that the idea could be dismissed, without ever studying the fossil record, and without conducting any experiments.

The simplicity and scope of this "opposing pressures" argument is breathtaking: theoretical population genetics tells us that, because mutation rates are small, internal orienting factors acting via mutation are ruled out, *completely*. Provine (1978) identifies this as one of the vital contributions of population genetics theory to the success of the Modern Synthesis, on the grounds that it eliminated the rival theory of Mendelian-mutationism. Indeed, the architects of the Modern Synthesis repeatedly denied mutational causes based on this argument, e.g., in addition to Fisher and Haldane, see relevant quotations from Huxley (item D.33), Simpson (item D.31), Mayr (items D.25, D.27), and Ford (item D.17). Many early authors such as Mivart (1871) or Eimer (1898) suggested that internal biases in variation could impart direction to evolution, thus the claim that there is no such mechanism, e.g., Simpson (item D.19), seems to indicate the influence of the "opposing pressures" argument.

Contemporary sources also cite Fisher's argument directly (e.g., Gould, item D.3), or draw on the argument that mutation is a "weak force" (see Freeman and Herron, 1998, Section 5.3; Sober, 1987, pp. 105–118; Maynard Smith et al., 1985). This argument even appears unexpectedly in the literature of molecular evolution (e.g., Li et al., 1985, p. 54). A white paper on "Evolution, Science and Society" endorsed by various professional organizations (Futuyma et al., 1988) describes the accomplishments of mathematical population genetics, claiming:

For instance, it is possible to say confidently that natural selection exerts so much stronger a force than mutation on many phenotypic characters that the direction and rate of evolution is ordinarily driven by selection even though mutation is ultimately necessary for any evolution to occur.

The influence of this argument may extend far beyond what is easily detectable. Here, the context (contributions of *mathematical theory*) makes it clear that the authors' confidence arises from the opposing pressures argument. In many other cases, confident claims about causes of direction are offered without providing a theoretical or empirical justification. For instance, Mayr (1980) sees neo-Darwinism as "the theory that selection is the only direction-giving factor in evolution" (item D.12). In the topic article on "natural selection" in the Oxford Encyclopedia of Evolutionary Biology, Ridley (2002, p. 799 to 800) states that:

In evolution by natural selection, the processes generating variation are not the same as the process that directs evolutionary change . . . What matters is that the mutations are undirected relative to the direction of evolution. Evolution is directed by the selective process . . .

Natural selection differs from most alternative theories of evolution in the independence between the processes that direct variation and that direct evolution . . . Darwin's theory is peculiar in that evolution is not an extension of the mutational process.

That is, Ridley imagines the possibility of "processes that direct variation" (analogous to Darwin's laws of variation) but does not assign such processes any leverage in determining the outcome of evolution. To imagine that the output of a process is unaffected by the qualities of its inputs seems odd. Yet this is precisely what is suggested, and the reasons may have much to do with the influence of the opposing forces argument.

6.4.3 Initiative and rate: the "gene pool" arguments

In this section we address arguments that invoke a "gene pool" to explain why mutation is not a source of initiative, dynamics, or direction in evolution. One might imagine that in a world of discrete inheritance, the introduction of a new mutation would be a key event that provides initiative for evolutionary change, and thus influences dynamics as well as direction. Following Shull (1936, p. 122), we might suppose that

If mutations are the material of evolution, as geneticists are convinced they are, it is obvious that evolution may be directed in two general ways: (1) by the occurrence of mutations of certain types, not of others, and (2) by the differential survival of these mutations or their differential spread through the population

and we might suppose that a new allele "produced twice by mutation has twice as good a chance to survive as if produced only once" (p. 140).

However, the architects of the Modern Synthesis invoked the "gene pool" theory to reject such possibilities (see Box 6.2). In this context, the term "gene pool" is not merely a synonym for population variation, but evokes the theory that (1) populations in a state of nature maintain abundant genetic variation, e.g., Stebbins writes that "a large 'gene pool' of genetically controlled variation exists at all times" (Stebbins, 1966, p. 12), and (2) this represents a dynamic buffer that shields evolution from effects of mutation. This theory, proposed by

Chetverikov in the 1920s (see Chetverikov, 1997) and popularized by Dobzhansky, holds that various features of genetics and population genetics—including recessivity, chromosome assortment, crossing over, sexual mixis, frequency-dependent selection, and heterosis—come together to create a dynamic in which variation is soaked up like a "sponge" and "maintained." This "gene pool" dynamic ensures that evolution is always a multi-factorial process in which selection never waits for a new mutation, but can shift the population to a completely new state based on readily available variation.

Thus, the "gene pool" theory could be used to argue against the early Mendelians in regard to (1) the source of initiative (i.e., an argument to the effect that episodes of evolutionary change do not begin with a new mutation), and (2) rates of evolution, which do not depend on mutation rates. A variety of "gene pool" claims are quoted in Appendix D, e.g., (Stebbins, 1959, p. 305) writes

Second, mutation neither directs evolution, as the early mutationists believed, nor even serves as the immediate source of variability on which selection may act. It is, rather, a reserve or potential source of variability which serves to replenish the gene pool as it becomes depleted through the action of selection . . . The factual evidence in support of these postulates, drawn from a wide variety of animals and plants, is now so extensive and firmly based upon observation and experiment that we who are familiar with it cannot imagine the appearance of new facts which will either overthrow any of them or seriously limit their validity.

In this speculative theory—never so strongly evidenced as Stebbins imagines—an episode of evolution begins when the environment changes, bringing on selection of variation in the gene pool, so that "mutation merely supplies the gene pool with genetic variation; it is selection that induces evolutionary change" (Mayr, 1963, p. 613). As Mayr explains at greater length,

In the early days of genetics it was believed that evolutionary trends are directed by mutation, or, as Dobzhansky (1959) recently phrased this view, "that evolution is due to occasional lucky mutants which happen to be useful rather than harmful." In contrast, it is held by contemporary geneticists that mutation pressure as such is of small immediate evolutionary consequence in sexual organisms, in view of the relatively far greater contribution of recombination and gene flow to the production of new genotypes and of the overwhelming role of selection in determining the change in the genetic composition of populations from generation to generation (Mayr, 1963, p. 101).

A second argument follows from this. If evolutionary change is not initiated by lucky events of mutation, then it seems to follow that the rate of evolution is not dependent on the rate of mutation. The architects of the Modern Synthesis repeatedly expressed this idea, e.g., Stebbins (1966, p. 29; my emphasis) writes that:

mutations are rarely if ever the direct source of variation upon which evolutionary change is based. Instead, they replenish the supply of variability in the gene pool which is constantly being reduced by selective elimination of unfavorable variants. Because in any one generation the amount of variation contributed to a population by mutation is tiny compared to that brought about by recombination of pre-existing genetic differences, even a doubling or trebling of the mutation rate will have very little effect upon the amount of genetic variability available to the action of natural selection. *Consequently, we should not expect to find any relationship between rate of mutation and rate of evolution. There is no evidence that such a relationship exists.*

Likewise, Dobzhansky et al. (1977), in their popular textbook, write:

The large number of variants arising in each generation by mutation represents only a small fraction of the total amount of genetic variability present in natural populations. . . . It follows that rates of evolution are not likely to be closely correlated with rates of mutation . . . Even if mutation rates would increase by a factor of 10, newly induced mutations would represent only a very small fraction of the variation present at any one time in populations of outcrossing, sexually reproducing organisms.

Again, multiple versions of this claim appear in Appendix D. No molecular evolutionist today would make this claim, because comparative data indicate clearly that the rate of evolution depends sensitively on the mutation rate. Nevertheless, the irrelevance of mutation rates follows from the shifting-gene-frequencies theory of the Modern Synthesis (see Box 6.2). Textbooks continue to repeat the idea that, because variation is typically abundant, "the rate of evolutionary change is not limited by the occurrence of new favorable mutations" (Stearns and Hoekstra, 2005, p. 99). Dawkins (2007), in his review of *The Edge of Evolution*, makes this the centerpiece of his critique:

If correct, Behe's calculations would at a stroke confound generations of mathematical geneticists, who have repeatedly shown that evolutionary rates are not limited by mutation. Single-handedly, Behe is taking on Ronald Fisher, Sewall Wright, J.B.S. Haldane, Theodosius Dobzhansky, Richard Lewontin, John Maynard Smith and hundreds of their talented co-workers and intellectual descendants. Notwithstanding the inconvenient existence of dogs, cabbages and pouter pigeons, the entire corpus of mathematical genetics, from 1930 to today, is flat wrong. Michael Behe, the disowned biochemist of Lehigh University, is the only one who has done his sums right. You think? The best way to find out is for Behe to submit a mathematical paper to *The Journal of Theoretical Biology*, say, or *The American Naturalist*, whose editors would send it to qualified referees.

Dawkins's views here serve as a useful time capsule, reflecting an understanding of the state of theoretical evolutionary genetics from about 1968 (see Box 8.2).

6.5 The methodological argument

As outlined in Box 3.1, Godfrey-Smith (2001) identifies three kinds of adaptationism: empirical, methodological, and explanatory. Likewise, we can articulate three flavors of the mutation-is-irrelevant claim. The empirical claim is that we know that mutation is irrelevant because we know that selection overwhelmingly determines what happens in evolution (e.g., initiative, creativity, rate, direction), and mutation does not. This seems to be the position of the architects of the Modern Synthesis, reflected in a variety of arguments given earlier to the effect that selection is important but mutation is not.

The idea of randomness as a null hypothesis discussed in Section 3.6, exemplified by Wagner (2012), is an example of a methodological position. Rather than treating mutation as irrelevant, Wagner (2012) describes a strategy that creates opportunities to reject randomness, and then incorporates the nonrandomness of mutation into ever-more-complex models.

A methodological position consistent with randomness as irrelevance would be that mutation, though perhaps consequential in some ways, is not a fruitful object of research, because mutations are unpredictable, or because they are difficult to control and thus not amenable to experimental manipulation, or for some other reason. This position is explicit in Mayr's statement quoted in Box 3.1. Mutation might be influential in evolution, but Mayr believes that this can never be discovered by the methods of science, because mutation is a "chance" factor. In Mayr's preferred approach, the evolutionist "must first attempt to explain biological phenomena and processes as the product of natural selection. Only after all attempts to do so have failed is he justified in designating the unexplained residue tentatively as a product of chance" (Mayr, 1983, p. 151).

Box 6.2 The shifting-gene-frequencies theory of evolution

As Orr (2005c) writes, "it's hard to think of evolution in the same way once you know that it has to obey the laws of Mendel's genetics." Indeed, the discovery of Mendelian genetics in 1900 undermined Darwin's three main theories of modification—"natural selection" by fluctuation-struggle-blending, effects of use and disuse (Lamarckism), and direct effects of the environment (Box 4.1). The early geneticists began to build a new understanding of evolution (Box 7.1), but offered no ruling principles. Their genetical framework had little power to generate predictions or meaningful explanations without information on mutation and inheritance inaccessible to paleontologists and field naturalists, who therefore had little to gain from integrating Mendelism into their evolutionary thinking.

Yet by 1959, Dobzhansky, Mayr, Stebbins, Simpson, and others were promoting a Mendelian justification of neo-Darwinism that excluded Lamarckism, orthogenesis, and mutationism. This theory carried all the power of neo-Darwinism (Box 6.1) to construct hypothetical explanations for any adaptation without regard to internal details—at the time, a theory to support evolutionary reasoning without knowledge of internal details had enormous appeal.

The touchstone of the theory was Castle's demonstration of what Provine (1971) calls the "effectiveness" or "efficacy" of selection. Castle and his colleagues bred nearly all-black or nearly all-white strains of rats from the same starting population of black-and-white rats, using just 20 generations of selection, which is not enough time for new mutations to play an appreciable role. Successively more extreme genotypes emerged, not by new mutations, but by recombination bringing together alleles from diverse loci. That is, the "effectiveness of selection" is the ability of selection to *create new types without mutation* ("Castle had been able to produce new types by selection," p. 114), and recombination is crucial to this effect.

This provided a basis to redefine "evolution" mechanistically, as a process of shifting the frequencies of pre-existing small-effect alleles (at many loci) from the previous optimal distribution to a new optimal distribution, in response to a change in conditions. Mutations, though ultimately necessary, are not directly involved, but merely supply the gene pool with variation: recombination is the proximate source of variation from which selection shapes adaptations. All of evolution follows from this process of microevolution (shifting gene frequencies) observed experimentally in animals and plants.

For instance, Stebbins appeals to this theory during a panel discussion at the 1959 Darwin centennial, when the new theory was being celebrated as an established orthodoxy:

> One very important point here is this: if we say that genetic recombination is necessary to generate new adaptive systems and then say that such highly adaptive and complex systems as the cell of an amoeba, or a euglena with its nucleus, chloroplasts, eyespots, flagella, etc., evolved without the aid of genetic recombination, we are contradicting ourselves. Even though we do not know that genetic recombination exists in these one-celled organisms, we must postulate its existence at the time they evolved. (Huxley et al., 1960, pp. 115 to 116).

Each part of this cohesive theory plays a role in justifying neo-Darwinism and excluding rival ideas. The notion that recombination, not mutation, is always the proximate source of variation precludes the mutationist idea that events of mutation initiate evolution: abundant pre-existing variation is always maintained in the "gene pool," thus the true source

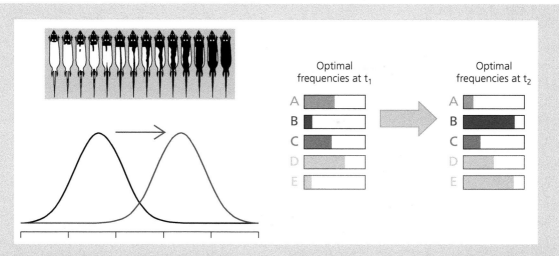

Figure 6.2 In the shifting-gene-frequencies theory, smooth change at the phenotypic level (left) corresponds to shifting gene frequencies at the genetic level (right), driven by selection.

of initiative is a change in conditions that brings on selection. Change must take place via small allelic effects, and it must always involve multiple loci simultaneously, otherwise it merely represents selection in the noncreative role of favoring one discrete type over another (as in Johannsen's experiments). As Stebbins (1966, p. 31) explains (see also Dobzhansky et al., 1977, p. 6, Mayr, 1963, p. 613, Simpson, 1964, p. 1536):

> Natural selection directs evolution not by accepting or rejecting mutations as they occur, but by sorting new adaptive combinations out of a gene pool of variability which has been built up through the combined action of mutation, gene recombination, and selection over many generations.

Thus, the shifting-gene-frequencies theory of evolution implicates a particular regime of population genetics, together with a set of precepts about explanations and causes, including both positive and negative claims. For instance, if the theory is correct, natural selection directs evolution and mutation cannot be a cause of direction, because the pressure of mutation is too weak to overcome selection. The shifting-gene-frequencies theory is correct to the extent that evolution works as Castle's experiment suggests—smooth phenotypic change driven by selection shifting the frequencies of many small-effect alleles present in the initial population.

The accomplishments claimed on behalf of the historic Modern Synthesis—restoring neo-Darwinism, repudiating all rival theories including mutationism and orthogenesis, and unifying evolutionary biology (Futuyma, 1988)—depend on the establishment of the shifting-gene-frequencies theory as the comprehensive explanation for organic evolution. Actually, the theory was never proven to be correct, but was advanced on the *plausibility argument* that it has the capacity to account for all of evolution using an experimentally validated mode of change.

6.6 The explanatory argument

The explanatory version of the mutation-is-irrelevant claim would be that, even if mutation is consequential in some ways, and even if its influence may be studied effectively via a suitable research program, to pursue such research would be a waste of time, because the things that matter are creativity, adaptation, and the direction of evolution, and mutation does not help us to understand these things. Explanations in terms of selection are the kind of explanations that satisfy us, whereas explanations in terms of mutation, though perhaps legitimate, would only apply to what Dawkins (1987, p. 303) calls "the boring parts" of evolution (see Box 3.1).

6.6.1 Darwin's architect

An example of the explanatory position would be the argument explaining "accidental" variability in the second of Darwin's volumes on variation (Darwin, 1868, Ch. 21), which relies on an extended architect analogy:

Throughout this chapter and elsewhere I have spoken of selection as the paramount power, yet its action absolutely depends on what we in our ignorance call spontaneous or accidental variability. Let an architect be compelled to build an edifice with uncut stones, fallen from a precipice. The shape of each fragment may be called accidental; yet the shape of each has been determined by the force of gravity, the nature of the rock, and the slope of the precipice—events and circumstances, all of which depend on natural laws; but there is no relation between these laws and the purpose for which each fragment is used by the builder. In the same manner the variations of each creature are determined by fixed and immutable laws; but these bear no relation to the living structure which is slowly built up through the power of selection, whether this be natural or artificial selection.

If our architect succeeded in rearing a noble edifice, using the rough wedge-shaped fragments for the arches, the longer stones for the lintels, and so forth, we should admire his skill even in a higher degree than if he had used stones shaped for the purpose. So it is with selection, whether applied by man or by nature; for though variability is indispensably necessary, yet, when we look at some highly complex and excellently adapted organism, variability sinks to a quite subordinate position in importance in comparison with selection, in the same manner as the shape of each fragment used by our supposed architect is unimportant in comparison with his skill.

To summarize, an architect or builder constructs a building from uncut stones that have fallen from a cliff. The stones are analogous to chance variations, and the builder or architect is analogous to selection, the agent who does the work. Darwin emphasizes that the shapes of the individual stones are not uncaused, but may reflect some law-like process, i.e., they are not spontaneous or uniform. Likewise, natural variations reflect the laws that govern the production of variation (Darwin suggested several such laws in the *Origin of Species*). However, the laws that shape the stones or variations "bear no relation" to the finished product, which is entirely due to the admirable skill of the architect, or of selection.

This analogy is often cited in discussions of "chance," as though it had a meaning such as spontaneous or independent (e.g., Sober, 2014; Merlin, 2010). However, this is simply not what Darwin emphasizes. The climax of the first paragraph is not the description of how the stones fall from the cliff, but the "bears no relation" claim, i.e., the laws that shape variations "bear no relation to the living structure which is slowly built up through the power of selection." One wonders how it can be that a process of building or evolution can have an input (stones, variations) that do not shape the outcome. Darwin's answer in the second paragraph is not (directly) an argument about causation, but an argument that some kinds of *explanations* or *attributions* are more deserving than others.

This point becomes clearer if we examine the account by Beatty (2010, 2014) of how Darwin developed the architect analogy over several years, to address the concerns of colleagues (Hooker, Gray, Lyell) uncomfortable with the role of accidental variation in his theory. As Beatty (2010) relates, Gray and others held that variation is not entirely accidental, but has been led along certain beneficial lines by divine intervention. Darwin argues that such thinking is not necessary. In a letter to Hooker in 1860, he writes

Squared stones, bricks or timber are indispensable for construction of a building; & their Nature will to certain extent influence character of building, but selection I look at, as the architect; & in admiring a well-contrived or splendid building one speaks of the architect alone & not of the brick-maker...

and in a letter to Lyell the same year he writes

As squared stone, or bricks, or timber, are the indispensable materials for a building, and influence its character, so is variability not only indispensable but influential. Yet in the same manner as the architect is the all important person in a building, so is selection with organic bodies (see Beatty, 2014).

That is, Darwin allows that the stones *are influential*, and yet at the same time, he represents the architect as "all important." This confirms that Darwin is making a claim about what kinds of attributions are appropriate, to the effect that assigning credit to architects, rather than to brick-makers or bricks, is simply better.

6.6.2 Later arguments

From Godfrey-Smith's account of adaptationism (Box 3.1), we might expect Dawkins to take the explanatory position on the randomness of mutation. This is indeed what Dawkins (1996, p. 80) does in the following passage:

But is this [criticism of Darwinism for relying on pure chance] one of those rare cases where it is really true that there is no smoke without fire? Darwinism is widely misunderstood as a theory of pure chance. Mustn't it have done something to provoke this canard? Well, yes, there is something behind the misunderstood rumour, a feeble basis to the distortion. One stage in the Darwinian process is indeed a chance process—mutation. Mutation is the process by which fresh genetic variation is offered up for selection and it is usually described as random. But Darwinians make the fuss they do about the 'randomness' of mutation only in order to contrast it to the non-randomness of selection. It is not necessary that mutation should be random for natural selection to work. Selection can still do its work whether mutation is directed or not. Emphasizing that mutation can be random is our way of calling attention to the crucial fact that, by contrast, selection is sublimely and quintessentially nonrandom [. . .]
Even mutations are, as a matter of fact, nonrandom in various senses, although these senses aren't relevant to our discussion because they don't contribute constructively to the improbable perfection of organisms.

That is, Dawkins admits that mutations might be nonrandom in some ways, and that they might contribute to evolution in some ways: they just do not contribute "constructively" to what is important for biologists to explain, which is "the improbable perfection of organisms," but instead only affect "the boring parts" of evolution (see Box 3.1).

Finally, a statement by Dobzhansky (1974, p. 323) about what is "meaningful," "interesting," and "fundamental" is worth quoting at length (emphasis mine):

The statement that the process required a supply of 'chance' mutations is true but **trivial**. What is far more **interesting** is that these genetic variants were not simply caught in a sieve, but were gradually compounded and arranged into adaptively coherent patterns, which went through millions of years and of generations of responses to the challenges of the environments. Viewed in the perspective of time, the process cannot **meaningfully** be attributed to the play of chance, any more than the construction of the Parthenon or of the Empire State Building could be ascribed to chance agglomeration of pieces of marble or of concrete. What is **fundamental** in all of these cases is that the construction process was **meaningful**. The meaning, the internal teleology, is imposed upon the evolutionary process by the blind and dumb engineer, natural selection. The 'meaning' in living creatures is as simple as it is basic—it is life instead of death.

In this theory, what is fundamental, interesting, and meaningful in evolution is contributed by selection and not variation.

6.7 Synopsis

To summarize, the architects of the Modern Synthesis invoked a number of arguments to the effect that mutation is inconsequential, irrelevant, or unimportant:

- Arguments from analogy and metaphysics
  - Mutation supplies "raw materials" only
  - Mutation is a material cause, whereas selection is an agent
  - Selection is a poet, painter or other artistic creator, whereas variation supplies words, pigments, etc
  - Mutation in an individual acts at the wrong "level," below that of determinative causes like selection.
- Empirical argument: Actual evolutionary changes are composed of many individually minor effects.
- Mechanistic arguments
  - Selection is creative because
    * variation is copious, undirected, and small in effect
    * it alone determines the direction of evolution
    * it brings together improbable combinations of alleles
  - Mutation is a weak force easily overcome by selection, therefore it cannot contribute to the direction of evolution
  - Because abundant variation is maintained in the gene pool.

* mutation is not a source of initiative in evolution; instead, change begins when a change in conditions brings on a selective response
* the rate of evolution is not linked in any precise way to the mutation rate.

- The methodological argument: Even if mutation has some effect on evolution, hypotheses of mutational effects cannot be proven.
- The explanatory argument: Even if mutation has some effects that we could study, doing so would be boring and a waste of time because what matters is the contribution of selection to the improbable perfection of organisms.

This pattern suggests an interpretation of the "mutation is random" claim that, relative to previous attempts to define "randomness," has more power to explain why evolutionary biologists say the things that they do. This interpretation depends on three points, one of which was established in previous chapters: the mutation-is-random claim is polymorphic, in the sense of taking on many forms. The second point, shown here, is that this polymorphic claim travels together with other claims of unimportance, inconsequentiality, and so on. The third point is that, like the randomness claim, these other claims rely on the same contrast case: natural selection.

This contrast is inescapable in the literature of the Modern Synthesis era. Evolution is described as a process that productively combines a random process of variation and a nonrandom process of selection (Darlington, 1958, p. 231). In his defense of the adaptationist research program, Mayr (1983) says that mutation is a "chance" process, but selection "is on the whole viewed rather deterministically: Selection is a non-chance process" (p. 150). Haldane (1959) writes that "a function of natural selection is to negate the randomness of mutation" (p. 109) and "Variation is in some sense random, but natural selection picks out variations in one direction, and not in another" (p. 147). In regard to Darwin's architect analogy (see Section 6.6.1), Beatty writes that "the main point of the architect analogy was to emphasize, more specifically, the importance of natural selection relative to the production of variation" (Beatty, 2014, p. 175).

Though often stated forcefully, this contrast reflects, not a well-founded conclusion, but the aspiration for a pure view in which selection governs evolution and variation is irrelevant (see Box 6.1). The architects of the Modern Synthesis developed and promulgated a view that denied importance to variation, even when this led them to rely, not on solid evidence, but on supposition and hopeful exaggeration. According to Beatty (2010), this aspirational approach goes back to Darwin: "I would even say that Darwin came to the notion of chance variation, as we now understand it, in large part through the realization that such a conception of variation would enhance the role of natural selection" (p. 22). Darwin himself said that

If selection consisted merely in separating some very distinct variety, and breeding from it, the principle would be so obvious as hardly to be worth notice; but its importance consists in the great effect produced by the accumulation in one direction, during successive generations, of differences absolutely inappreciable by an uneducated eye (Darwin, 1872, Ch. 1).

The fact that this kind of theory is conjectural does not mean that it lacks a rational basis. The scientific motivation for proposing a theory in which selection governs evolution, with no saltations or internal dispositions, and no idiosyncratic dependence on which mutations occur, is the belief that evolution must be governed by selection, because (the argument assumes) organisms are observed to be exquisitely adapted, down to the finest detail, i.e., the premise of empirical adaptationism (Box 3.1).

In this broader context, the "mutation is random" claim emerges as one of many ways to proclaim the importance of selection and deny the importance of variation; it represents one of many ways of stating a deeper "mutation is irrelevant" doctrine in which variation must be made subservient to selection, to account for the supposed level of adaptation in nature. The mutation-is-irrelevant doctrine implies that the details of mutation do not have any influence on the important features of evolution. The doctrine tells us that, if we set aside the boring parts of evolution (e.g., sequence composition) and focus on understanding important things like adaptation, mutation will never be the answer to the scientific questions we are asking. The doctrine exists to support the assignment of roles in the topic article on "natural selection" (Ridley, 2002) in the *Oxford Encyclopedia of Evolutionary Biology*:

In evolution by natural selection, the processes generating variation are not the same as the process that directs evolutionary change. . . What matters is that the mutations are undirected relative to the direction of evolution. Evolution is directed by the selective process.

This interpretation—randomness means irrelevance—explains a number of things that are otherwise confusing. The first is that, because the doctrine is not primarily about the nature of mutation, but about how to frame its subservient role in evolution, evolutionary biologists feel comfortable deciding the issue without spending years studying the molecular biology of mutation. No ordinary definition of randomness cleanly distinguishes one biological process from another, thus a special "evolutionary" definition is brought forth, the only criterion for which is that *it must apply to mutation but not to selection*. Indeed, the definition may refer merely to the idea that mutation, unlike selection, does not invariably increase fitness (Section 4.2). Eble (1999) makes this explicit: "Patterns are evolutionarily random whenever selection and adaptation are not directly involved."

The second matter clarified by this interpretation is that adherence to the mutation-is-random doctrine is (to some extent) a matter of research programs and scientific identity-politics. The scientists who insist that mutation is random are particularly the ones sympathetic to neo-Darwinism. For supporters of neo-Darwinism, the causes of evolution are external: the notion of internal causes acting by way of variation is a heresy, treated with suspicion as an appeal to vitalism. To pour scorn on internalist thinking while asserting that mutation is random is to show one's allegiance to neo-Darwinism. Scientists either defend the doctrine, or refuse to defend it, depending on whether they identify with, or reject, the characteristic concerns of neo-Darwinism.

Furthermore, this interpretation explains the polymorphism of justifications, and the dispensability of any specific justification. The doctrine that mutation is insignificant might be defended on the grounds that mutation has no propensities whatsoever; on the grounds that propensities of mutation, whatever their nature, have no influence; on the grounds that the influence of mutation on evolution, however real, is inaccessible to rigorous investigation; or finally, on the grounds that mutation, whatever its demonstrable influence, is *boring*. If one justification fails, another will suffice. In fact, we have seen all of these claims, often mixed together, e.g., in the passage from Dawkins (1996) cited earlier (Section 6.6.2). Arthur (2001) perfectly captures this polymorphism when he writes

Since the earliest days of evolutionary theory there has been a school of thought in which the introduction of new variants has been considered to be incapable of playing a role in determining the direction of evolutionary change, and, related to that, has been deemed uninteresting and unproductive as a focus of evolutionary study. This view can be traced back to Wallace (1870), who thought that variation was always present in all phenotypic directions, without any bias one way or another (that is, the variation lacked structure). Later versions of this school of thought took the form of vehement denials of any role for mutation in influencing the direction of evolution (Fisher 1930; Ford 1971).

In the space of three sentences, this passage invokes a lack of propensities or biases, a lack of influence on the direction of evolution, a lack of explanatory value ("uninteresting"), and a lack of methodological value ("unproductive"), associating all four claims with a single "school of thought." What unifies these claims is that they all point in the same direction: the practical irrelevance of mutation to what matters most.

The mutation-is-irrelevant doctrine is slippery but not entirely immune to refutation. Biologists have been offering adaptive explanations for the features of organisms and the patterns of nature for centuries. That is, we could make a very long list of things that biologists find important or meaningful enough to consider as candidates for adaptive explanations. To the extent that mutational dispositions can explain aspects of these patterns, an alternative research program focused on mutational causes is a threat to the irrelevance doctrine. For instance, biologists felt in the past that genome size and genome composition were important enough to be given adaptive explanations, but these adaptive explanations have not fared well, and instead, models that depend on mutation biases are often considered.

CHAPTER 7

The problem of variation

In evolution, selection may decide the winner of a given game, but development non-randomly defines the players.

Pere Alberch (1980)

7.1 Introduction

Analyzing the "mutation is random" doctrine leads in two directions, depending on how one interprets the doctrine. The most obvious interpretation is that the thing called "mutation" has a property of "randomness." The best way to justify this type of claim is to argue that "randomness" means something very specific: the incipient fitness effects of intermediate mutational states do not influence their emergence as heritable mutations in such a way as to result in a positive correlation. This is a conjecture of unknown generality, not a principle of biology. Furthermore, this conjecture has limited utility: it does nothing to guarantee that the occurrence of mutations is uncorrelated with fitness effects, because mutagenesis is directly subject to processes and conditions related to fitness (e.g., gene expression), and mutation-generating systems can evolve so as to favor useful mutations.

A completely different way to interpret the randomness doctrine is to link it to an explanatory irrelevance claim about variation that goes back to Darwin. Rather than being a principle of biology, explanatory irrelevance is more of a guiding assumption for a research program or school of thought, like the assumption of behaviorists that animals have no internal mental states, or the assumption of free-market economists that consumers behave as rational agents. For a behaviorist, the statement "animals have no internal mental states" does not signify "the evidence convinces me

that animals lack internal mental states," but rather "I am a behaviorist committed to assuming that animals have no internal mental states." Likewise, the correct translation of "mutation is random" (as a claim of irrelevance) is something like "I identify with a neo-Darwinian view in which the details of mutation are irrelevant to anything of genuine biological interest; I accept that selection is the ultimate source of meaning and explanation in biology."

No amount of evidence on *the nature of mutation* will affect the status of this doctrine, which is not about how mutation works, but about *the role assigned to mutation in evolutionary explanations*. The only way to challenge the doctrine directly is to interrogate the role of mutation in evolution, and to discover that we cannot explain evolution satisfactorily by calling on selection to do all the important work, while treating variation merely as raw material.

Perhaps most evolutionary biologists today would agree on rejecting the irrelevance doctrine, on the basis of familiar kinds of evidence. However, the case has not yet been made, if only because the issue has not been framed in precisely the way we have framed it for the purposes of this book. The work of this chapter is to explain the significance of the irrelevance doctrine for evolutionary theory, what kinds of familiar evidence cast doubt on the doctrine, and what are the general implications of rejecting it.

Mutation, Randomness, and Evolution. Arlin Stoltzfus, Oxford University Press (2021). © Arlin Stoltzfus. DOI: 10.1093/oso/9780198844457.003.0007

7.2 The power of the morphotron

The morphotron shown in Fig. 7.1 is an imaginary device for changing morphologies by adjusting a large number of dials. Every feature that one might wish to change has a dial. By moving the dials, any form can be morphed into any other in a graduated manner. For instance, by adjusting the dials, we can change an alligator into a horse, a horse into a rabbit, and so on. The shape is controlled ultimately by the hand that moves the dials. In Darwin's theory, the "struggle for life" is the invisible hand that moves the dials on the morphotron, and the whole process of shifting in response to conditions is called "natural selection."

The advantage of the morphotron, and of neo-Darwinism more generally (see Box 6.1), is that it frees us from having to worry about internal details. To understand evolution, or to predict its course, we only need to understand the hand that controls the dials, the hand of selection. What is happening inside the morphotron is irrelevant: so long as every trait shows infinitesimal variation—so long as every dial may be moved one way or the other—every change is possible. The laws operating inside

the black box "bear no relation" to the outcome of evolution, which is determined by the architect, selection (see Section 6.2).

Thus, the morphotron is helpful for visualizing the implications of the neo-Darwinian theory for conducting research and formulating hypotheses, and likewise, for considering arguments against this theory. Bateson (1894, p. 80) objected that one cannot simply add or subtract indefinitely from every attribute:

For the crude belief that living beings are plastic conglomerates of miscellaneous attributes, and that order of form or symmetry have been impressed upon this medley by selection alone; and that by variation any of these attributes may be subtracted or any other attribute added in indefinite proportion, is a fancy which the study of variation does not support.

Accordingly, Bateson believed that the first task of the evolutionist was to make a catalog of variations, to understand their character. By contrast, Mayr (1959a) asserts that "infinitely variable natural populations are of such evolutionary plasticity that natural selection can mold them into almost any shape." When selection is the invisible hand on the

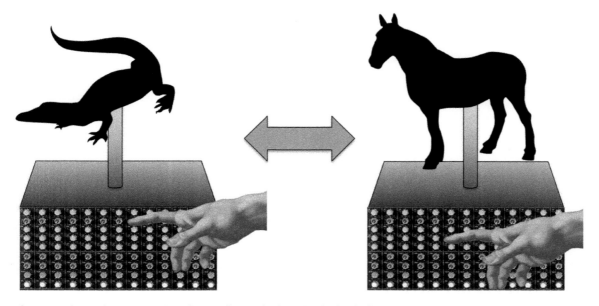

Figure 7.1 The morphotron, an imaginary device to illustrate the theory that the details of genetics and development are irrelevant to the outcome of evolution, which is governed by the hand of selection.

dials of the morphotron, "the ultimate source of explanation in biology is the principle of natural selection" (Ayala, 1970).

Indeed, the morphotron's black box recalls the contemporary criticism, common in the evo-devo literature, that the "Modern Synthesis" treats development literally as a "black box" (e.g., Palopoli and Patel, 1996; for an early version, see Hamburger, 1980). For instance, Hendrikse et al. (2007, p. 393) write that

One of the chief complaints about the Modern Synthesis is that it black-boxes development in such a way that the variation needed for selection to act on is just assumed to be generated in sufficient quantities and with sufficient qualities that adaptations can be adequately understood solely in terms of selection.

Perhaps few scientists today would defend the morphotron as a realistic theory. Whenever results show that we cannot predict evolution from phenotypes alone, without knowing the internal details, the morphotron is contradicted. We review some results of this type later in the chapter.

In addition to these empirical arguments, one may consider a general theoretical argument against the morphotron that follows from the multivariate generalization of quantitative genetics (Lande and Arnold, 1983). This theory describes how various quantitative traits change together under selection, assuming the kind of abundant infinitesimal variation that, under the neo-Darwinian theory, would provide infinite plasticity for selection to mold any shape. It is the closest thing we have to a mathematical formalization of neo-Darwinism. Yet, as explained in Section 7.5.1, this mathematical formalization actually shows that the morphotron is an impossible machine, in the same sense that a perpetual motion machine is an impossible machine. The hand of selection may be on the dials, but turning one dial causes another dial to turn in a way that cannot be predicted without knowing what is inside the black box, with the result that selection cannot possibly be the governing force previously imagined.

Nevertheless, the logic of the morphotron is too powerful and convenient to be ignored. Facile

theories are at work in our thinking, whether we like it or not, and sometimes we can expel them only by conscious effort. The morphotron is at work implicitly whenever the role assigned to selection is a teleological force that molds features, an approach that remains common in the scholarly literature, and is pervasive in popular books about evolution (e.g., Coyne, 2010). As noted previously (Section 5.7), the research literature occasionally gives serious attention to the possibility that selection rules the evolution of quantitative traits in the long term, in spite of any short-term effects of the internal details.

7.3 Source laws and consequence laws

Suppose that we want to develop a theory of evolution that is, in principle, sufficient to account for the immense scope and complexity of evolution. The inputs to this theory will be various principles of biology, population genetics, and so on, and when we combine the theory with salient facts, the outputs will include predictions of the expected course of evolution, and explanations of the observed course. Among the principles that we must consider in our theory of evolution are principles that relate to variation.

The morphotron provides a simple solution to this challenge. If we are willing to make *one rather large assumption*, we can have a complete theory of evolution *without having any substantive theory of biological form and its variation*. That is, our theory of evolution will not need to say anything about how bodies are built or how genetic circuits are constructed. As long as each phenotypic feature varies infinitesimally, the theory says that selection will govern the outcome, with variation playing the role of raw material. Yet, as we will discover, the morphotron is both theoretically impossible, and is contradicted by results from diverse areas of evolutionary research. We must reject it.

What happens if we reject the morphotron, and return to the problem of developing a theory that is genuinely sufficient to account for evolution? The outcome of evolution now depends, in some unknown way, on the processes by which variants emerge in biological systems. To have a sufficient theory, we need a theory of form and variation. For instance, Melo et al. (2016) begin their treatment of modularity by stating that "Modularity has emerged as a central concept for evolutionary biology, providing the field with a theory of organismal structure and variation," implying the lack of any such theory previously.

What kinds of principles are needed to achieve sufficiency? As a first approximation, these will fall into two categories, the first about understanding the *generation of variation*. For instance, in Chapter 4, we used a vector of rates **u** to characterize mutation, with one value u_i for each possible mutation i. Being able to assign values to **u** would mean that we have a very good empirical understanding of mutation. Given a theory for the process of mutation, we might be able to say something about patterns in **u**, i.e., we might be able to explain why some types of molecular or morphological variations are more likely to emerge. Note that we concluded earlier that mutations do not occur independently of the environment, which means we must represent mutation as a function of the environment. Thus, one of the tasks of post-morphotron evolutionary biology is to develop an understanding of the process of variation. Any fact about the process of variation is potentially relevant.

Medawar (1967) recognized the need to understand the causes of variation when he said that "The main weakness of modern evolutionary theory is its lack of a fully worked out theory of variation, that is, of candidature for evolution, of the forms in which genetic variants are proffered for selection" (p. 104). However, the evolutionary theory of Medawar's time also lacked *causal principles that relate tendencies of variation to evolutionary tendencies*. Given a bias of magnitude B in candidature for evolution, what is the effect on the outcome of evolution? How does the degree of effect depend on demographic conditions?

These two categories correspond to the distinction of *source laws* and *consequence laws* that Sober (1984) applies to selection (see his section 1.5). The *consequence laws* of selection tell us the consequences of differences in fitness, e.g., they tell us what will happen if A and B differ in fitness by 2 % (given some set of background conditions including a scheme of inheritance). An early example would be Norton's allelic selection model, presented by Punnett (1915). The consequence laws of selection are generally in the province of population genetics. As Orr (2005b) points out, models in population genetics "begin with selection coefficients, but say nothing whatever about where the selection coefficients come from" (p. 4). For that, we need the

---

Box 7.1 The Mendelians in history and in Synthesis Historiography

Upon the discovery of Mendelian rules of inheritance in 1900, the early geneticists or "Mendelians" began to rethink old evolutionary ideas, and to offer new ones, from the perspective of genetics. For instance, Bateson and Saunders (1902, p. 131) described the idea later known as the Hardy-Weinberg equilibrium, and foresaw its uses in research:

It will be of great interest to study the statistics of such a [Mendelian] population in nature. If the degree of dominance can be experimentally determined, or the heterozygote recognised, and we can suppose that all forms mate together with equal freedom and fertility, and that there is no natural selection in respect of the allelomorphs, it should be possible to predict the proportions of the several components of the population with some accuracy. Conversely, departures from the calculated result would then throw no little light on the influence of disturbing factors, selection, and the like.

Similarly, Bateson (1902, p. 31) explained how a model with multiple Mendelian factors can account for a smooth bell curve ("chance distribution") for a quantitative trait, i.e., the multiple-factor theory (see also Bateson and Saunders, 1902; Punnett, 1905, pp. 52–53). Punnett urged his more mathematically gifted colleagues to solve the Hardy-Weinberg equilibrium, as well as the allelic selection model (Punnett, 1915) showing how selection can drive a beneficial allele from low to high frequency (see Provine, 1971). Morgan (1916, p. 187–189) articulated the concept of a differential probability of acceptance for beneficial, neutral, or deleterious mutations (see also Punnett, 1911, p. 142). Punnett (1911, p. 151) argued that species must be defined by interfertility rather than by museum specimens or visible characters, i.e., the biological species concept.

In general, the Mendelians entertained any idea that could be reconciled with genetics (for review, see Stoltzfus and Cable, 2014), including beneficial changes by selection of single mutations (de Vries, Morgan, Punnett), neutral change (Morgan), superficially smooth change based on multiple factors (Bateson, Johannsen, Punnett), one-step speciation (de Vries), the origin of Mendelian incompatibilities (the Bateson–Dobzhansky–Muller model), parallelisms based on parallel mutations (Morgan, Vavilov), effects of mutation rates on the chances of evolution (Morgan, Punnett, Shull), and the possibility of trends (orthogenesis) due to mutational tendencies (Bateson, Morgan).

The Mendelians put the new-found principles of heredity first, even if this meant discarding traditional thinking. They argued against effects of use and disuse (Lamarckism) and direct environmental effects on heredity, the factors emphasized by neo-Lamarckians (e.g., Delage and Goldsmith, 1913). Likewise, they rejected Darwin's fluctuation-struggle-blending theory, which Johannsen had refuted (Roll-Hansen, 1989; Stoltzfus and Cable, 2014). At the time, historian Erik Nordenskiöld (1929) wrote that the proposition that natural selection "does not operate in the form imagined by Darwin must certainly be taken as proved."

Yet, in the typical version of history repeated in the evolutionary literature, the Mendelians are depicted, not as the pioneers of genetical thinking about evolution, but as bumbling "mutationists" who reject selection and gradual change, imagining evolution by dramatic species-creating mutations alone (e.g., Dawkins, 1987, p. 305; Cronin, 1991, p. 47; Ayala and Fitch, 1997; Eldredge, 2001, p. 67; Segerstråle, 2002; Charlesworth and Charlesworth, 2009). The period of conceptual innovation from 1900 to the 1920s is depicted literally as an "Eclipse"—evidently a period of darkness when the light of Darwin was absent.

This treatment of the Mendelians is part of a larger pattern that historians call "Synthesis Historiography" (Amundson, 2005), i.e., telling history in ways that make things turn out right for the Modern Synthesis. Synthesis Historiography features narratives such as the Eclipse, the Mutationism Story, and the Essentialism Story ("fabricated" by Ernst Mayr, in the words of historian Mary Winsor, 2006). In these stories, opponents of neo-Darwinism behave irrationally and hold views with obvious flaws, while Darwin's followers use reason and evidence to make foundational discoveries. For instance, Synthesis Historiography credits Fisher (1918) with reconciling Mendelism and smooth change. In reality, as historian Jean Gayon (1998) writes, "the fundamental doctrines of quantitative genetics were developed early in the century, long before the publication of Fisher's canonical article of 1918" and "it was not the mathematical formalism that convinced scientists of the plausibility of the multifactorial hypothesis" (p. 316).

Remarkably, Synthesis Historiography does not even give the early geneticists credit for bringing genetics into evolutionary thinking! Instead, the credit for rejecting the old developmental view of heredity shared by Darwin and Lamarck, and for introducing discrete inheritance, is assigned to the nineteenth-century physiologist August Weismann, though historians object that this story does not reflect what Weismann actually believed (Winther, 2001; Griesemer and Wimsatt, 1989).

In Synthesis Historiography, the rejection of Darwin's theory a century ago is depicted as a case of irrational prejudice, e.g., both Ridley (1993, p. 13) and Charlesworth and Charlesworth (2009) cite Nordenskiöld's comment as though this indicates a world gone mad with anti-Darwinian animus. As if to refute Nordenskiöld, Charlesworth and Charlesworth (2009) write

> This contrasts with 3253 articles mentioning natural selection and evolution in 2008 in the Web of Science database. For a detailed discussion of anti-Darwinian evolutionary ideas, see Bowler (1983) and Gayon (1998).

Actually, Gayon (1998) cites precisely the same passage from Nordenskiöld, explaining that "the decline of Darwinism was virtually always attributed to the experimental refutation of the hypothesis of 'natural selection' in the highly restrictive sense that Darwin had intended" (p. 2). Gayon's main thesis is that Darwin's mistaken views of heredity created a "crisis" resolved by "the most important event in the history of Darwinism: the Mendelian reconstruction of the principle of selection" (p. 289). That is, the 3,253 articles refer to the Mendelian conception of selection as an increase in frequency of one discretely inherited type at the expense of another (see Gayon, 1998, pp. 181–182).

source laws of selection, which account for the origins of differences in fitness, based on an understanding of physiology and ecology and so on. A simple example of a source law is that overall fitness is a function of both survival and reproduction.

Likewise, to develop a sufficient theory of evolution without assuming the morphotron, we need source laws and consequence laws of variation. The source laws of variation would emerge from an understanding of mutation and development, and would tell us what kinds of variations tend to occur in what circumstances. The consequence laws, which tell us how such tendencies will influence evolution, are again (as a first approximation) a matter of evolutionary genetics. Note that this formula of source laws and consequence laws is not a complete framework for transcending the morphotron, because it focuses only on first-order effects on evolutionary trajectories, rather than higher-order effects, e.g., the effects that play an important role in discussions of evolvability (see Chapter 5).

7.4 The Mendelian challenge

With the distinction of source laws and consequence laws in mind, let us turn briefly to the historical development of evolutionary thought in the twentieth century. How did the Mendelians approach the problem of variation?

The discovery of genetics in 1900 brought on a major challenge to the neo-Darwinian view that variation merely serves as raw materials or passive clay shaped by selection to fit conditions. The early geneticists—de Vries, Bateson, Johannsen, Punnett, Morgan, and others—argued that the mechanistic basis of evolutionary change is discrete, and the introduction of new variants likewise occurs by discrete events of mutation with discrete consequences (Stoltzfus and Cable, 2014). Because of this, they argued, one must rethink the roles of both selection and variation. Selection can no longer be seen as a creative force, because the outcome of evolution depends importantly on discrete events of mutation that introduce variation.

Even before the genetical revolution, Bateson (1894) spent years compiling *Materials for the Study of Variation*, a compendium of 886 cases of discontinuous variation. He planned a second volume on continuous variation, but the need for a volume based on his "method of miscellaneous collection" was obviated by subsequent systematic work on quantitative trait distributions (see Bateson, 1909b, p. 287).

Did any source laws emerge from this? One point that Bateson and others (e.g., Davenport, 1909) emphasized was that variant forms were not mere aberrations or monstrosities, but could exhibit a kind of completeness or developmental stability. In historical sources going back at least to Darwin, this property is referenced with the adjective "definite" and contrasted with "indefinite" variability: indefinite variability is fuzzy nondirectional noise, whereas a definite variant represents a distinctly different type. Bateson illustrated this point with the example of a cockroach species typically showing five-jointed legs, but sometimes found with four-jointed legs. For both types of legs, the lengths of segments show a normal distribution of variation among individuals, leading Bateson (1894, p. 65, my emphasis) to suggest that

> . . . the final choice between these two may have been made by Selection, yet it cannot be supposed that the accuracy and completeness with which either condition is assumed is the work of Selection, for the 'sport' is as *definite* as the normal.

Morgan (1916, p. 67) pointed out, in regard to an eyeless mutant fly, that "formerly we were taught that eyeless animals arose in caves. This case shows that they may also arise suddenly in glass milk bottles, by a change in a single factor."

Morgan frequently suggested that characters do not vary, and could not evolve, independently of each other, e.g., he suggests this indirectly in Morgan (1904, p. 63) and more clearly in:

> I am inclined to think that an overstatement to the effect that each factor may affect the entire body, is less likely to do harm than to state that each factor affects only a particular character. The reckless use of the phrase 'unit character' has done much to mislead the uninitiated as to the effects that a single change in the germ plasm may produce on the organism. (Morgan, 1916, p. 72; see also p. 117).

In addition to the generalization that variants of one genetic factor may affect many traits ("pleiotropy"), is the generalization that variation in many factors may affect one trait ("polygeny"), e.g.,

We have seen that there are thirty mutant factors at least that have an influence on eye color; and it is probable that there are at least as many normal factors that are involved in the production of the red eye of the wild fly (Morgan, 1916, p. 87).

Another generalization frequently stated by the early geneticists is that mutational effects do not necessarily differ in either character or size from nonheritable fluctuations (what we would call "environmental variation" today): breeding experiments were the only way to sort out whether observed phenotypic variations (including small ones) were actual genetic variations (e.g., Shull, 1907; Ortmann, 1907; see Stoltzfus and Cable, 2014). The contemporary analog of this notion is Cheverud's conjecture on the correlation of environmental and genetic variation (Cheverud, 1988).

Finally, in regard to source laws, the early geneticists drew attention to nonuniformity in mutation, and wished to understand the underlying causes, though with little success. Vavilov (1922) argued that "Variation does not take place in all directions, by chance and without order, but in distinct systems and classes analogous to those of crystallography and chemistry" (p. 84). Bateson (1909a) said that "the greatest advance that we can foresee will be made when it is found possible to connect the geometrical phenomena of development with the chemical" (p. 97). Morgan (1923) writes that

Nevertheless, the discovery that the same mutation happens over and over again, not only within the same species, but in different species, is, I think, one of the most interesting discoveries in recent genetic work. It means that certain kinds of changes in the germ material are more likely to occur than are others. If we adopt the Galton metaphor of the equilibrium polygon, these changes might be interpreted to be the more stable conditions of the genes.

What did the early geneticists suggest in regard to *consequence laws* outlining the evolutionary role of properties of mutation? They clearly believed that non-uniformities in rates of mutation would be influential. For instance, Morgan considered it important that some mutations happen repeatedly, and particularly that parallel mutations, e.g., toward melanism or albinism in the case of mammals, could happen in different species (Morgan, 1923). Morgan (1925, p. 142) writes

There is another result, clearly established by the genetic work on *Drosophila*, that is favorable to the final establishment of a new type of character if it is beneficial. Most, perhaps all, of the mutations appear more than once. This improves their chances of becoming incorporated in the species, and if the mutation produces a character that favors survival the chance of its becoming established is still further increased.

Thus Morgan suggests a positive relationship between the rate of occurrence of a mutation and the chance of becoming established, and he does this in a context of dual causation in which the chance of becoming established is also affected by whether the mutation has a favorable effect on survival.

Likewise Shull (1936, p. 140) writes that

Probably the most effective aid in establishing new genes lies in their repeated production by independent mutations. A gene produced twice by mutation has twice as good a chance to survive as if produced only once.

Here Shull is making a more exact statement. Whereas Morgan says that the chance of becoming established merely increases with occurrence, Shull implies that it increases linearly. Note that Shull (1936, p. 122) also invokes dual causation of the direction of evolution "(1) by the occurrence of mutations of certain types, not of others, and (2) by the differential survival of these mutations or their differential spread through the population." In Chapter 8, we develop a more exact consequence law of this type, and formalize the idea of dual causation.

The Russian geneticist Nikolai Vavilov, famous today for his pioneering efforts to preserve biodiversity through seed banks (and for his death as a political prisoner), was the most extreme advocate of parallel evolution via parallel variation. He argued for a "law of homologous variation" in parallel evolution. Indeed, he found that the same varieties or polymorphisms often occur in parallel, even in distantly related species in the same genus or family. As an example illustrating parallelism and dual causation, Vavilov (1922) reports that lentils (*Ervum lens*), a food crop, and vetch (*Vicia sativa*), a weed, have many homologous variations, and notes that vetches sometimes mimic lentils so closely in cultivated fields that their seeds cannot be separated by mechanical sorters:

...the role of natural selection in this case is quite clear. Man unconsciously, year after year, by his sorting machines separated varieties of vetches similar to lentils in size and form of seeds, and ripening simultaneously with lentils. The same varieties certainly existed long before selection itself, and the appearance of their series, irrespective of any selection, was in accordance with the laws of variation.

Punnett (1915) held so strongly to the role of parallel variation that he argued, quite implausibly, that some alleged cases of polymorphic mimicry were not actually mimetic in the sense of being maintained by selection.

Morgan made several suggestions about genetic factors that would facilitate the explanation of apparent trends, one of which is merely a reference to what theoretical population geneticists today would call a "shift model," in which the chance that mutation results in a trait value *t* is not absolute, but is just a matter of how far *t* is from the parental value for the trait. Morgan (1919, p. 269) explains this by saying that "A rolling snowball that already weighs 10 pounds is more likely to reach 15 than is another that has just begun to roll." However, Morgan (1910, p. 208) makes the further suggestion that, if the system happens to show a tendency to vary in a particular direction, and this direction turns out to be useful, then "After the first step, which was undirected, i.e., not purposeful, the subsequent events are rendered more probable; for the dice are loaded."

In summary, the early Mendelians explored various ideas that went beyond the neo-Darwinian dichotomy of the potter and the clay. They found ways to combine selection and variation in a theory of dual causation, and recognized some consequences of the nonuniformity of mutation. Some of these ideas are recognizable today, though not due to any direct influence. The development of evolutionary thinking subsequently went in a different direction, documented in Chapter 6, and

the major contributions of the Mendelians to the conceptual foundations of genetical thinking about evolution were erased (see Box 7.1).

7.5 The contemporary challenge

The discovery of genetics initially threatened to undermine both Darwinism and Lamarckism, and led to attempts to rethink the role of variation within a genetical framework. However, the Mendelian-mutationist challenge to neo-Darwinism was beaten back, and defenders of Darwin's view began to dominate the mainstream of evolutionary discourse with their doctrines of selection as the ultimate source of explanation, and variation as random raw materials.

A second challenge came in the 1960s, when results of protein sequence comparisons evoked a world so inescapably discrete that no direct reconciliation with neo-Darwinism was possible. Accordingly, Mayr and Simpson responded by declaring that "molecular evolution" occurs in a different domain or on a different level, providing a superficial window on the process of change, scarcely relatable to the true causes of evolution (see Box 9.2). That is, the second challenge was deflected by creating a two-tiered system, with "evolution" referring to the selection-driven evolution of visible phenotypes and behavior, and allowing for a different level of "molecular evolution" to cover the unimportant internal details, which might follow a different set of rules.

The third major challenge, which began with evo-devo in the 1980s, is ongoing. Like the original Mendelian challenge, this one is taking a long time to reach a resolution. Furthermore, just as the original Mendelian challenge was multi-faceted (Stoltzfus and Cable, 2014), the contemporary challenge is multi-faceted, including key data and arguments from evo-devo, comparative genomics, molecular evolution, and evolutionary genetics. The contemporary challenge takes place at a time when scientists often are focused on aspects of biology unimagined by our predecessors, conducting research on systems and phenomena that were largely or completely foreign to them, e.g., lateral gene transfer, endosymbiogenesis, genetic conflicts. The historical disparity in the objects and methods of evolutionary research makes it difficult to reconcile contemporary findings with past theories. This difficulty is compounded by the surprising popularity of a form of "Synthesis" apologetics that makes these past theories a "moving target" (Smocovitis, 1996).

The remainder of this chapter is a brief review of some theoretical and empirical challenges to the neo-Darwinian dichotomy of the potter and the clay, based on results that are widely known. Then, in Chapter 8 and Chapter 9, we focus on a different challenge involving the role of biases in the introduction of variation, featuring analyses that are not widely known.

7.5.1 The G matrix as predictor

If a quantitative trait such as height is influenced by allelic variation at infinitely many Mendelian loci, each with an infinitesimal effect on the value of the trait, then phenotypic variation will be smooth and normally distributed. The expected change in the trait value z under directional selection follows the classic "breeder's equation" $\Delta z = h^2 s$, which is the product of two terms, one relating trait values to fitness, and the other indicating the amount of heritable variation (specifically, additive genetic variation, relative to total variation). The resulting change in the trait mean is called the "response to selection."

The model does not have any term for mutation, but begins with an abundance of infinitesimal variation in an implicit gene pool, with a term for the *amount* of variation in a trait, but not its direction—the mean effect of variation on the trait is 0. Masses of variations provide fuel, but each individual variation is infinitesimal and thus unimportant. Selection provides impetus and direction. The role of variation is strictly that of a raw material, used in bulk.

This model of quantitative genetics is the closest that mathematical theories get to the verbal theory of neo-Darwinism.

Yet, the extension of this model from one trait to many—the multivariate generalization of Lande and Arnold (1983) that emerged long after the Modern Synthesis orthodoxy—introduces something subversive. To represent two traits, such as height and weight, requires a positive value for the variance in height, another positive value for

the variance in weight, and a covariance of height and weight, which has both a sign and a magnitude. The meaning of the canonical equation:

$$\Delta \bar{z} = G\beta \tag{7.1}$$

is easier to understand if we look at a specific example with three traits:

$$\begin{pmatrix} \Delta \bar{z}_1 \\ \Delta \bar{z}_2 \\ \Delta \bar{z}_3 \end{pmatrix} = \begin{pmatrix} G_{11} & G_{12} & G_{13} \\ G_{21} & G_{22} & G_{23} \\ G_{31} & G_{32} & G_{33} \end{pmatrix} \times \begin{pmatrix} \beta_1 \\ \beta_2 \\ \beta_3 \end{pmatrix} \tag{7.2}$$

Here the $\Delta \bar{z}$ vector of changes in mean trait values is the product of the **G** matrix of variances (diagonal elements) and covariances (off-diagonal elements) and a vector of selection gradients for each trait. Multiplication here means, for instance, $\Delta \bar{z}_1 = G_{11}\beta_1 + G_{12}\beta_2 + G_{13}\beta_3$, where the latter two terms represent *the effect that selection on traits 2 and 3 has on trait 1*, due to correlations.

Although the mean effect of variation in every trait is still zero, the variances and covariances will differ. This makes the **G** matrix of variation something more than passive clay: as Steppan et al. (2002) say, "Together with natural selection (the adaptive landscape) it determines the direction and rate of evolution." For instance, there is an axis in the multidimensional space of characters with the most variance, **G**$_{max}$: the greatest prospects for evolutionary change are in directions that align with this axis. The expected correspondence between evolutionary divergence and **G**$_{max}$ has been observed repeatedly (Schluter, 1996; McGuigan, 2006; Bégin and Roff, 2003).

Importantly, this is *not* a model of side effects. That is, the model does *not* say that selection governs important traits, and exerts side effects on unimportant traits due to correlated variation, as in the theory of Sturtevant (1924), or Darwin's theory of correlated growth. Instead, the model says that the trajectories for *all traits*, including important traits, are linked together by a complex, structured factor that represents standing variation. Standing variation, in turn, reflects the generative process (represented by the **M** matrix) together with recent population history.

In other words, this model tells us that one of the core syllogisms of neo-Darwinism, explained again and again by leading thinkers, is incorrect. We have been told, in assorted ways (see Sections 6.2.1, 6.2.2, 6.2.3, 6.4.1, 6.4.2, 6.6.1), that variation merely supplies the raw materials in evolution and is not a difference-maker, source of direction, or shaping factor. Darwin's architect analogy (see Section 6.6.1) tells us that the laws of variation "bear no relation" to the finished structure, and that we must give full credit to the architect, selection. As Gould (1977) explains:

The essence of Darwinism lies in its claim that natural selection creates the fit. Variation is ubiquitous and random in direction. It supplies the raw material only. Natural selection directs the course of evolutionary change.

Yet, the argument fails, even if we allow the kind of abundant infinitesimal variation that Mayr (1959a) believed would provide infinite plasticity for selection to mold any shape. Even under ideal conditions, we cannot justify saying that the role of variation in evolution is merely to supply fuel or raw materials, with no dispositional influence. We stated this previously, relative to the morphotron, by saying that the hand of selection may be on the dials of the morphotron, yet turning one dial causes another dial to turn in an unsolicited way that depends on what is inside the black box.

In summary, the most neo-Darwinian mathematical model that we can possibly devise fails to uphold the assignment of roles that is central to neo-Darwinism. This failure tells us that *there is no sufficient theory of evolution without a theory of how the black box works.* Note that this is a different argument from the *causal completeness* argument (see Section 7.5.4), which holds merely that we have to include the detailed inner workings of development somewhere in a complete causal account of evolution, *even if these are inconsequential for determining the course of phenotypic change.* Instead, the mathematical theory of quantitative genetics shows that the inner workings of development must be included in evolutionary accounts *precisely because they can not be inconsequential.*

7.5.2 The challenge to gradualism

Darwin employed both a bottom-up and a top-down argument for gradualism. The top-down

argument (from empirical adaptationism: see Box 3.1) is that organisms are too perfectly adapted for the process of evolution to involve any large changes (e.g., see the passage in Chapter 2 of *The Origin of Species* that begins "Almost every part of every organic being is so beautifully related to its complex conditions of life . . ."). The bottom-up argument is that species often fall into perfectly graduated series; their differences appear to be composed of infinitesimal effects, and never correspond to the effects of sports, which may play a role in domestication but which would be monstrous and unworkable in a state of nature.

The architects of the Modern Synthesis offered both an empirical and a theoretical argument to justify infinitesimalism (gradualism). The empirical argument is that species differences inevitably resolve to an accumulation of many tiny effects. Orr and Coyne (1992) review this argument and conclude, not only that large-effect loci are sometimes involved in species differences, but that the original conclusion was not justified based on the evidence available at the time, e.g., they say that "Dobzhansky (1937) and Muller (1940) seem to have based their support of micromutationism on almost no data at all" (p. 731). They pointed out several kinds of historic bias in the treatment of evidence, and cite several sources of evidence for the role of large-effect alleles. The conclusion that large-effect alleles are sometimes important was only strengthened with subsequent work (Orr, 2001, 2005a).

The other argument is a bottom-up theoretical argument, known as "Fisher's geometric model," which holds that adaptive change will occur preferentially in infinitesimal increments instead of any larger increments. Suppose that P_O represents the optimal combination of values for phenotypes 1 and 2, and that fitness decreases in every direction similarly, such that any point the same distance from P_O has the same fitness (Fig. 7.2). What Fisher's model shows is that the chance of moving in a beneficial direction approaches 0.5 for the smallest possible changes.

Fisher's argument actually refers to a many-dimensional space, rather than a two-dimensional space. However, the conclusion is the same: the smallest changes are the most likely to be beneficial.

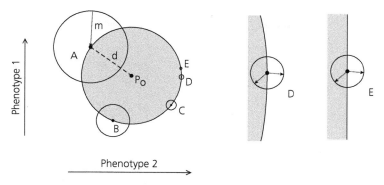

Figure 7.2 Fisher's geometric model addresses the chance that a change in a quantitative trait is beneficial, as a function of the size of the change. Different sizes are illustrated by the circles around points A through E. Changes of size m from point A form a circle around point A: less than half of the changes move the system into the shaded circle, nearer to the optimum P_O. The circles around points B through E represent successively smaller changes. The figures to the right magnify the areas around points D and E, suggesting that the chance of a beneficial effect will approach 0.5 as the size becomes infinitesimally small.

Orr (2005a) suggests that Fisher's infinitesimal model stifled research for decades by "assuming away" the problem. This criticism recalls Bateson's sarcastic critique from a century earlier (which Orr quotes):

By suggesting that the steps through which an adaptive mechanism arose were indefinite and insensible, all further trouble is spared. While it could be said that species arise by an insensible and imperceptible process of variation, there was clearly no use in tiring ourselves by trying to perceive that process. (Bateson, 1909a, p. 99).

Indeed, little further work was done on the topic until Kimura (1983) pointed out that, if one treats prospective changes as new mutations with a chance of acceptance proportional to the selection coefficient—which will be proportional to the size of effect—evolution will not favor the smallest possible changes, but changes of *intermediate* sizes, which will have the highest joint probability of both being beneficial and of reaching fixation. Assuming a stochastic population, as the size of a mutation becomes arbitrarily small, the chance of being incorporated in evolution does not increase to a maximum, but decreases until it converges on the chance of neutral fixation.

In more recent years, a variety of theoretical models have appeared using Fisher's geometric framework (Weinreich and Knies, 2013; Blanquart et al., 2014; Matuszewski et al., 2014). For our purposes, what is important to understand is that

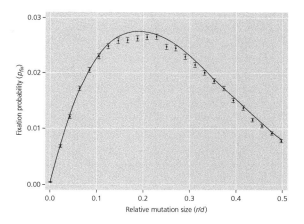

Figure 7.3 The chances of substitution in the Fisher–Kimura geometric model, showing calculated values (line) and simulated ones (dots with bars) from Ram and Hadany (2015). The relative size of a mutation is shown as the ratio of its size r to twice the distance to the optimum, i.e., d is analogous to the diameter of the grey circle in the previous figure (image credit: Yoav Ram, CC-BY-SA).

Fisher's formalism is no longer used to argue for *natura non facit saltum*, i.e., it is not used to argue that infinitesimal changes are definitively favored by the evolutionary process. Instead, the geometric formalism is integrated into an implicit or explicit origin-fixation model of evolutionary dynamics (see 8.6) that yields a frequency distribution of effect-sizes from small to large, like that shown in Fig. 7.3 (see Ram and Hadany, 2015; also Orr, 1999; Blanquart et al., 2014).

Thus, like the empirical case for gradualism, the theoretical case for gradualism has been undermined, and is no longer advanced in its original form.

7.5.3 The new genetics of adaptation

A minor renaissance of interest in the population-genetic theory of adaptation has occurred in the past 25 years. Tenaillon (2014) identifies three findings that gave reasons to revisit the topic and reconsider the classic neo-Darwinian view of gradual shifts in phenotype driven by mass selection on standing variation. First, empirical results reviewed earlier (7.5.2) implicated large-effect changes in species divergence. Second, rich data on experimental adaptation of laboratory cultures of microorganisms began to emerge, including from the famous long-term evolution experiment (LTEE) of Lenski and colleagues, with results suggesting that large-effect mutations were, according to Tenaillon (2014), "the drivers of adaptation." Third, Tenaillon (2014) notes the genomic evidence for selective sweeps in which a new mutation arises and then sweeps to fixation, becoming the predominant allele. By contrast, in the shifting-gene-frequencies theory, evolution is a shift from one multi-locus frequency distribution to another, based on abundantly available variation (see Box 6.2).

Thus, this newly expanded body of theory on adaptation evokes the discretized view that re-emerged in the 1960s, in which the unit step in evolution is the fixation of a mutation (by selection or drift), resulting in a discrete genetic change. Evolution is understood in terms of discrete steps, and a central question is how each step is chosen.

Some of this theory builds on Fisher's geometric model, discussed in the previous section. Yet this work follows Kimura in considering Fisher's framework in the context of an origin-fixation process (see Section 8.6). The shortcoming of Kimura's original argument was that it inferred a size distribution as though there were only one step involved. If we are considering multi-step walks, an initial intermediate change will tend to be followed by smaller changes to the extent that the distance to the optimum decreases, a point emphasized by Orr (1999). From this kind of analysis, one expects an exponential distribution of effect-sizes—a distribution in which the chances of occurrence decrease with size, so that larger effects are increasingly rare, yet frequent enough that they make an important contribution to the total change (Orr, 2005b). In particular, one expects changes with larger effects early in an adaptive walk.

Further innovations in the genetic theory of adaptation emerged along a different route, from models that attempted to build an explicit molecular framework for change. These models use a key locality assumption introduced in an early form by Maynard Smith (1970). The assumption is simply that, when considering the evolution of a sequence from the current state to some alternative state, only the "neighbor" states—the alternative states that differ at one site in the sequence—are important, on the grounds that multiple mutations (or mutations changing multiple sites) are vanishingly rare. For instance, a nucleotide sequence of length L has $3L$ one-mutant neighbors. When the sequence changes, the accessible neighborhood shifts slightly and must be recalculated, but this is still an enormous simplification relative to considering the entire set of $4^L - 1$ alternatives to the current sequence.

In passing, note that, before the molecular revolution, the genetic space of evolution was conceptualized relative to a set of ordered lists of loci (ordered via chromosomal location), numbering in the thousands, each with a small unordered list of alternatives (variant alleles). In the simplest diploid case, each locus has two alleles and three genotypes (one heterozygote and two homozygotes), so that the size of the space is 3^L, where L is the number of loci. This is the conception of genetic space that underlies various classic arguments regarding the totipotent gene pool and the creative power of selection (Section 6.4).

The next issue is which neighbor to choose. In the model of Orr (2002), the choice is again based on the probability of fixation, using Haldane's classic result that the probability of fixation of a new beneficial allele is approximately $2s$, where s is the selection coefficient (Haldane, 1927).

Where do the selection coefficients come from? This is a huge unknown. One possibility is to draw from some distribution. A crucial innovation in contemporary models of adaptation is to apply extreme value theory (EVT) to generate useful assumptions about selection coefficients, without making restrictive assumptions about the underlying shape of the distribution. If we can assume that the current genotype is already highly adapted, then its fitness is in the tail end of the fitness distribution. EVT applies to the tails of distributions, even if we do not know the overall shape of the distribution, which is very useful. Specifically, EVT tells us that the intervals between the top-ranked values are exponential random variables such that, if the expected value of the first interval (between the top-ranked value and the next one) is d, the expected value of the second interval is $d/2$, the expected value of the third is $d/3$, and so on. In the case of ranked fitness values, these intervals are the selection coefficients relative to the next lower rank.

Given a probability of fixation proportional to s, this leads to specific predictions about the probability of jumping from rank r to some higher rank r', including a fascinating invariance property where the expected upward jump in the fitness ranking is the same no matter where we are in the ranking. Namely, if the rank of the current genotype is j (i.e., $j-1$ genotypes are better), we will jump to rank $(j+2)/4$.

This suggests the possibility of obtaining results for extended adaptive walks, if one can solve a problem that goes back to the neighborhood concept. Each time the sequence changes, the neighborhood changes, e.g., some sequences that were neighbors are now two steps away, and some sequences that were previously two steps away are now neighbors. Gillespie (1983a) articulated this problem and later addressed it with simulations (Gillespie, 1984); analytical results did not appear until the work of Jain and Seetharaman (2011). Orr (2002) likewise addresses the issue using simulations. Adaptive walks under his model contradict the expectations of gradualism, in that the first step of an adaptive walk contributes 30 % of the total effect on fitness, on average, and the largest step contributes 50 %.

Box 7.2 When "Darwinian adaptation" is neither

Today, a process involving the fixation of new mutations with beneficial effects (i.e., the lucky mutant view) is universally called "adaptation" or "Darwinian adaptation" (e.g., Orr, 2005b; Weinreich et al., 2006; Jain and Seetharaman, 2011). Previous generations of evolutionists would have objected to this terminology. Classically, "adaptation" refers to a stimulus-response dynamic: conditions change, then a biotic system adjusts (adapts) in a graduated manner, depending on the strength and duration of the stimulus, representing either short-term (physiological) or long-term (evolutionary) adaptation. Colloquial usage has the same connotation: when we admire someone's adaptability, we mean to draw attention to how they respond to new or changing conditions, not to how they suddenly have great new ideas, or how they continually improve in a constant environment.

Our intellectual progenitors used a different term for the lucky mutant view. For instance, Shull (1936, p. 254) states, in regard to mutation-initiated evolution, "the whole process as Goldschmidt conceives it is one that was sponsored by Cuénot and Davenport and by the former called preadaptation. First a change occurred, fitting individuals for a new

situation, and then sometimes the new situation was either near at hand or was accidentally reached" (Charles Davenport and Lucien Cuénot were early geneticists known for their evolutionary views). When Cavalli-Sforza and Lederberg (1956) set out to show that mutations conferring antibiotic resistance occur prior to the onset of selective conditions, i.e., spontaneously in the Luria–Delbrück sense (see Section 2.7), their article was entitled "Isolation of Pre-Adaptive Mutants in Bacteria by Sib Selection," and it referred to the pre-adaptation theory.

Likewise, the architects of the Modern Synthesis used "pre-adaptation" or "random pre-adaptation" to distinguish the lucky mutant view from genuine adaptation (e.g., Simpson, 1967, pp. 157, 236, 257; Mayr, 1963, p. 121). Dobzhansky (1974, p. 325) gives DDT resistance in insects as an example of "preadaptation," arguing that resistant flies must have been present in the population prior to exposure. Referring to the early geneticists, Simpson (1967) complained that "the problem of adaptation was, in their opinion, solved by abolishing it: they proclaimed that there is no adaptation, only random pre-adaptation" (p. 276).

Continued

Box 7.2 *Continued*

Darwin would have agreed. Ever since he proposed "natural selection" based on indefinite variability (fluctuation) and blending, some readers have misconstrued it as a theory of pre-adaptation based on discretely inherited mutations. Because this misinterpretation happened in Darwin's time, no speculation is required to know his response. When challenged that his theory was "not a theory of the Origin of Species at all, but only a theory on the causes which lead to the relative success and failure of such new forms as may be born into the world," Darwin's response was "that may be a very good theory, but it is not mine" (see Poulton, 1909, p. 45).

That is, not only did the architects of the Modern Synthesis have a different term for the adventitious fixation of ben-

eficial mutations, they downplayed the importance of this process, which they associated with the Mendelians (Box 7.1). The lucky mutant models referred to as "Darwinian adaptation" are actually models of *aptation* that fit more comfortably with the thinking of de Vries or Morgan than of Darwin. In contemporary evolutionary discourse, "Darwinian" and "adaptive" are contrasted with "neutral" and "nonadaptive," with the result that Mendelian selection is called "Darwinian" selection, de Vriesian aptation is called "Darwinian adaptation," and mutation as well as anything else that is not selection may be called a "nonadaptive process."

7.5.4 Evo-devo

Evo-devo began to assert itself in the 1980s with a verbal theory of "constraints" intended to contrast an idealized version of neo-Darwinism (see Box 6.1) with a less-than-ideal version in which selection still governs evolution, just not as well as one might have imagined, due to the numerous constraints under which it acts. Thus the agenda of the "constraints" advocate in the 1980s was to identify factors that prevent selection from achieving perfection. Over the past 20 years, the "constraints" agenda has given way to more positive ideas such as evolvability (Hendrikse et al., 2007), "facilitated variation" (Kirschner and Gerhart, 2005), and "innovability" (Wagner, 2014) (for a general discussion of this shift, see Brigandt, 2015; one still finds the "constraints" terminology in the quantitative genetics literature, e.g., Bolstad et al., 2014).

The ideas of evo-devo may be difficult to grasp. Ever since the architects of the Modern Synthesis popularized the notion that "the mechanisms of evolution constitute problems of population genetics" (Dobzhansky, 1937, p. 11), the conventional expectation is that causal theories in evolutionary biology will call on population-genetic forces to account for observed patterns. For instance, the neutral theory proposed that random fixation of selectively neutral (or nearly neutral) alleles was much more important than previously imagined, and that this explains patterns such as the tendency

for unimportant parts of molecules to change faster than important parts.

By contrast, evo-devo arguments typically have not put forth broad falsifiable claims about the causes of evolution, but focus on explanatory concepts that give developmental biology an important place. Indeed, a constraint is not a cause. Likewise, the focus on evolvability does not reflect a belief that there is a new population-genetic force called "evolvability," but rather, the belief that the observed ability of organisms to evolve cannot be understood without consulting principles of development, and that understanding the evolvability of organisms in these terms is a major task of evolutionary biology (see Section 5.7). To use the language of philosophers, the major *explanandum* (thing to be explained) in evo-devo is that organisms are so "evolvable" or "innovable." The focus is on "natural principles that accelerate life's ability to innovate" (Wagner, 2014).

Note that, whereas the evo-devoists are seeking to explain evolvability, they typically are *not* seeking an adaptationist "why" that would attribute evolvability to a process of evolutionary adaptation. Instead they focus more on a mechano-structuralist explanation of how evolvability works, an explanation in terms of developmental principles like modularity, weak linkage, compartmentation, and so on. That is, their main focus is on what we defined earlier (Box 5.1) as E2 claims that link internal organizational principles with evolutionary tendencies or capacities.

With this brief introduction to evo-devo, let us return to the problem of variation stated earlier. A complete evolutionary theory must either (1) include a theory of form and its variation, or (2) assume the morphotron. In explanations of the history of evo-devo, one often hears the complaint that development was treated as a black box and left out of the "Modern Synthesis." This reflects a way of thinking that mixes up the issues of theoretical unification and disciplinary integration. If the morphotron assumption is valid, we do not need to understand development to understand why evolution turns out the way that it does. The theory underlying the claims of the original Modern Synthesis does not "leave out" or "ignore" development: it takes full account of the role of development in evolution, which is that development pertains to unimportant details inside the black box, and selection determines important phenotypic features outside the black box.

Scholl and Pigliucci (2015) make a parallel argument in their analysis of Mayr's proximate-ultimate distinction. Mayr cannot be accused of ignoring development, as he so often insisted that selection acts on phenotypes and not genotypes: instead, his dismissal of the relevance of developmental *explanations* for evolution is based on the theory that, because variation is abundant, infinitesimal, and undirected, developmental processes are not explanatorily relevant to evolutionary outcomes (i.e., they are not difference-makers).

How do the ideas of evo-devo address the problem of variation, helping us to understand its source laws and consequence laws? Most of the arguments offered by evo-devo apparently are *not* attempts to solve the problem of variation as I have posed it above. A frequent feature of evo-devo advocacy is what Amundson (2001, p. 176) calls the "causal completeness" argument—that evolution cannot be explained without explaining phenotypic forms, which cannot be explained without developmental genetics. That is, population genetics cannot be a complete theory of evolution: development has to be included somewhere.

In practice, the causal completeness argument has led to a plea for including an alternative "nar-

rative" of developmental causation in evolution, rather than re-thinking conventional narratives of population-genetic causation. This way of thinking implies that explanations for episodes of evolution have two parts (e.g., Wilkins, 1998), a dry population-genetics account of the evolutionary forces at work, and a wet biological narrative of changes in development that accompany evolutionary transitions. A full account of evolution would have both narratives. This position mistakenly accepts all the flawed arguments in Chapter 6 regarding causes, levels, and so on.

In some cases (e.g., West-Eberhard, 2003), the ideas of evo-devo are broader and more sophisticated than what we can understand easily within the dichotomy of source laws and consequence laws.

Yet, *some* evo-devo arguments clearly align with the need to solve the problem of variation with adequate source laws and consequence laws. Many advocates of evo-devo, e.g., Thomson (1985), Arthur (2004), Hendrikse et al. (2007), Psujek and Beer (2008), Braendle et al. (2010), and Sears (2014), see the challenge posed by development as a matter of understanding the impact of biases in variation on the course of evolution. Phenotypic variation exhibits biases that are important in evolution—"nonrandom variation produces preferred trajectories in phenotype space" (Hendrikse et al., 2007)—and the study of development is the means to understand these biases. This is not merely an explanatory claim, nor is it a plea for an alternative narrative to satisfy the causal completeness criterion. Instead, this is a claim that the course of evolution importantly reflects nonuniformities in the direction and amount of variation that are developmental in origin. As Arthur (2004, pp. 65–66) writes

No reasonable person is going to object to me or any other biologist saying that our theory of evolution would be more complete if it took on board not just the selective reasons for a bird's beak getting bigger but also the nature of the developmental system that produced the ancestor's particular beak and the ways in which it got modified . . . But many people may object if I say, as I will, that these very developmental changes are in part responsible for the directions that evolution takes.

Indeed, Brookfield (2005), in his review of Arthur (2004), finds it "disturbing" to suggest that mutational or developmental biases exert a dispositional influence on evolution that is not adaptive. If there is such an influence, mediated by variation, there must be some set of developmental source laws governing the production of biased variation, and there must be some set of consequence laws governing their influence on evolution. We will consider some of these laws in Chapter 8.

7.5.5 Molecular evolution: the case of codon usage bias

The genetic code (see Fig. 8.8) has 61 codons for 20 amino acids. The codons for the same amino acid are called "synonymous codons," and the general many-to-one relationship of codons to amino acids is called "degeneracy." For instance, GGT, GGC, GGA, and GGG are synonymous codons encoding glycine (Gly). Since the 1980s, it has been clear that synonymous codons for Gly and the 17 other amino acids with multiple codons (all but Met and Trp) are not used equally. Molecular evolutionists have been discussing the causes of codon usage bias since the 1980s. Most explanations implicate the combination of mutation biases together with selection favoring some synonymous codons over others, possibly for reasons of translational efficiency that correlate with tRNA abundance (Hanson and Coller, 2018).

Over time, the analysis of codon usage has matured to the point that some robust conclusions are possible. Modern high-throughput methods enable a genome-wide assessment of genes, tRNA abundances, mRNA abundances, and even translation rates. Shah and Gilchrist (2011) integrate all of this information in a stochastic model of codon bias in which fitness is based on a biophysical model of translational efficiency.

The results of applying this approach to *S. cerevisiae* are shown in Fig. 7.4. Each panel shows the frequency of synonymous codons in yeast genes, where the genes have been binned according to expression level along the horizontal axis, with high expression at the right. The first ten panels show the cases of 2-codon blocks, which are the easiest to interpret. In a 2-codon block, there is always one codon ending in G or C (red) and one ending in A or T (blue). Mutation-accumulation experiments indicate a modest (roughly 2-fold) AT bias in yeast (Zhu et al. 2014). Figure 7.4 shows that the AT-ending (blue) codons, which are mutationally preferred, predominate in most genes. In highly expressed genes, however, the codon favored by translational selection increases in frequency. The blue and red lines cross when the mutationally favored codon is not the translationally favored one, which is the case for seven out of ten blocks.

The remaining nine blocks in the figure include amino acids with more than two codons (note that serine is split into a 2-codon block and a discontiguous 4-codon block). The overall pattern is roughly the same, in the sense that translationally favored codons become strongly favored for log expression levels above 0.

Inferred effects of mutation and selection are both strong. Mutation apparently is not just a material cause, but shapes the codon composition of genomes. To make more precise statements about the magnitude of effects, one must consider the histogram at lower right, which shows that the vast majority of genes are not highly expressed. For the 90 % of yeast genes below −0.5 in log expression level, i.e., the vast majority, the pattern of codon use is dominated by mutational preferences. Translational efficiency dominates codon use only in the top 10 % of genes in terms of expression. However, the bias can be extremely strong: as one moves into the top 5 % of genes (above −0.1 in log expression), nonpreferred codons are typically held to a frequency below 10 %; and in the top 1 % of genes (above 0.5 log expression), the nonpreferred codons are *nearly absent* for the cases of Gln, Glu, Tyr, Ile, Ala, Gly, Thr, Pro, Val, Leu, and Arg, indicating powerful selection.

More generally, the literature of molecular evolution has featured claims of mutational effects on composition ever since Sueoka (1962) and Freese (1962) suggested that differences in genome composition might reflect simple mutation biases. Unfortunately, the vast majority of claims for mutational effects in the literature of molecular evolution are weak for the same reason that the Sueoka–Freese

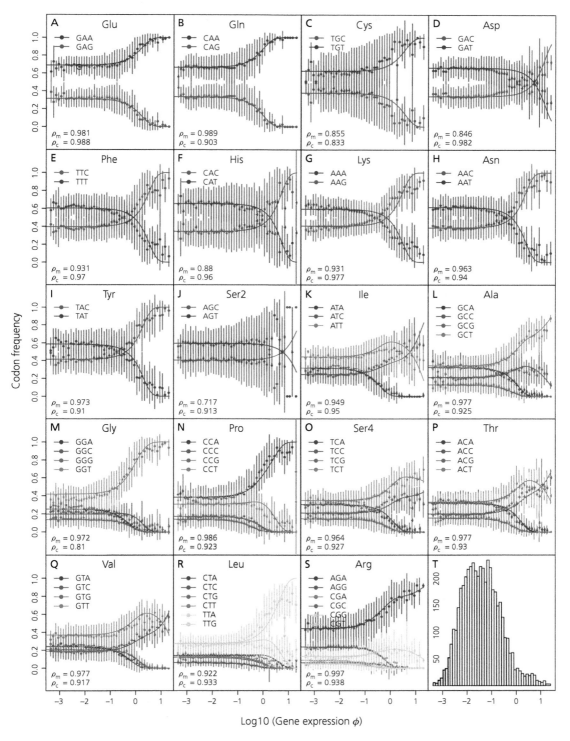

Figure 7.4 An analysis of codon usage, mutation bias, and gene expression level in yeast (Shah and Gilchrist, 2011). Abundance for mutationally preferred (blue) and nonpreferred (red) codons as a function of expression level. Mutationally preferred codons tend to predominate unless they are translationally nonpreferred and the gene is highly expressed. Figure kindly provided by Dr. Premal Shah.

argument was weak: there is no experimental measurement of mutation biases, nor independent verification. Instead, a mutational source law is simply called into existence to explain an observed pattern. The assumption was that high-GC (high-AT) organisms had, at some time in their past, evolved under a regime of GC-biased (AT-biased) mutation, but no attempt was made to verify the postulated direction and magnitude of the biases experimentally, which is extremely difficult.

The dangers with this way of thinking were exposed over a decade ago when it became clear that GC-biased gene conversion can have a significant effect on GC composition (Duret and Arndt, 2008). Prior to this, it was assumed that genome composition would have an explanation in terms of either mutation bias or selection, and then suddenly, there was a third factor—not previously considered—that, in principle, could render void all previous analyses, because these analyses were not based on validated quantitative models of mutational effects or selective effects, but were merely inferences created out of thin air to fit a model.

In fact, because biased gene conversion depends on heterozygosity, its potential impact on genome composition in prokaryotes is limited, nonetheless its potential role in eukaryotic GC content evolution shows the danger of accepting inferred mutation biases without any experimental verification. This danger was exposed yet again when Hershberg and Petrov (2010) reported that, for five different prokaryotic species, actual mutation biases favored AT, even for species from clades with GC-rich genomes.

Though claims of adaptive effects in molecular evolution are often little better (i.e., fitness effects are postulated without any detailed verification), the study by Shah and Gilchrist (2011) is again an exception in that fitness is based on a model in which some parameters represent biological properties whose values are fixed by physical measurements (tRNA concentrations, gene expression) and some parameters are free to vary, with values inferred from observations (codon usage) using a model. The biophysical model represents a source law for fitness effects of codon choices.

Box 7.3 Taking neo-Darwinism seriously: a practical example

A side effect of the perverse tendency to equate "neo-Darwinism" (or "Modern Synthesis") with whatever is widely accepted is that it is now regularly assaulted on grounds that have nothing to do with genuine neo-Darwinism. For instance, the attack by Noble (2015) on the "conceptual framework of neo-Darwinism, including the concepts of 'gene,' 'selfish,' 'code,' 'program,' 'blueprint,' 'book of life,' 'replicator' and 'vehicle'" is almost entirely a critique of late-twentieth-century genetic reductionism à la Dawkins, and addresses neither neo-Darwinism, nor the conceptual framework developed in the 1950s by Dobzhansky and colleagues. Clearly this framework minimizes the role of development and certain concepts of emergence, yet it relies distinctively on *other kinds of emergence* (the gene pool, population-level forces) in the service of a theory in which selection acts as a governing principle—hardly a reductionist premise!

This disciplinary amnesia is unfortunate. Genuine neo-Darwinism has been influential in evolutionary thinking precisely because it is *a powerful theory*, literally capable of explaining any adaptive outcome as the selection-guided accumulation of infinitesimals. The shifting-gene-frequencies theory, and neo-Darwinism more generally, are useful theories even when wrong, just as neutral models are useful even when wrong. One cannot carry out *modus tollens* reasoning about evolution without such models (see Box 1.1). Being able to specify a neutral model, e.g., as a null model, is an important skill for the evolutionary biologist. Likewise, an important skill is to specify a neo-Darwinian hypothesis in which selection is a creative governing force and mutation is merely the supplier of raw materials.

Consider the case of lateral gene transfer. For lateral gene transfer to occur in a manner conformant with the shifting-gene-frequencies theory, the entire process of change must take place gradually by shifting allele frequencies, it must involve may loci at once, it must be fueled by recombination, and it must be governed by selection.

How can we satisfy these criteria with some scheme of lateral transfer? Let us begin with the imaginary case of transferring a 5-bp segment denoted "TRANS" from species X to species Y. If we can accomplish this small transfer, then we can scale up from 5 bp to 500 or 5,000 bp. We begin with a species X genome with the sequence xxxxxxxxxTRANSxxxx and a species Y genome with the sequence yyyyyyyyyyyyyyy. To be successful, the transfer process must give rise to an altered Y genome yyyyyyyyyyTRANSyyyy that did not exist previously.

Figure 7.5 shows how this could happen in a neo-Darwinian manner roughly consistent with the shifting-gene-frequencies theory. Before the start of the process (before "evolution" starts), micromutational introgression from species X fills up the gene pool of species Y, so that many Y genomes have infinitesimal pieces of the TRANS segment. A sample of six Y genomes is shown on the left of Fig. 7.5. With abundant variation available in the gene pool, the selection-driven creative process of assembling a complex combination of features can begin.

From the beginning, critics of Darwin's thinking have objected that selection would not be able to distinguish the incipient stages of useful structures, e.g., the rudimentary beginnings of a wing, or in this case, the rudimentary beginnings of a gene. However, the response of Darwin and his defenders has always been that one does not need to propose large initial steps (saltations), because selection can leverage even the smallest effects.

Therefore, in our model of lateral gene transfer, we simply take neo-Darwinian thinking seriously and assume that selection would recognize the rudimentary beginnings of a new gene with partial function, bringing together new combinations and increasing the frequencies of subsequences like TRA and NS. Over time, recombination and selection would bring together a useful TRANS gene in the Y genome. In this process, introgression was merely a source of raw materials, and not a creative force, consistent with the shifting-gene-frequencies theory. Instead, selection is the creative factor that brings together unlikely combinations from the abundance of variation in the gene pool.

Thus, neo-Darwinian lateral gene transfer is possible in principle. Taking a disciplined approach to applying the theory leads to a specific set of predictions, and sharpens our understanding of how the evolutionary process conflicts with neo-Darwinism: lateral gene transfers are macromutations that transfer an entire gene, or a set of them, at once.

Likewise, the evolutionary emergence of a mitochondrion or chloroplast is not by itself contrary to neo-Darwinism: there is no conceptual difficulty in imagining a mitochondrion being built up gradually by thousands of microsteps. At the level of phenotypes, there is no conceptual difficulty in imagining something like the gain or loss of a digit or a vertebra one cell (or one molecule) at a time. There is no conceptual difficulty in imagining a gene being built up (or laterally transferred, or intercompartmentally transferred) one nucleotide at a time, via a gene pool full of abundant variation.

Neo-Darwinism is a fully general theory in the sense that it can generate a potential explanation for any feature *considered as an end-state*. However, neo-Darwinism does not accommodate every possible *series of successor states*, because it allows only infinitesimal shifts that are beneficial. The theory fails for the examples listed above because our inference of history implicates saltations or discrete jumps. This failure induces the need for other theories that address the size and character of variations (what we called "source laws"), and how these properties of variation influence evolution.

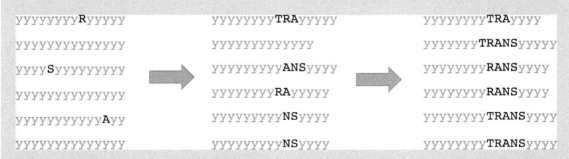

Figure 7.5 A neo-Darwinian model for lateral gene transfer of a segment.

7.5.6 The genomic challenge to adaptationism

Entire books have been written about genome evolution (Gregory, 2005b; Koonin, 2011; Lynch, 2007b), a topic much too large to consider in depth here. Below, I will simply sketch the phenomenology of genome diversity based on these sources, then offer brief comments on what is known or suspected about the underlying causes, relating that to the problem of variation. The title of this section follows Sarkar (2014), who addresses a narrower range of issues in a more detailed way.

If we simply survey the landscape of genomes of cellular organisms (setting aside the viruses arbitrarily), and attempt to understand their diversity abstractly, certain *dimensions of diversity* rise to prominence, even before trying to decode the sequences: the vast range of sizes, compositional diversity, repeat structure, and so on. More dimensions of diversity emerge by mapping genes and genetic elements onto the genome, and by comparing the genes within and across genomes, considering their relationships in the context of a branching process of evolution.

A million-fold range of genome sizes The diversity of genome sizes ranges nearly a million-fold from 0.6 Mb for the smallest free-living microbe (the endosymbiont *Carsonella ruddii* has an even smaller genome of 0.16 Mb) to > 150,000 Mb in *Paris japonica* (which is the largest sequenced genome but perhaps not the largest genome), with only a 100-fold range of gene numbers.

Vast within-species variation in genome size Most prokaryotic species have genome sizes in the same narrow range of 2 to 5 Mb, yet genome size often varies intra-specifically by large amounts (e.g., up to 30 % in *E. coli*). By contrast, for eukaryotes, there is a 10^5-fold range of genome sizes, but within-species variation is typically very small except in the case of karyotypic variation.

GC content diversity and heterogeneity The main dimension of diversity in genome composition is the frequency of the G and C nucleotides, which ranges from 15 % to over 75 %. Within genomes, GC content may be very constant, or may vary widely on a scale of kilobases to megabases. In mammals, a 1 Mb region can vary from 40 % to 60 % GC.

Prokaryotic strand bias In prokaryotes, the GC skew measure $(G - C)/(G + C)$ and the AT skew measure $(A - T)/(A + T)$ show a disparity between strands that peaks at the origin of replication and (typically) diminishes symmetrically going out from the origin.

Repeat content Prokaryotic genomes typically have very little repetition, whereas many eukaryotic genomes have long arrays of tandem repeats, or repeats dispersed throughout the genome. In some plants, over half of the genome is devoted to repetitive DNA, mostly mobile elements but also tandem repeats.

Gene number diversity Cellular genomes differ in the number of genes from 525 in *Mycoplasma genitalium*—or 180 in the endosymbiont *Carsonella ruddii*—to 50,000 in many plants.

Split genes Eukaryotic protein-coding genes are typically split into introns and exons, whereas prokaryotic genes typically are not. Intron density ranges from < 0.01 per gene in yeast and some other unicellular eukaryotes, to ≈ 6 per gene in mammals and vascular plants. The introns are so numerous and long in mammals and vascular plants that genes are > 10-fold longer, on average, than those in prokaryotes, even though the length distribution of proteins varies little between species of cellular organisms.

Pseudogenes Dead, nonfunctional copies of genes called "pseudogenes" are rare in prokaryotes but common in eukaryotes, which also have retroposed pseudogenes that entered the genome via an RNA intermediate.

Scrambling of genetic maps Changes in gene order occur at a wide range of rates. Often the genetic maps of two species are extensively scrambled in 100 MY, but in some cases, maps remain constant over this time scale, or become scrambled on a much shorter time scale.

Genome doublings and segmental duplications Modern genomes of plants, animals, and fungi contain signs of ancient genome doublings, as well as smaller segmental duplications. These processes are ongoing today, though they occur considerably more often in plants.

Differential gene family evolution by birth and death When genes are assigned to families, the families have widely varying numbers of members and rates of evolution.

Mobile element invasion and propagation Nearly all genomes are infested with mobile elements, and some house a zoo of diverse elements. Plant genome evolution is subject to explosive bursts of transposition so extreme they significantly effect overall genome composition and even genome size.

Intercompartmental transfer Nearly all eukaryotes have compartmentalized mitochondrial or plastid genomes with evolutionary dynamics distinct from that of the nuclear genome, and with a gene composition that reflects a long-term trend of gene transfer to the nucleus.

Lateral gene transfer and the pangenome The gene composition of many prokaryotic species is so fluid that the concept of a genome with a specific structure and a predetermined gene complement simply does not apply. Instead, the core genes (present in all members of a species) and dispensable genes (present only in some) make up a "pan-genome" that may have as many genes as a large eukaryotic genome.

Now, let us consider this genomic diversity of living organisms from the perspective of the problem of variation. Genome evolution, in general, might be understood as the selection-guided adaptation of genomes to changing conditions, with abundant micromutations providing the raw material. When conditions dictate adaptive increases in genome size, the genome would grow gradually; when conditions dictate an adaptive shift in sequence composition, it would change accordingly. The features of genomes might have been built gradually by the hand of selection on the genomic morphotron.

Genomes are discrete, but this discreteness is very fine grained and easy to understand, because the smallest heritable unit of change in a double-stranded DNA genome is obviously the base-pair. The smallest possible variations—those that allow the most precise adaptation and are most likely to be beneficial, by Fisher's argument—are to insert, delete, or substitute a single base-pair. Any

sequence can be converted to any other sequence with this basic repertoire of operations. If we expand this repertoire with the ability *to break a sequence* into two parts, or *to join two ends*, then we can change the number of chromosomes in a genome, and their topology (circular or linear).

Thus, a set of five simple micromutational operations provides the universality to change any known genomic set of DNA molecules into any other, i.e., to account for *all of the diversity described earlier*. This point is important to repeat: in principle, we can account for all known genomic diversity with infinitesimal changes. For instance, a series of about 6×10^9 micromutational changes would be sufficient to convert the single circular chromosome of a hypothetical prokaryotic ancestor into the 23 chromosome pairs of the 6×10^9 bp diploid human genome. That is, the shortest path of micromutational change, assuming no detours or reversals, would have 6×10^9 micromutational steps. If we imagine following this path over 3×10^9 years of evolution, the average rate of change is two micromutational changes per year. In a neo-Darwinian theory, these 6×10^9 steps would be selection-driven shifts.

Of course, *real genome evolution looks nothing like this*. The first thing one notices, when faced with the literature on genome evolution, is that a dominant adaptive explanation is often absent. Lynch's reference to "the frailty of adaptive hypotheses" (Lynch, 2007a) might seem shocking in some other field, but the diversity and complexity of genomes do not look particularly adaptive, and attempts to explain them as organismal adaptations have repeatedly fallen apart. For instance, it was long supposed that genomic differences in GC content would reflect thermal adaptation, until systematic data were gathered relating the genomic GC content of an organism to its optimal growth temperature, with the result was that there was no correlation (Hurst and Merchant, 2001). Ideas about genome size and organismal complexity have a long history, mostly full of wonderment about why there does not seem to be much of a correlation (Gregory, 2005a).

Selection is often assigned a negative role. When assigned a positive role, selection may be called upon to account for the propagation of sub-genomic elements that are harmless or slightly harmful to the host—selfish DNA such as transposons that adapt to maintain an infestation of the genome, and occasionally to infect new hosts. Genomic components that are not inherited in exactly the same way, such as organellar and nuclear genomes, can come in conflict with each other (McLaughlin and Malik, 2017). The concept of "the unity of the genotype" (Mayr, 1963, Ch. 10) has no apparent influence in contemporary evolutionary genomics. Eukaryotic genomes, in particular, are a mess.

Repeatedly, evolutionary thinkers of the past have promoted the intuitive claim that evolution requires macromutations, and that these macromutations, in contrast to numberless indistinct micromutations, must play a distinctive and important role. Whether or not this intuition was ever justified in the past, it is justified in genome evolution. Genome doublings, segmental duplications, lateral gene transfers, rearrangements of all sorts, transposon insertions—these are all macromutations that play distinctive roles in the evolution of genomic novelty. They are macromutations not only in the sense of being orders of magnitude larger than the smallest possible changes, but also in the important sense that they have a distinctive character that influences any further steps. For instance, a 10-kb gene duplication mutation is not just a bucket of base-pairs with 10,000 times more raw material than one base-pair. Instead, it springs forth as a complex and highly ordered sequence, ready for operation.

7.6 Synopsis

The problem of variation can be defined in an abstract way, by reference to the morphotron. If variation is sufficiently abundant, infinitesimal, and undirected, then perhaps selection is the guiding hand that shapes the features of organisms. The biological world might be a world in which variation is like clay, merely supplying raw materials. In such a world, we would not need a substantive theory of form and variation to have a fully general theory of evolution, supporting complete explanations and predictions.

The problem of variation arises because this view breaks down both in theory and in practice. Even under the most forgiving conditions, the proposi-

tion that selection is the ultimate source of explanation in biology fails. There is no possible world in which selection governs evolution as a ruling principle, and variation merely supplies raw materials. Empirical results from quantitative genetics, molecular evolution, and evo-devo all show that the outcome of evolution depends on peculiarities, discontinuities, and regularities of variation, giving the generation of variation a dispositional role.

The preceding chapters all serve to provide depth to our appreciation of this problem—the problem of variation.

First, we have seen historically how the need to rethink the role of variation emerges from the peculiar circumstances of the early twentieth century. An empirical and computational revolution that has been accelerating for several decades supports the capacity (today) to gather massive amounts of systematic data and to infer detailed retrospective accounts of evolutionary changes. Yet, this revolution was still decades out of reach in 1930. The neo-Darwinian conception of the role of variation made a virtue of necessity: it was convenient to assume that what was unknown did not matter. Today, we may look back at the open-ended Mendelian synthesis—mutation, selection, and inheritance, without the neo-Darwinian ideology—and find it more appealing. Yet, a century ago, the Mendelian view was utterly lacking in power, because it offered no governing forces, only an open-ended framework that (to be applied) required detailed knowledge that simply was unavailable.

Second, we have learned how to identify the specific technical and rhetorical implications of this historic response, as embodied (for example) in the shifting-gene-frequencies theory and the randomness doctrine. We have discerned, in the writings of the architects of the Modern Synthesis, a coherent theory—the shifting-gene-frequencies theory of evolution—that brings together gradualism, the gene pool, and population-genetic forces, so as to justify the neo-Darwinian dichotomy of the potter and the clay. In addition, we have learned to recognize the vast arsenal of argument used by the architects of the Modern Synthesis to promote the general irrelevance doctrine—arguments covered in Chapter 6 about proximate causes, forces, creativity, raw materials, and so on.

This issue is not merely historic, of course. Today, the neo-Darwinian response to the problem of variation remains deeply embedded in evolutionary discourse. For instance, popular and educational treatments of evolution continue to cast selection as the ultimate source of explanation or meaning in biology, with variation as merely a source of raw materials. Indeed, the "raw materials" metaphor continues to be used in the research literature, even in the evo-devo literature, where it conflicts with the basic premise of internalism. Clearly, rethinking the role of variation for the twenty-first century will be difficult.

Yet, third, we can recognize multiple ways in which this process of rethinking the role of variation has already begun, with some threads—from molecular evolution, evo-devo, evolvability, and quantitative genetics—that stretch back decades.

As a first approximation, we can make use of the distinction of source laws and consequence laws of variation, which together allow predictions about variational effects on evolution. This distinction helps us to formulate questions for theoretical or empirical research, and to identify common concerns in different parts of the evolutionary literature focused on molecular evolution, evo-devo, and evolvability. Much more could be said in this regard.

In Chapter 8 and Chapter 9, we use this framework to confront a major issue from a narrow perspective. The major issue is the problem of variation. The narrow perspective is to understand a few very basic source laws and consequence laws of variation, mainly from a molecular perspective, but also extending this thinking in some ways to include conventional phenotypes.

CHAPTER 8

Climbing Mount Probable

In this important class of cases of causation, one cause never, properly speaking, defeats or frustrates another; both have their full effect. If a body is propelled in two directions by two forces, one tending to drive it to the north, and the other to the east, it is caused to move in a given time exactly as far in *both* directions as the two forces would separately have carried it; and is left precisely where it would have arrived if it had been acted upon first by one of the two forces, and afterwards by the other.

John Stuart Mill, *A System of Logic*

8.1 Introduction

The notion that *mutation supplies the raw materials that selection shapes into adaptations* is so familiar, one might read this claim as though it were a bland truism suitable for a textbook. However, the previous chapters have established that, in the context of a century of evolutionary discourse, this claim represents a *useful, provocative, and falsifiable theory* positing specific and contrasting roles for selection and variation—the potter and the clay. Historic advocates of this theory ridiculed the idea that internal tendencies of variation could act as dispositional causes, e.g., Simpson (1967, p. 159) refers to "the vagueness of inherent tendencies, vital urges, or cosmic goals, without known mechanism."

Importantly, the historic advocates of neo-Darwinism did not dismiss internal causes merely on the basis of ideological prejudice. Instead, they worked hard to establish the irrelevance of variation, attacking the idea with the diverse arsenal of arguments covered in Chapter 6, summarized in Section 6.7. Some of these arguments appeal to *basic principles*, e.g., the "opposing pressures" argument and the "wrong level" argument appeal to the idea of evolution as a population-level process controlled by "forces"—forces that are alleged today to be

fundamental to evolutionary biology (e.g., Lynch, 2007a; Futuyma, 2017).

And yet, we began this book with a brief review of results showing a dispositional role of the generation of variation (Section 1.1), contradicting this historical arsenal of arguments. The "opposing pressures" argument, the "wrong level" argument, and so on, are mistaken. This mistaken position defines the work of this chapter, in both (1) a negative sense of articulating what is wrong in these arguments, and (2) a positive sense of formulating alternatives that allow us to grasp the potential for a dispositional role of variation in evolution, in a practical way that allows us to formulate hypotheses and propose explanations.

Note that a part of this work has already been done in the first half of this book, where we learned to dismiss simplistic ideas and appreciate the complexity of mutation as a process that exhibits propensities.

The work that remains is to build up a body of theory with which to understand the implications of such propensities for evolutionary change, considering the types of biases that are possible (including developmental ones), population-genetic models of the influence of biases on one-step adaptation and on adaptive walks, concepts of dual

Mutation, Randomness, and Evolution. Arlin Stoltzfus, Oxford University Press (2021). © Arlin Stoltzfus. DOI: 10.1093/oso/9780198844457.003.0008

causation, the forces theory, and the special case of parallel evolution.

8.2 Climbing Mount Probable

Consider, as an analogy for evolution, a climber operating on a rugged mountain landscape like that shown in Fig. 8.1. A human climber would develop a plan to reach the peak, but an analogy for evolution requires a kind of movement that follows only local rules, without planning; therefore let us imagine a blind robotic climber. The robot will move by a simple proposal-acceptance mechanism. In the proposal (introduction) phase, the robot climber reaches out with one of its limbs to sample a nearby handhold or foothold. Each time this happens, with some nonzero probability, the robot may employ the handhold or foothold to shift its position.

Biasing acceptance, such that relatively higher points of leverage have relatively higher probabilities of acceptance, causes the climber to ascend, resulting in a mechanism, not just for moving, but for climbing.

What happens if a bias is imposed on the proposal process? Imagine that the robotic climber (perhaps by virtue of longer or more active limbs on one side) samples more points on the left than on the right. The probability of proposal is greater on the left, therefore the joint probability of proposal-and-acceptance is greater (on average) on the left, so the trajectory of the climber will be biased, not just upwards, but to the left as well. If the landscape is rough (as in Fig. 8.1), the climber will tend to get stuck on a local peak that is upwards and to the left of its starting point. In this case, the lateral bias in the proposal step causes a bias in both (1) the path and (2) the final destination. If the landscape is smooth, the climber will spiral to the left, ascending to the single summit, resulting in a bias in the path, but not in the final destiation.

This analogy of "Climbing Mount Probable" helps us to conceptualize the possible influence of a bias in the introduction of novelty in evolution. The analogy seems promising, but analogies and metaphors are only useful if they are not misleading. Therefore, it is important to understand whether, and under what conditions, the metaphor of Climbing Mount Probable accurately depicts evolutionary behavior. To find out, we need to evaluate the idea in the terms of population genetics.

Figure 8.1 The rugged landscape of the Aiguille Verte (Green Needle) in the French Alps (modified from original image by John Johnston, reproduced under Creative Commons Attribution CC-BY-2.0 license).

8.3 One-step adaptive walks under mutation bias

To establish a principle, one begins with the simplest case. In the simplest case of Climbing Mount Probable, the climber faces just two different steps, one going up and to the left, and the other going up and to the right. If the leftward path both offers a greater ascent *and* is favored by a proposal bias, then obviously the climber will tend to go to the left, and likewise if both biases favor the rightward path, the climber will tend to go to the right. That much is obvious. The nontrivial case to explore is when the proposal step favors going left and the acceptance step favors going right, as in Fig. 8.2 (left).

Now, having articulated the simplest case, let us render it in the terms of population genetics, following Yampolsky and Stoltzfus (2001). From the starting point, we must be able to access two different alternatives, each with a single mutation. In the simplest possible model, an <u>ab</u> genotype may mutate to <u>Ab</u> or <u>aB</u>, i.e., the simplest case is a 2-locus, 2-allele model as shown in Fig. 8.2 (right).

Arbitrarily, we will choose <u>aB</u> as the alternative favored more strongly by mutation ($\mu_2 > \mu_1$), and <u>Ab</u> as the alternative that is more advantageous ($s_1 > s_2$). Let B represent the bias in mutation favoring <u>aB</u>, $B = \mu_2/\mu_1$, and let K represent the bias in selection coefficients favoring <u>Ab</u>, $K = s_1/s_2$. By making t smaller than s_2, short-term evolution will proceed either to <u>aB</u> or <u>Ab</u>. The kind of question we wish to ask is whether B will ever be influential, and if so, how influential will it be? How will the outcome depend on K, or on the strength of selection?

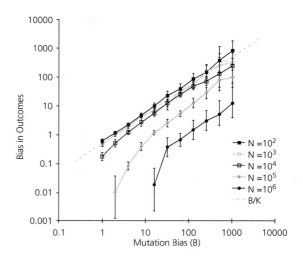

Figure 8.3 Effect of population size in the Yampolsky–Stoltzfus model.

To simulate evolution, we start with a population of <u>ab</u> individuals that may evolve either to be predominantly <u>Ab</u> or predominantly <u>aB</u>. Population size (N) will vary from 10^2 to 10^6, and B will vary from 1 to 1,000 by setting $\mu_2 = 10^{-5}$ per generation, and allowing the other mutation rate to vary; meanwhile, $s_1 = 0.02$ and $s_2 = 0.01$, so that $K = 2$. For each set of conditions, we will repeat the simulation hundreds of times and plot the results.

Simulation results from Yampolsky and Stoltzfus (2001) are shown in Fig. 8.3, with the bias in outcomes—the ratio of <u>aB</u> outcomes to <u>Ab</u> outcomes—as a function of mutation bias B. In all but the largest populations, most of the time is spent waiting for a beneficial allele to arise and

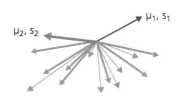
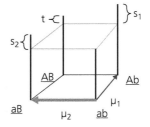

Figure 8.2 The Yampolsky–Stoltzfus model. The left panel shows an abstract set of choices faced by the climbing robot, presenting just two upward steps, one of which is more likely to be proposed (thick green arrow), and the other of which is more likely to be accepted due to being higher (blue arrow). The right panel shows a 2-locus, 2-allele model that captures this difference, used by Yampolsky and Stoltzfus (2001).

reach sufficient numbers that loss becomes unlikely, after which it quickly rises to a high frequency, thus the time required for a simulation varies inversely with N, on the order of 10^3 to 10^5 generations.

The first thing to notice is that all of the lines are going up, i.e., the bias in outcomes increases with the bias in mutation. For $N = 100$ or $N = 1000$, the bias in outcomes is proportional to B and inversely proportional to K, thus B/K (short grey dashes). As explained in Section 8.6, B/K is the exact expectation for a population in the regime of origin-fixation dynamics. For larger population sizes, we depart from the origin-fixation regime, but mutation bias continues to influence the outcome of evolution.

Now, let us consider the behavior of this model according to the common view that evolution is a matter of applying the various forces of population genetics, including selection, drift, mutation, and recombination (in older sources, migration is often listed as a force). The forces each have a magnitude: s for selection, u or μ for mutation, and something like $1/N_e$ for drift. Population geneticists describe the forces as being weak or strong depending on their magnitude, e.g., the opposing-pressures argument from Fisher and Haldane (6.4.2) holds that mutation is a weak force relative to selection, and therefore cannot influence outcomes. This is why population geneticists (following Gillespie,

1983b) sometimes refer to part of the origin-fixation regime (McCandlish and Stoltzfus, 2014) as "strong selection, weak mutation" (SSWM), namely the part where selection coefficients are far from neutrality, and mutation rates are such that $\mu N << 1$.

The behavior of the Yampolsky–Stoltzfus model seems to violate the opposing pressures argument, allowing for internal tendencies of variation to influence the outcome of evolution—something that was held to be impossible. Why?

Is it simply a matter of selection not being strong enough to overcome mutation? We can probe this possibility by varying the magnitude of selection without changing anything else. The results in the left panel of Fig. 8.4 are based on varying s_1 (the higher selection coefficient) 200-fold from 0.001 to 0.2, using $B = 16$ (triangles), $B = 4$ (squares), or $B = 1$ (diamonds), with $N = 1,000$ and other parameters as before ($\mu_2 = 10^{-5}$, $K = 2$). The lines showing the bias in outcomes are flat. That is, relative to the conditions in Fig. 8.3, we have varied the magnitude of selection from 10-fold weaker to 20-folder stronger, without seeing a significant change in the bias in outcomes.

Similarly, we can ask whether the biasing effect is dependent on high mutation rates. That is, let us consider in detail what happens as mutation rate gets smaller, shown in the right panel of Fig. 8.4. For

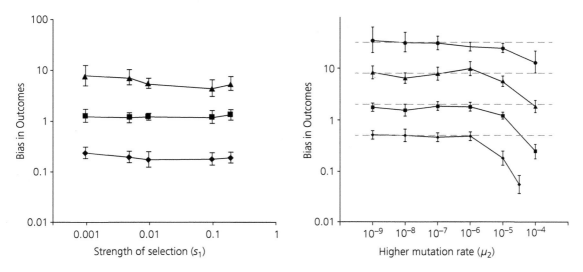

Figure 8.4 Effect of strength of selection (left), and convergence on origin-fixation expectations (right) in the Yampolsky–Stoltzfus model.

Box 8.1 Dual causation: lessons from chemotaxis

Many prokaryotic cells are motile by means of rotary flagella that act as propellers, pushing the cell from behind. Such motile cells often exhibit *chemotaxis*: moving toward attractants (e.g., nutrients) and away from repellents.

The system of chemotaxis in *E. coli*, which has been studied in detail (Parkinson et al., 2015), relies on the alternation of two phases, tumbling and swimming. When the flagella are all turning counter-clockwise they combine (on one end of the cell) into a single bundle that propels the cell; when the flagellar motors are reversed, the bundle unwinds and the flagella continue to spin separately, pointing in multiple directions, causing the cell to tumble. If the bacterium is swimming up a gradient of attractant (or down a gradient of repellent), tumbling is suppressed and swimming is prolonged. That is, the cell simply swims longer when it is swimming in a favorable direction. *The combination of re-orientation via tumbling and prolongation of swimming results in a biased random walk.*

Humans sometimes use a similar procedure: our noses cannot easily detect the orientation of an odorous gradient, but we can sense changes in intensity over time, thus we can detect whether movement in a given direction brings us closer to, or further from, the source of an odor.

The process is easy to simulate in a computer. In the simulations represented in Fig. 8.5, all trajectories begin at the origin and end when the cell has traveled four units to the

right, up the gradient of attractant (each trajectory is plotted in a series of colors from red at the start, to green at the end). The left figure shows chemotactic movement based on the model for *E. coli*, with unbiased tumbling, whereas the right panel shows an *imaginary* kind of chemotaxis in which tumbling is biased in the vertical dimension (specifically, a tumble upward is three times more likely than a downward one).

In both the left and right panels, all cells move at least four units to the right. In the left panel, some paths go up and others go down on the vertical axis, but the mean trajectory is roughly flat. In the right panel, by contrast, the mean trajectory is biased upwards by several units.

These examples illustrate several concepts useful in our discussion about evolutionary causes—causal completeness, dual causation, orthogonality, and the composition of causes.

First, bacterial chemotaxis is analogous to evolution under the neo-Darwinian theory, in that it combines one process that is responsive to external conditions (swimming) and another process that scrambles orientation (tumbling).

The dual process is adaptive, and we could call this "dual causation," but it is a rather unequal duality. Tumbling is clearly a part of the process, and so a complete mechanistic account of chemotaxis (satisfying causal completeness) must include tumbling. Yet, tumbling is not the source of the key adaptive feature of chemotaxis, which is biased movement relative to the gradient (i.e., biased along the horizontal axis).

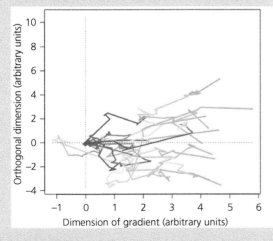 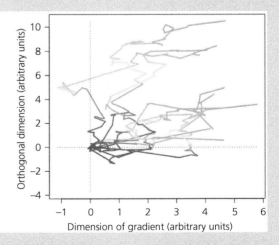

Figure 8.5 Simulated trajectories of ten chemotactic cells responding to a gradient of attractant increasing along the horizontal axis. Tumbling is unbiased in the horizontal dimension, and is either unbiased (left) or biased (right) in the vertical dimension.

Continued

Box 8.1 *Continued*

In the same way, variation is required in a neo-Darwinian mode of evolution, but plays no dispositional role.

When tumbling is biased (right panel), the trajectories exhibit two different biases, a rightward adaptive bias due to the prolongation of swimming, and an upward bias due to tumbling. The process is still composed of the alternation of two phases, tumbling and swimming, but now tumbling contributes a bias to directionality, though not to adaptive movement.

This bias in tumbling, which is orthogonal to the gradient of attractant, does not prevent chemotaxis and does not incur a cost or penalty. We could call the tumbling bias a "constraint" or "limit" on chemotaxis, but this usage is problematic. If the tumbling bias were not perfectly orthogonal to x, but aligned partially with x, or against

x, then this would sometimes improve chemotaxis and sometimes hinder it, depending on the direction of the gradient.

Finally, we can imagine more complex scenarios for movement in more than two dimensions. In fact, the simulations shown above are three-dimensional, with both plots showing the same set of ten walks projected onto different planes. Specifically, tumbling is biased only in the z dimension, and the left and right figures are projections onto the xy and xz planes, respectively. This means that the left panel fails to reveal a tumbling bias *because it does not show the dimension in which tumbling is biased*, the z dimension. Likewise, if our analysis of evolution focuses only on fitness, rather than movement in phenotype space, we may fail to see biases that are actually present.

$N = 1,000$ and other conditions the same as above, the bias in outcomes is shown as a function of the rate of mutation (the higher rate, μ_2), for values of B of 64 (circles), 16 (triangles), 4 (squares) and 1 (diamonds). Reading the figure *in reverse* from right to left, i.e., as the mutation rate decreases, the bias in outcomes converges on B/K (dashed grey lines) for each set of conditions. That is, *weaker* mutation gives a *stronger* effect that converges on B/K. Thus, contrary to what is suggested by the opposing pressures argument (see Section 6.4.2), the effect of mutation in imposing a bias on evolution depends, not on the largeness of the mutation rate, but on its smallness.

8.4 Extended adaptive walks under mutation bias

The model in Section 8.3 depicts a one-step adaptive walk, in which evolution ends after one move. What will happen in a multi-step walk? Are long-term trends possible?

This question has two parts: one involving source laws, and the other involving consequence laws. The question about source laws is whether the biological world includes variational biases that persist, and whose effects may accumulate. For instance, if a GC (or AT) bias persists and is influential, the result will be a shift in composition

toward GC (or AT), until some balance is reached. Some other kinds of biases do not have this property, as discussed in Section 8.9.

The consequence-law questions are about the conditions under which the effects of a directional bias in variation will accumulate. What conditions will allow such an effect? What factors affect the dynamics?

To address this issue, i.e., to allow a multi-step adaptive walk analogous to the path of the mountain-climbing robot we imagined previously (8.2), we must consider some larger space of possibilities for which fitness can be defined (i.e., a fitness landscape). The "NK" formalism of Kauffman (1993) represents N loci or sites, each with some number of possibilities (minimally 2), and K interactions per site. This model allows a fitness landscape with "tunable" ruggedness, in the sense that we can use $K = 0$ for a smooth landscape, and use higher values for an increasingly rugged landscape (explained later).

Here we will consider a simplified NK model from Stoltzfus (2006a), before using this type of model to represent a genetically encoded protein (next section). Suppose that each site in a sequence may have one of two arbitrary states designated "F" or "G." We could think of this as a protein sequence composed of two kinds of amino acids, e.g., hydrophobic and hydrophilic. If there is a mutational bias favoring either F or G, then this may influence the composition of the sequence that results from an adaptive walk.

For each adaptive walk, beginning with a randomly generated starting sequence of length $N = 100$, we will iterate a process of (1) choosing a mutation from the mutational neighborhood of the sequence (i.e., the set of one-mutant neighbors), then (2) either accepting or rejecting it depending on its fitness effects. Because we are interested only in the composition of the sequence (the relative proportions of F and G), we will average many walks, and the uniqueness of the individual walks will not be visible. To combine individual walks of different lengths, we simply stretch them to the same length and plot them relative to the mean length (for details, see Stoltzfus, 2006a).

First, consider the special case of a random walk in which every mutation that is proposed, whether $F \rightarrow G$ or $G \rightarrow F$, is accepted with equal probability. Over time, we expect the composition to approach a level that directly reflects B, the bias in mutation. That is, a sequence will either wander in the direction of more F, or more G, depending on the bias. Figure 8.6 (left) shows this process, by which each sequence simply converges on a ratio of F to G that depends on the mutation bias.

Next, let us consider a special kind of adaptation—greedy adaptation—in which only the best possible step is chosen at each opportunity. That is, for a given sequence of F and G states, there is precisely one alternative sequence (in the mutational neighborhood) with the highest fitness, and this sequence is chosen as the next step in the adaptive walk.

Whereas evolution continues forever under the condition of no selection, here the adaptive walks are short—typically, after just two dozen steps, no higher-fitness mutational neighbors are available.

Under this Panglossian form of adaptation, an adaptive walk proceeds deterministically: evolution from a given starting sequence always follows the same course. Because this course is determined fully by effects of fitness, mutation bias has no effect, as shown at right in Fig. 8.6.

A more natural rule for an adaptive walk would be that a mutation, if beneficial, is accepted with a probability $2s$, and otherwise is rejected. The $2s$ rule used in the model of Orr (2002) reflects the classic theoretical result that the probability of fixation is approximately $2s$ for a newly introduced beneficial allele (Haldane, 1927).

Before applying this rule, let us consider what K means. In the NK model, the contribution made by each site to fitness depends on the state (the residue or allele) at that site, and the states at K interacting sites. The fitness of the entire sequence is the sum of all the contributions of each site. Suppose that each site interacts with $K = 1$ other sites, and the map of interactions is such that each site interacts with the site that follows it. Then, the fitness contribution of site 23 (for instance) will depend on which state, F or G, is present at (1) site 23, the focal site, and (2) site 24, the interacting site. Thus, to compute the fitness under an NK model with $K = 1$, we need to specify values of the component of fitness for FF, FG, GF, and GG for site 23, and for every other site. In the model described here, these values are drawn from a uniform distribution (Stoltzfus, 2006a).

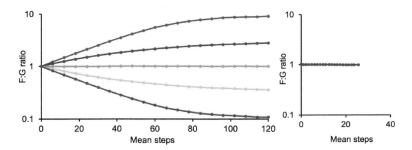

Figure 8.6. Special cases of neutral (left) and greedy (right) evolutionary walks on an NK landscape under mutation bias. Each series represents a different mutation bias, namely $B = 10$ (blue), $B = 3$ (red), $B = 1$ (green), $B = 1/3$ (orange), and $B = 1/10$ (brown). In the neutral case, the ratio of F to G simply approaches 10, 3, 1, 1/3, or 1/10. Under greedy adaptation (right), mutation bias is irrelevant. Here, $K = 4$, $N = 100$.

If the state at site 23 changes, this will change the fitness component both for site 23 and for site 22, which interacts with site 23. By contrast, when $K = 0$, contributions of only one site are changed by a mutation, and all of the other fitness components remain the same. Therefore, changing one state does not change the local landscape much when K is small. That is, when we take one step on a relatively smooth landscape, the topology of the neighborhood still looks very much the same. At the other extreme, when $K = N - 1$, each site interacts with every other site, thus a single change, at just one site, scrambles all the fitness contributions: the landscape is so rugged that the fitness of a sequence is completely uncorrelated with that of its one-mutant neighbors.

Now, with this understanding of the NK model, let us consider evolution using the more natural $2s$ rule for acceptance, but with the unnatural case of $K = 0$. The expected behavior is easy to work out from the starting conditions. Given that $K = 0$, each site has a preferred state (F or G) independent of any other site, i.e., each site has a universally preferred state. The optimal sequence is simply the sequence with the preferred state at each site. If the starting sequence has x sites with the nonpreferred state, then the adaptive walk will consist of all the changes from the nonpreferred state to the preferred state at the x sites. If we simulate 100

sites, then a randomly chosen starting sequence will have the nonpreferred state at 50 sites, on average, and the adaptive walk will consist of 50 steps. On average, half of the steps will involve changing an F to a G, and the other half will change a G to an F. What will adaptive walks look like under mutation bias?

Figure 8.7 (left) shows the evolution of the ratio of F to G under these conditions. To understand these dynamics, we only need to understand that, though the set of steps required to change from the starting sequence to the optimal sequence is determined by the fitness landscape alone, and (on average) has equal numbers of $F \rightarrow G$ and $G \rightarrow F$ changes, the *order* of these steps depends on mutation bias. Again, any walk is a roughly equal mixture of $F \rightarrow G$ or $G \rightarrow F$ changes, but when the mutational bias favors $F \rightarrow G$ (for instance), then $F \rightarrow G$ changes tend to happen first, which is why the composition initially moves in the direction favored by the mutation bias. However, all paths are converging on a peak that, on average, has the mean composition, so the trajectory of composition bias returns to the same central value.

This recalls the metaphor of Climbing Mount Probable. When the proposal step is biased to the left or to the right, and the landscape is a perfectly smooth hemisphere or cone, the climber simply spirals up and to the left, or up and to the right

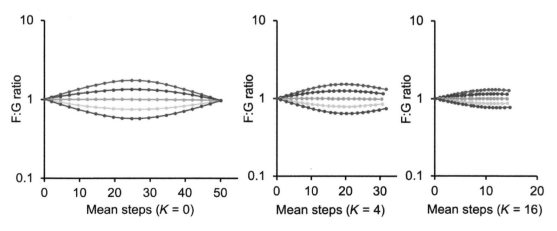

Figure 8.7 Adaptives walks on a simplified NK landscape under mutation bias with three different values of K. As before, each series represents a different mutation bias, namely $B = 10$ (blue), $B = 3$ (red), $B = 1$ (green), $B = 1/3$ (orange), and $B = 1/10$ (brown). The effect of a bias on the initial trajectory is roughly the same regardless of K, but the lengths of walks and the final bias differ greatly (see text for explanation).

(respectively) until the final destination—always the same—is reached.

In passing, note that the order of steps is also biased by fitness effects. The chance of acceptance is proportional to s; therefore, the early steps tend to be the ones with the larger fitness advantages, so the increase in fitness is rapid initially, and then tapers off, an effect noted previously in the discussion of models of adaptation (7.5.3).

What will happen in the more natural case in which there is some roughness in the fitness landscape, i.e., $K > 0$? Adaptive walks for $K = 4$ and $K = 16$ are shown in the center and right panels of Fig. 8.7. For the most extreme biases, the rate of change in composition is 3 % to 4 % per accepted step in the first few steps, the same as in the neutral case or the case of $K = 0$. Since the adaptive walks are shorter as K increases—only 62 % as long when $K = 4$, and only 29 % as long when $K = 16$—the maximum change in composition is not as great as when $K = 0$. This effect is offset somewhat by the fact that the trajectory does not recurve as far for larger values of K.

8.5 Protein adaptation under mutation bias

Based on the previous section, we know something of the behavior of abstract adaptive walks under mutation bias, and these results tend to confirm the intuition suggested by Climbing Mount Probable.

Now, let us consider a more naturalistic case: the effect of a GC bias on the composition of an evolving protein. The meaning of a GC bias in mutation (typically) is that the total rate of mutations from A or T to G or C is greater than the reverse rate. An AT bias would work in the opposite direction. For instance, mutation-accumulation experiments indicate that yeast has a 2-fold bias toward AT (Zhu et al., 2014). If such a bias is effective in biasing evolution, one result will be to change the frequency of amino acids, because some of them have GC-rich codons, and others have AT-rich codons.

In Fig. 8.8, the blue codons encode amino acids with GC-rich codons, and the pink ones encode amino acids with AT-rich codons. Sometimes the ratio of AT-rich to GC-rich amino acids is labeled the FYMINK:GARP ratio (Singer and Hickey, 2000), based on the single-letter codes for amino acids

F	TTT TTC	S	TCT TCC	Y	TAT TAC	C	TGT TGC
L	TTA TTG		TCA TCG	*	TAA	*	TGA
				*	TAG	W	TGG
L	CTT CTC CTA CTG	P	CCT CCC CCA CCG	H	CAT CAC	R	CGT CGC CGA CGG
				Q	CAA CAG		
I	ATT ATC ATA	T	ACT ACC ACA ACG	N	AAT AAC	S	AGT AGC
M	ATG			K	AAA AAG	R	AGA AGG
V	GTT GTC GTA GTG	A	GCT GCC GCA GCG	D	GAT GAC	G	GGT GGC GGA GGG
				E	GAA GAG		

Figure 8.8 The canonical genetic code showing amino acids with AT-rich codons (pink) and GC-rich codons (blue).

(FYMINK for Phe-Tyr-Met-Ile-Asn-Lys, and GARP for Gly-Ala-Arg-Pro).

To understand the effect of a GC or AT bias on protein composition, we will simulate a sequence of 100 amino acid residues, encoded by a gene of 300 nucleotides, following Stoltzfus (2006a).

The use of a genetic code makes the behavior of this model more complex. In the previous section, we considered two mutationally interchangeable states (F and G). However, the 20 natural amino acids cannot change so freely. Given the canonical genetic code in Fig. 8.8, and allowing only single-nucleotide mutations, each codon for one amino acid can mutate to encode four to seven alternative amino acids (TCA is the least connected codon; AAY, CAY, GAY, TGY, and ATY are the most connected). This means that, even when $K = 0$ for the *protein* landscape, the space of *protein-coding genes* has isolated peaks, because some locations that are adjacent in the space of protein sequences are not adjacent in the space of nucleotide sequences. For instance, Ala (GCN) can change to Pro (CCN) and to Gly (GGN), but Gly and Pro codons are not mutationally accessible to each other (by single mutations). Because of this, we can imagine a case in which, for a given site in the protein with Ala, the alternatives Pro and Gly represent isolated fitness peaks accessible from Ala, but not from each other.

Figure 8.9 shows simulated effects of a mutation bias during an adaptive walk on a protein landscape with $K = 0$, with a range of GC bias from $B = 1/10$

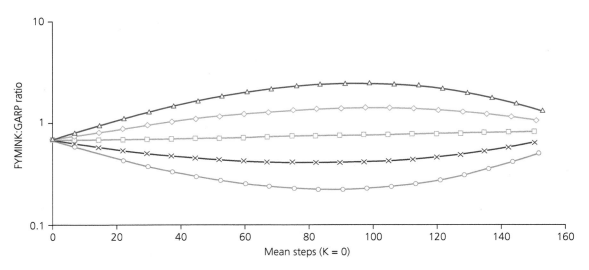

Figure 8.9 Adaptive walks on a protein NK landscape under GC mutation bias, with $K = 0$. Values of GC bias are $B = 1/10$ (red), $B = 1/3$ (orange), $B = 1$ (green), $B = 3$ (brown), and $B = 10$ (blue). Although $K = 0$ at the protein level, the evolving gene sequence experiences a somewhat rough landscape because a codon for one amino acid cannot mutate to access all 19 amino acid alternatives.

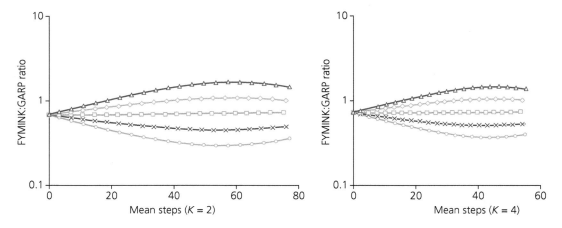

Figure 8.10 Adaptive walks on a rugged protein NK landscape under mutation bias.

to $B = 10$. As before, each series shows the average behavior of hundreds of walks under the same conditions. Because the genetic code has fewer AT-rich FYMINK codons (pink in Fig. 8.8), the average FYMINK:GARP ratio in initial sequences is not 1, but about 0.67. Even though $K = 0$, the trajectories do not all return to the same composition, for the reason just explained.

The two panels for Fig. 8.10 show simulations for the cases of $K = 2$ and $K = 4$. The scale of effects is important to notice. For $K = 2$, the ratio increases

from 0.67 to 1.4 when the bias is 10-fold in favor of AT (red series), and decreases from 0.67 to 0.35 when the mutational bias is 10-fold against (blue series). For the more modest biases of 3 and 1/3, the effect is less, but still substantial.

Note again that these are adaptive walks including only beneficial changes. Evolution here is a climbing algorithm with *a trajectory subject to dual causation* (see Box 8.1): the system climbs upward, but it also shows a lateral bias due to biases in the introduction of variation. In this way, the results confirm the intuitive implications of the analogy of Climbing Mount Probable.

8.6 Origin-fixation dynamics

By the mid-1960s, biochemists comparing proteins had begun to conceive of evolution as a Markov chain of discrete amino acid changes, each determined by a mutation (e.g., Zuckerkandl and Pauling, 1962; Margoliash, 1963; Zuckerkandl and Pauling, 1965). Natural selection was said to "accept" or "reject" mutations, and was often depicted as a filter preventing harmful changes (e.g., Eck and Dayhoff, 1966, p. 161, p. 200). This new way of looking at evolution quickly proved useful in the analysis of sequences (see Box 9.2).

What the new field of "molecular" evolution needed in the 1960s was a theory that generalized the concept of a rate of change and related it to causal factors, including mutation. The origin-fixation formalism emerged in this climate, having been proposed independently by Kimura and Maruyama (1969) and by King and Jukes (1969). Origin-fixation models represent evolutionary change as a simple two-step process of (1) the introduction of a new allele by mutation, and (2) its fixation or loss (for review, see McCandlish and Stoltzfus, 2014). The most familiar version of this formula is $K = 4Nus$, where $2Nu$ is the rate of mutational introduction (in a diploid population with $2N$ alleles) of beneficial alleles with selection coefficient s, and $2s$ is the probability of fixation for such alleles. For the case of neutral alleles, the probability of fixation is $1/(2N)$ in a diploid

population, so the origin-fixation rate of neutral evolution is $K = 2Nu/(2N) = u$.

This radical simplification skips over most of what we usually think of as "population genetics," e.g., it does not include any allele frequencies. Figure 8.11 illustrates how origin-fixation models relate to population genetics. A population is depicted as a single particle that periodically jumps from one state to another.

Origin-fixation models have distinctive implications regarding the rate of evolution, the direction of evolution, and the role of selection (McCandlish and Stoltzfus, 2014). The first implication is that the effect of selection is represented as a stochastic sieve or filter that accepts or rejects new mutations one at a time, depending on their fitness effects (and the population size).

The second implication is that the rate of evolution is directly proportional to the mutation rate, e.g., doubling the rate of mutation doubles the rate of evolution.

The third implication, and the most important implication for our purposes, is that, in the regime of origin-fixation dynamics, biases in mutation rates can bias the course of evolution, including adaptive evolution. For example, consider a model in which a population is currently fixed at allele i and can mutate to a variety of other alleles. Using μ for a mutation rate and π for a probability of fixation (dependent on a selection coefficient s and a population size N), the odds that the population will next become fixed at allele j rather than allele k are given by Eqn. 8.1:

$$\frac{P_{ij}}{P_{ik}} = \frac{2N\mu_{ij}\pi(s_{ij}, N)}{2N\mu_{ik}\pi(s_{ik}, N)} = \frac{\mu_{ij}}{\mu_{ik}} \times \frac{\pi(s_{ij}, N)}{\pi(s_{ik}, N)} \qquad (8.1)$$

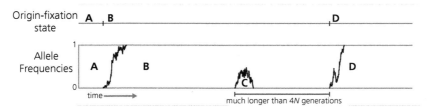

Figure 8.11 The origin-fixation model. The lower figure shows the frequency of alleles at a locus that begins with allele A. After an interval of time, allele B is introduced by mutation and rises to fixation. At some later point, allele C is introduced and eventually goes extinct. Later, an allele D arises and replaces B. In the origin-fixation formalism in the upper figure, this process is depicted as a series of shifts in state, as if the entire population were a single particle that jumps from state A to B and then to D.

This can be described as the product of two factors: a ratio of mutation rates, and a ratio of fixation probabilities. For the case of beneficial alleles, we can use the probability of fixation $2s$ (Haldane, 1927), and this reduces to Eqn. 8.2 (Yampolsky and Stoltzfus, 2001).

$$\frac{P_{ij}}{P_{ik}} = \frac{\mu_{ij}}{\mu_{ik}} \times \frac{s_{ij}}{s_{ik}} \qquad (8.2)$$

Note that, in these equations, we have switched our terms of reference from rates to probabilities; the switch is accomplished mathematically by converting rates to waiting times by inversion, then finding the chance of a smaller waiting time ($\frac{1/r_2}{1/r_1}$) rather than computing a ratio of rates (r_1/r_2).

In the Yampolsky–Stoltzfus model presented earlier (see Section 8.3), we learned that, for some population sizes, the bias in outcomes is not different from B/K. This is simply an instance of the relation in Eqn. 8.2, i.e., $B = \mu_{ij}/\mu_{ik}$ and $1/K = s_{ij}/s_{ik}$. Not only does a bias in mutation have a direct, proportional effect on the odds of one outcome relative to another, this effect is neither weaker nor stronger, but exactly the same as, a bias in fixation of the same magnitude. That is, in the origin-fixation regime, doubling the mutation rate to an allele or doubling its probability of fixation both have the same effect on the odds of being the next allele to be fixed.

Equation 8.1 (or 8.2) provides a useful framework for comparing the impact of biases in the origination process to the impact of biases in the fixation process (Farlow et al., 2011; Stoltzfus and Yampolsky, 2009; Streisfeld and Rausher, 2011) or modeling adaptation under the influence of mutation bias and fitness differences (Rokyta et al., 2005). We review some of these cases in Chapter 9.

Gillespie (1984) introduced a "mutational landscape model" that has become important in the more recent literature due to the work of Orr and others (see 7.5.3; Orr, 2005a; McCandlish and Stoltzfus, 2014). In this model, each genotype has a fixed number of mutational neighbors, and evolution continues until a local peak is reached (i.e., all of the mutational neighbors are less fit). Typically this kind of model uses an origin-fixation move-rule as the basis for assigning a probability to each of the possible steps. The chance that the next step is to alternative j, normalized to the full set of alternatives k, and assuming Haldane's $2s$, is given in Eqn. 8.3:

$$\frac{P_{ij}}{\sum_k P_{ik}} = \frac{2N\mu_{ij}2s_{ij}}{\sum_k 2N\mu_{ik}2s_{ik}} = \frac{\mu_{ij}s_{ij}}{\sum_k \mu_{ik}s_{ik}} \qquad (8.3)$$

For instance, the protein-NK model that we described in the previous section has these dynamics: the chance of each step is proportional to the rate of mutational introduction and to the selection coefficient. The model of Orr (2002) makes the further assumption of mutational homogeneity such that μ is always the same, leading to a simpler form, Eqn. 8.4,

$$\frac{P_{ij}}{\sum_k P_{ik}} = \frac{\mu_{ij}s_{ij}}{\sum_k \mu_{ik}s_{ik}} = \frac{s_{ij}}{\sum_k s_{ik}} \qquad (8.4)$$

in which s is the only factor (we show in Section 9.3 that this simplification goes too far).

Two very general points about origin-fixation models are of interest. First, origin-fixation models have developed into an important branch of theory with applications to the dynamics of adaptation, properties of evolutionary rates (e.g., their dispersion), the evolution of codon bias, co-evolving sites, analyzing Fisher's geometric model, the evolution of binding sites, and phylogenetic inference (McCandlish and Stoltzfus, 2014). Second, the wide use of origin-fixation models has not been accompanied by much critical attention to their applicability or their broader evolutionary significance. Indeed, origin-fixation models have been applied to systems in which one simply cannot justify the underlying assumptions (McCandlish and Stoltzfus, 2014). When multiple alleles are segregating in a population, the probability of fixation is no longer a simple function, and indeed, the concept that an allele has a single unchanging selection coefficient can become unworkable (Gillespie, 2004).

8.7 The sushi conveyor and the buffet

In Chapter 6, we viewed an array of arguments that appeared to rule out, or render unlikely, a dispositional role for the generation of variation. Some of these arguments appealed to population genetics. Yet, in the previous sections, we used population genetics to show how biases in variation may impose tendencies on evolution, including adaptive evolution. Do the rules of population genetics prohibit variational trends, as argued by the founders of theoretical population genetics and the architects of the Modern Synthesis? Or do the rules of population genetics support the possibility of variational trends?

In fact, the classic opposing-pressures argument, and the argument of Yampolsky and Stoltzfus (2001), reach different conclusions because they rely on two different conceptions of evolutionary genetics, both of which can be understood as special cases of a broader conception of evolutionary genetics. To understand the difference intuitively, let us consider two different styles of self-service restaurant, one equipped with a buffet, and the other, with a sushi conveyor (Fig. 8.12). The customer of the buffet begins with access to an abundance of choices, and fills a plate with the desired amount of each dish. By contrast, from the sushi conveyor, the customer iteratively makes a yes-or-no choice of the chef's latest creation as it passes by. In either case, the customer exercises choice, and may obtain a satisfying meal.

In the shifting-gene-frequencies theory of the Modern Synthesis, evolution occurs according to the buffet model, based on abundant pre-existing variation in the "gene pool." Just as the staff who tend the buffet will keep it stocked with a variety of choices sufficient to satisfy every customer, the gene pool "maintains" abundant variation sufficient to meet any adaptive challenge (selection never has

to wait for a new mutation). Adaptation happens when the customer gets hungry and proceeds to select a platter of food from the abundance of choices, each one ready at hand, choosing just the right amount of each ingredient to make a well-balanced meal. The customer provides the initiative and makes all the choices.

Out of many kinds of buffets, the most appropriate for this analogy is a large and well-stocked salad bar, because the ingredients of a salad bar are often raw unrefined parts (not carefully prepared, finished dishes), consistent with classic view of the role of mutation as supplying unfinished pieces, rather than fully formed features. To be true to the "gene pool" concept, we have to imagine a salad bar with an extravagant, carefully maintained abundance of ingredients.

The thinking of molecular evolutionists (and the early geneticists) frequently corresponds to the sushi conveyor model, which presents us with a simple proposal-acceptance process reminiscent of the mutationist mantra "mutation proposes, selection disposes" (decides). The customer exercises choice, but does not control *what is offered or when*: instead, the choice consists of accepting or rejecting each dish that passes by. Even though the customer has decided on each dish, initiative and creativity belong largely to the chef.

The effect of a bias in the choices offered to customers (analogous to a bias in variation) depends on which model applies. Let us suppose that the buffet has four apple pies and one cherry pie. For

Figure 8.12 The buffet and the sushi conveyor, two styles of self-service restaurant that illustrate some differences between the "gene pool" regime depicted in the shifting-gene-frequencies theory, and the mutation-fixation view of molecular evolutionists (buffet image by Helder Ribeiro, sushi conveyor image by NipponBill, reproduced under Creative Commons Attribution Share-alike licenses).

most customers, this quantitative bias will make no difference. The customer who prefers cherry pie will choose it every time. The only kinds of biases that are relevant are *absolute constraints*—the complete lack of some possible dish. A customer who prefers cherry pie will choose apple pie only if cherry pie is not available.

At the sushi conveyor, the effect of a bias will be different. Let us suppose that occasionally a plate of sushi appears, with a 4 to 1 ratio of salmon to tuna. And let us suppose that a particular customer prefers tuna. Even a customer who would prefer tuna in a side-by-side comparison may choose salmon more frequently, because a side-by-side comparison simply is not part of the process. In our hypothetical scenario of a 4:1 bias for salmon rather than tuna, a customer with a 2-fold preference for tuna will choose salmon more often.

Now, just as we previously explored the analogy of Climbing Mount Probable with a precise model, let us consider how this difference-by-analogy maps to a difference in evolutionary genetics. We have already seen sushi-conveyor dynamics, both in simulations of the Yampolsky-Stoltzfus model (Section 8.3), and in origin-fixation models of the sequential-fixations type (Section 8.6).

What about the buffet regime? The key condition in a hypothetical buffet regime is that the variants relevant to the outcome of evolution are already present in the gene pool: evolution is merely a process of shifting gene frequencies to a new optimum distribution. Therefore, let us consider what is the effect of initial variation in the Yampolsky–Stoltzfus model presented earlier (see Section 8.3). We will simulate the model exactly as before, except that, instead of a pure ab starting population, we will add the alternative Ab and aB genotypes.

The results, shown in Fig. 8.13, are simple and compelling. The upper series (closed triangles) represents the same results for $N = 1000$ shown earlier. In the lower two series, the initial population contains aB and Ab at frequencies of 0.005 (open circles) or 0.01 (open squares). Even when the initial frequencies of the alternatives are just 0.5 %, a 1,000-fold difference in mutation bias has almost no effect, as shown by the flat lines. Thus, the distinction between the buffet and the sushi conveyor helps

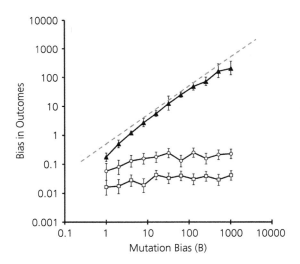

Figure 8.13 Effect of initial variation in the Yampolsky–Stoltzfus model.

us to understand a genuine distinction between regimes of population genetics.

Why is selection so powerful in the buffet regime? The reason is that, if two alternative alleles are established in a population, having reached sufficient numbers to make random loss unlikely, selection nearly always establishes the more fit alternative, even if it starts out at a lower frequency. But the same difference in selection coefficients has a more modest effect in the sushi conveyor regime. Given a probability of fixation of $2s$, a 2-fold difference between $s_{Ab} = 0.02$ and $s_{aB} = 0.01$ represents merely a 2-fold preference for Ab.

Why is the mutation bias so influential in the sushi-conveyor regime, yet negligible in the buffet regime? In the sushi-conveyor regime, a *B*-fold bias in introduction is a *B*-fold bias on evolution. In the buffet regime, this kind of causation is not just weakened, but absent. This is because the bias operating so effectively in the origin-fixation regime is a bias in the *introduction process*, i.e., a bias in the rate of introduction of new alleles caused by the mutation bias. This kind of bias is a profoundly important effect that directly impacts the course of evolution. But in the buffet regime, there is no bias in the introduction process, because *there is no introduction process*—all relevant alleles are present already.

To summarize, biological processes of reproduction and mutation may have quite different implications depending on the regime of population genetics. At one extreme of origin-fixation dynamics (the sushi conveyor), biases in mutation act as biases in the introduction of novelty, which are a powerful influence on the outcome of evolution. At the other extreme of the buffet, in the shifting-gene-frequencies theory of evolution, there is no introduction process, and biases in mutation can act only by shifting allele frequencies, which makes them largely ineffectual.

Between these two extremes are intermediate regimes with intermediate behavior, as indicated by the cases in Fig. 8.3 where the effect of a B-fold bias is less than B.

Much more work is needed to clarify this intermediate behavior. These results are from simulations. The behavior of the Yampolsky–Stoltzfus model does not have obvious mathematical solutions except in two limiting cases. In the limiting case as $\mu N \to 0$ with N fixed (where $\mu = \mu_1 + \mu_2$), we may consider an absorbing Markov chain whose behavior is readily understood with origin-fixation dynamics. In the limiting case as $\mu N \to \infty$ with μ fixed, we may consider a set of coupled differential equations whose behavior dictates that the fittest type asymptotically approaches a frequency near 1. In the intermediate regime, the problem has not been solved, but a solution may be possible using methods developed for the case

of "concurrent mutation" (Desai and Fisher, 2007; Weissman et al., 2009). Indeed, new work on this issue by Gomez et al. (2020) suggests that effects of mutation bias are not limited to the origin-fixation regime, but extend into the concurrent mutations regime.

8.8 Why the theory of forces fails

In the shifting-gene-frequencies theory of evolution (Box 6.2), change at the phenotypic level is a smooth shift to a new adaptive optimum. Underlying this smooth shift in phenotypes is a process of shifting gene frequencies, in which small-effect alleles at multiple loci are shifted from their previous frequencies to a new optimal distribution of frequencies.

Figure 8.14 represents this process for three loci, each represented by a dimension in a three-dimensional space. This conception of evolution, in which genetic change is movement in an allele-frequency space, provides the framework for understanding the theory of population-genetic forces: a process that can displace a population in allele-frequency space qualifies to be a "force" of population genetics. For instance, in the classic allelic selection model, the force of selection raises the frequency of a rare beneficial allele from a low initial value, to a high value approaching $f = 1$ (fixation). In this case, the allele-frequency space in which selection acts has a single dimension.

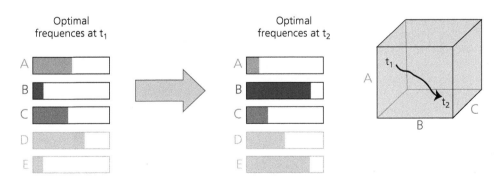

Figure 8.14 In the shifting-gene-frequencies theory of evolution, change in trait values is caused by population-genetic "forces" that shift frequencies. Here, an episode of evolution from t_1 to t_2 consists of shifts in frequency at multiple loci, shown at the left. On the right, this change is depicted in an allele-frequency space with three dimensions corresponding to the first three loci, A, B, and C (additional dimensions for D and E are not shown).

Likewise, in the mutation-selection balance model, with two different forces, mutation pushes the frequency of a deleterious allele upward, and selection pushes it downward (see Stephens, 2004). The resulting equilibrium frequency ($f \approx \mu/s$ in the simplest case) is typically much closer to 0 than to 1, a result explained by saying that the force of selection is typically much stronger than the force of mutation. Note that, as Sober (1984) explains, "In evolutionary theory, mutation and selection are treated as *deterministic* forces of evolution," whereas "random genetic drift is the source of the stochastic element in evolution." The forces are not all equal in strength. Selection is typically the most powerful, and mutation the weakest.

The concept of a force is deliberately general, allowing multiple forces and multiple dimensions. Consider the front face of the cube in in Fig. 8.14, representing the A and B dimensions. Let the position in the A dimension from 0 (origin) to 1 (top) be the frequency of allele A2, as distinct from A1, and likewise for B2 and B1 in the B dimension. In a population of 1000, a small shift from the center of this area (0.500, 0.500) to (0.506, 0.506) would mean adding six more A2 and six more B2 individuals. We can imagine such a shift happening by drift, selection, or mutation, although in practice, a shift of $\sim 1\%$ is too much for mutation to accomplish in a generation: it is indeed a weak force.

We could change the details of this example in many ways without changing the conclusion, and that is the power of the force analogy: a force is a force because it has a *generic ability to shift a frequency* from f to $f + \delta$, where δ is an infinitesimal. Stated differently, the *common currency of causation* for population-genetic forces is a shift in allele frequencies. By way of this common currency, the forces have a property of comparability and, in ideal cases, are combinable, as in the mutation-selection balance equation earlier. When considering more than two dimensions (loci), the evolution of a population becomes a path or traverse through a multidimensional frequency-space, and this path can be broken down into infinitesimal shifts that, in principle, could be caused by any force. This is how the theory of forces achieves generality.

Yet, this generality only holds in the *topological interior* of an allele-frequency space, e.g., the interior of the cube in Fig. 8.14, but not its surfaces or edges. This is why the theory fails for systems subject to the dynamics of the introduction process. We described these failures earlier in regard to the Yampolsky–Stoltzfus model (Section 8.3): neither increasing the magnitude of selection coefficients (which increases the force of selection), nor decreasing the magnitude of mutation rates (which decreases the force of mutation), causes selective preferences to overpower mutational ones (see Fig. 8.4).

The introduction process is a different kind of cause with *a different currency of causation*: the ability to change a frequency from nothing to something, from 0 to $1/N$ (or $\frac{1}{2N}$). Mutation has this ability, but not selection or drift. For instance, considering two dimensions A and B as before, the shift from an initial point $P_0 = (0.500, 0.500)$ to $P_1 = (0.506, 0.506)$ is mathematically identical to the shift from $P_0 = (0.000, 0.000)$ to $P_1 = (0.006, 0.006)$, but it is not evolutionarily identical: the latter shift absolutely requires the involvement of mutation. A mutation that introduces a new allele allows the system to jump off of the periphery into the interior of an allele-frequency space, where the forces of selection and drift operate.

The assumption that we can understand evolution as a process that takes place in the interior of an allele-frequency space is the key assumption of the shifting-gene-frequencies theory of the Modern Synthesis (Box 6.2). Under this assumption, mutation is truly a weak force, so that it becomes possible to *construct models of evolution that do not include mutation rates*, as explained in Box 8.2. If we accept this theory, we can accept the claim of Dobzhansky (1974) that "It is not on the level of mutation, but rather on that of natural selection and other genetic processes that take place in living populations, that the creativity of evolution manifests itself" (p. 330). That is, high-level forces are responsible for displacing the system in allele-frequency space. If we define evolution as a process whose outcome depends on alleles already present, then the introduction process is literally not part of evolution, but something that happens to establish a precondition for evolution (population variation).

Yet, at the same time, we must now reject this argument, along with many of the arguments of Chapter 6 about "proximate causes," "raw materials," and so on, as being inapplicable when evolutionary change occurs in the sushi conveyor regime, or any regime that depends significantly on the dynamics of the introduction process, e.g., the concurrent mutations regime of Gomez et al. (2020). The arguments about causation from the Modern Synthesis canon are not universal principles, but principles specific to the shifting-gene-frequencies theory of evolution.

Interestingly, the "levels" argument (see Section 6.2.3) could be reinterpreted to yield a different conclusion. Population-level forces have the character of emergent statistical laws, like those of statistical physics. Accordingly, the architects of the Modern Synthesis argued that the process of mutation (or development) in an individual is not a population-level force. In the same way, the introduction process is not a force, because it does not have the character of a statistical law. Yet, one could argue that the introduction process *is emergent at the population level*, in the precise sense that we cannot determine whether an event of mutation (e.g., from A1 to A2) in an individual is an introduction event, without examining the population: the concept of "introduction" has no meaning for individuals, but emerges at the population level.

One might object to this critique on the grounds that the forces theory is really an invention of philosophers and that actual scientists would never make the kind of mistakes induced by a literal application of the forces theory. This objection is undermined by a rich documentary record of scientists reasoning about evolution under the guidance of the forces theory. For instance, the canonical founders of theoretical population genetics, and the architects of the Modern Synthesis, all relied on the opposing-pressures argument (Section 6.4.2) to dismiss the prospects for variation-biased evolution. Likewise, consider the seminal "developmental constraints" article by Maynard Smith et al. (1985), whose list of authors included three eminent theoreticians. They cite the opposing pressures argument, which presented a major obstacle:

[The issue] is whether biases on the production of variant phenotypes (i.e., developmental constraints) such as those just illustrated, cause evolutionary trends or patterns. Since the classic work of Fisher (1930) and Haldane (1932) established the weakness of directional mutation as compared to selection, it has been generally held that directional bias in variation will not produce evolutionary change in the face of opposing selection. This position deserves reexamination.

Though Maynard Smith et al. (1985) called for a re-examination, they offered no resolution (other than the unsatisfactory suggestion of neutral evolution). Thus, in a critical response, Reeve and Sherman (1993) simply called on the opposing-pressures argument, claiming that Maynard Smith et al. (1985) had failed to provide any valid evolutionary mechanism.

Box 8.2 Population genetics without the introduction process

How important was the shifting-gene-frequencies theory of evolution? Though largely forgotten today, the theory strongly shaped ideas of causation and approaches to modeling, with ongoing impacts. In particular, the commitment of theoreticians to a view in which all the variants relevant to the outcome of evolution are present initially in the "gene pool" (Box 6.3) is evident in a large body of theory in which evolution is *literally defined as the sorting out of variation in an initial population*. Under these conditions, the influence of mutation is merely to shift the relative amounts of the alleles, an effect of little importance due to the smallness of mutation rates. As Lewontin (1974) noted, "There is virtually no qualitative or gross quantitative conclusion about the genetic structure of populations in deterministic theory that is sensitive to small values of migration, or any that depends on mutation rates" (p. 267).

Continued

Box 8.2 *Continued*

Thus, it was considered reasonable in some classical theoretical studies to omit mutation, on the grounds that its effects are trivial and can be ignored, e.g., the treatment of the mathematical foundations of population genetics by Edwards (1977) has no terms for mutation in its hundreds of equations; the word "mutation" occurs only on page 3, in the sentence "All genes will be assumed stable, and mutation will not be taken into account."

The need for a theory relating the rate of evolution to the rate of mutation became apparent in the 1960s, due to the molecular revolution. Origin-fixation models emerged in this context, in 1969 (McCandlish and Stoltzfus, 2014). However, for decades, these models were used mainly to address the fate of neutral or slightly deleterious mutations (McCandlish and Stoltzfus, 2014). Meanwhile, outside of molecular evolution, theoretical population genetics still relied heavily on the shifting-gene-frequencies theory. Theoreticians began to notice this restriction about 20 years ago, e.g., Yedid and Bell (2002) write

> In the short term, natural selection merely sorts the variation already present in a population, whereas in the longer term genotypes quite different from any that were initially present evolve through the cumulation of new mutations. The first process is described by the mathematical theory of population genetics. However, this theory begins by defining a fixed set of genotypes and cannot provide a satisfactory analysis of the second process because it does not permit any genuinely new type to arise.

Likewise, Hartl and Taubes (1998) write

> Almost every theoretical model in population genetics can be classified into one of two major types. In one type of model, mutations with stipulated selective effects are assumed to be present in the population as an initial condition . . . The second major type of models does allow mutations to occur at random intervals of time, but the mutations are assumed to be selectively neutral or nearly neutral.

A similar distinction is made by Eshel and Feldman (2001, p. 182):

We call short-term evolution the process by which natural selection, combined with reproduction (including recombination in the multilocus context), changes the relative frequencies among a fixed set of genotypes, resulting in a stable equilibrium, a cycle, or even chaotic behavior. Long-term evolution is the process of trial and error whereby the mutations that occur are tested, and if successful, invade the population, renewing the process of short-term evolution toward a new stable equilibrium, cycle, or state of chaos.

They conclude that

> Since the time of Fisher, an implicit working assumption in the quantitative study of evolutionary dynamics is that qualitative laws governing long-term evolution can be extrapolated from results obtained for the short-term process. We maintain that this extrapolation is not accurate. The two processes are qualitatively different from each other. (p. 163)

Thus, the influence of the shifting-gene-frequencies theory resulted in a blind-spot in modeling, such that theoreticians were not prepared to recognize a role for biases in the introduction of variation.

In 1969, a half-century after theoretical population genetics emerged as a discipline, the introduction process began to appear as an unnamed technical feature of certain models (origin-fixation models as per McCandlish and Stoltzfus, 2014), though not as a feature of contending theories, which focused on selection, drift, dominance, heterosis, sex, and so on (e.g., see Crow, 2008).

After several more decades, theoreticians began to abandon the extrapolationist doctrine of the original Modern Synthesis—the doctrine that all of evolution (macroevolution) follows from the short-term process of shifting gene frequencies (microevolution) seen in experimental populations of animals and plants. A central argument of this book is that recognizing the introduction process is not a minor technical detail, but a major innovation that challenges previous thinking about how evolution works, opening new avenues for theoretical and empirical research, including research on biases in the introduction process.

8.9 The sources and forms of biases

This chapter has thus far considered models with abstract differences in mutation rates, whereas in the early chapters of the book, we considered various concrete examples of mutational processes and their spectra. What is the relationship between the two? What kinds of mutation biases might be relevant to these abstract models? Are all mutation processes the same with respect to their potential influences on evolution?

In this section and in the next two sections, we consider some intermediate concepts to make sense of where biases come from, what forms they take, and how they may influence evolution. We find that differences in rates of introduction are not all the

Figure 8.15 Transition–transversion bias (left) and GC:AT bias (right) are different kinds of biases with different implications for evolution.

same with respect to their causes and their possible consequences. Some biases are developmental rather than mutational. Different kinds of nonuniformity have different long-term implications for evolution.

For instance, earlier in the chapter we considered effects on protein composition of a mutation bias toward (or against) G and C nucleotides. In Section 7.5.5, we found that yeast shows a roughly 2-fold bias toward AT that is important for understanding codon usage bias. Transition-transversion bias is another type of bias (left in Fig. 8.15) commonly invoked in studies of molecular evolution.

Interestingly, even if the magnitude of bias is the same, the possible evolutionary effects of these two kinds of biases are different. Both could be the cause for an increase in parallelism, but only the GC:AT bias could cause a long-term trend. That is, a consistent bias toward GC (or AT) has a predictable effect on the state of the system, moving it toward more GC (or AT), until an equilibrium is reached. Other examples of directional biases are an insertion:deletion bias, a bias toward duplications of repetitive regions, a bias toward insertions in heterochromatin, etc.

By contrast, a transition–transversion bias is not by itself directional: an excess of transitions does not lead to a state of increased transition-ness. Other things being equal, it will cause some kinds of *changes* more often than others, without making any kinds of *states* more frequent than others.

The analogous phenomenon for a continuous phenotypic trait would be the case in quantitative genetics where the population exhibits greater additive genetic variation for one trait than another. Suppose trait 1 has value t_1 and variance V_1, and trait 2 has value t_2 and variance $V_2 > V_1$. Because $V_2 > V_1$, we expect more change in trait 2, but the

expected value of trait 2 in the absence of any other effect is still t_2.

A second important distinction is that some biases are properly "mutational" in the sense of arising from local mechanistic properties of the pathways that generate mutation, whereas others reflect the location of the focal system within an abstract possibility-space (Stoltzfus, 1999). To determine if a bias is mutational, we must specify each mutation as the change from the starting genome to a fully specified alternative. If one mutation takes place at a different rate than a second mutation, this is properly a mutational bias.

Though asymmetries in locally accessible possibility-spaces are not precisely the same thing as mutation biases, their effects on evolution can be exactly the same. That is, consider a form of Eqn. 8.1 for the case in which we are aggregating across two classes of possible changes to determine the odds of getting a change of type j rather than k (Eqn. 8.5):

$$\frac{P_{ij}}{P_{ik}} = \frac{\sum_j 2N\mu_{ij}\pi(s_{ij}, N)}{\sum_k 2N\mu_{ik}\pi(s_{ik}, N)} \qquad (8.5)$$

If mutation rates and selection coefficients are uniform, there will still be a bias when $j \neq k$. For instance, if we have categorized all the locally accessible alternatives as to whether they increase or decrease complexity (as in Xue et al., 2015), then any asymmetry will represent a prior bias on the outcome of evolution.

For a more extended example, consider Fig. 8.16, which depicts a model of "constructive neutral evolution" applied to the evolution of gene-scrambling in ciliates (from Stoltzfus, 2012). The genes in question are split into segments by DNA-based intron-analogs that are developmentally excised from somatic copies of the genes prior to expression: this process is proposed to render innocuous the ordinarily disastrous reordering and inversion of segments (for reasons having to do with the specific mechanism of developmental excision). To the extent that this condition is met, the system may be free to wander into a morass of scrambled possibilities like the one shown for a three-segment gene in Fig. 8.16.

The complexity of this type of landscape explodes rather quickly and becomes difficult to visualize.

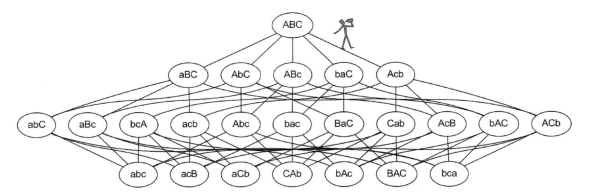

Figure 8.16 A drunkard's walk through a space of scrambled and unscrambled gene configurations. Here a lower-case letter indicates an inverted segment, thus an inversion including the last 2 segments changes ABC to Acb. A random walk from the unscrambled state ABC will tend to wander into the much larger space of scrambled possibilities.

However, the landscape involving just three segments in Fig. 8.16 is complex enough to show why a process of neutral mutation and random fixation will tend to lead the system into scrambling. From ABC with scrambling rank 0 in the top row, five paths lead to configurations with scrambling rank 1. Each individual configuration at rank 1 has one path to rank 0, and four paths to rank 2, thus there is a 4-fold bias. Each configuration at rank 2 has three or four paths to rank 3, and only two or one paths back to rank 1. This difference in degrees of connectivity to neighboring ranks will be manifested as a bias in mutational introduction of alternatives, even when all individual rates of mutation are the same (i.e., no mutation bias)

The case of transitions and transversions, though familiar, is a bit complicated, because it combines two effects: (1) there are twice as many transversions (see Fig. 8.15), and (2) individual transition mutations typically take place at a higher rate. For instance, a T residue can mutate to C, A or G. The $T \rightarrow C$ mutation, a transition, typically takes place at a higher rate than $T \rightarrow A$, a transversion. This is properly a mutation bias. When we group together two different transversions ($T \rightarrow A$ and $T \rightarrow G$) and compare them with a single transitions ($T \rightarrow C$), the result is a 2-fold bias in neighboring possibilities (alternative nucleotides) that favors transversions. The overall tendency for transition mutations to be more common indicates that the mutation bias is typically > 2-fold, overcoming the bias in adjacent possibilities.

Earlier we raised the question of where biases come from. What causes mutation bias? Given the way that we have defined mutational bias, it is a manifestation of mechanisms of mutation. For some very general types of mutation biases, this means that the cause of the bias is heterogeneous, e.g., many pathways lead to nucleotide mutations, and therefore an effect such as GC:AT bias will represent the aggregate influence of many pathways. Types of mutations that occur only in very specific ways, e.g., insertions of transposon Tn1, may have biases with a more homogenous causal explanation, in terms of the action of specific proteins.

What causes biases in the local neighborhood of mutational possibilities? In the sense that these kinds of biases are formal or structural, they do not have mechanistic causes, but formal or historical explanations. A number of useful examples are given in Stoltzfus (1999), where they are called "systemic" biases. In several examples, the systemic bias is presumed to be attributable to a history of selection, because the system has been driven, under the influence of selection, into a highly unusual region of its state-space. For instance, the splitting of a self-splicing intron into separate pieces that reassemble to restore splicing depends on the fact that the original intron evolved so as to have extensive regions of complementarity conferring folding stability. Likewise, the gene duplication model of Stoltzfus (1999), analogous to the DDC model proposed by Force et al. (1999), depends on the fact that mutations that compromise activity or expression are much more likely than those that augment it, because a history of selection has driven the system to a state where activity is unusually high, so that mutational changes tend to result in lower activity.

8.10 Understanding developmental biases as evolutionary causes

A thorough discussion of the role of development in evolution is outside the scope of this book (and beyond my expertise). The aim of this section is simply to develop an argument from first principles to the effect that developmental biases in the introduction of variation may act as genuine causes of direction or orientation in evolution.

The most familiar *molecular* model of development is represented by the genetic code, illustrated earlier in Fig. 8.8. The genetic code is a genotype–phenotype (GP) map relating codon genotypes to amino acid phenotypes. This mapping is realized by the developmental process called "translation." Translation is a complex, multi-step process and, as for other developmental processes, there is not a 1:1 mapping of phenotypes to genotypes: most amino acid phenotypes can be encoded in multiple ways. Unlike many other developmental processes, translation results in discrete phenotypes. More distinctly, translation results in discrete phenotypes in such a highly canalized way that the developmental realization of the modal phenotype occurs over 99.9 % of the time, i.e., the error rate for translation is typically less than 10^{-3} (Ellis and Gallant, 1982).

Yet, neither of these differences is so vital as to exclude translation as a developmental process. Many developmental processes result in discrete phenotypes, e.g., limb development in chordates results in an autopod with a discrete set of digits, with a modal number of 5 for mammals. Canalization of cranial development is so strong that nearly every mammal is born with precisely one head and not two, three, or more. Rare cases indicate that double-headedness is developmentally possible (typically with duplication of the neck and upper torso), yet the chance of occurrence is extraordinarily low, less than that of a translation error.

Thus, translation is a developmental process, and may be used as an example of developmental processes, with the understanding that many other developmental processes show greater plasticity, a phenomenon with a variety of evolutionary implications (West-Eberhard, 2003).

The genetic code is a GP map that induces asymmetries of the type indicated in Fig. 8.17a. Consider an evolving system that starts with the Asp (aspartate) phenotype, encoded by GAT (it also could be encoded by GAC). Here the dotted lines represent developmental expression of a genotype. From the initial state, twice as many mutational paths lead to Glu (glutamate) genotypes as to Val (valine) genotypes. If all individual mutation rates are the same, the introduction of Glu phenotypes will be favored over Val phenotypes by a 2-fold developmental bias in the introduction of variation.

Note that, in Section 8.9, we distinguished biases that are properly mutational from biases that arise

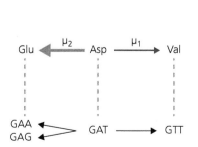
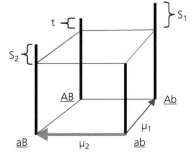

Figure 8.17 Understanding developmental bias induced at the molecular level by the genetic code, a GP map. A 2-fold bias induced by the genetic code (left) has the same implications as a 2-fold mutation bias in the Yampolsky–Stoltzfus model (right).

from formal ways of dividing up the space of adjacent possibilities, so as to implicate different sets of mutations. The kind of developmental bias illustrated in Fig. 8.17 (left) results from dividing up possibilities in genotype-space according to their mapping to phenotype-space.

From the perspective of population genetics, the potential influence of this kind of bias is not any different from that of a mutational bias of the same magnitude. We can map this precisely to the Yampolsky–Stoltzfus model (Fig. 8.17, right), and all of the same results of that model will apply. Specifically, positions 2 and 3 of the Asp codon map to the $\underline{A}$ and $\underline{B}$ loci, respectively. The $\underline{ab} \rightarrow \underline{Ab}$ mutation is a nucleotide change from GAT to GTT, introducing Val at the phenotypic level. The $\underline{ab} \rightarrow \underline{aB}$ mutation is a nucleotide change from GAT to GAA or to GAG, introducing Glu at the phenotypic level. In the latter case, $\underline{B}$ represents a compound allele comprising allelic states A and G at codon position 3.

Given that we can map this case to the Yampolsky–Stoltzfus model, all of the same implications emerge, e.g., the two-fold bias favoring the introduction of Glu over Val exerts a 2-fold bias on the outcome of evolution in the origin-fixation regime.

The asymmetry in the density of codon mutation paths depicted in Fig. 8.17 is not different from a conceivable form of developmental bias mentioned repeatedly in the evo-devo literature, namely the idea that some phenotypes are more likely to arise by mutation than others (e.g., Emlen, 2000). This is represented abstractly in Fig. 8.18: from the starting genotype encoding phenotype P_0, four times as many mutations lead to P_2 as to P_1. In the language of genetics, the mutants at the four different loci that generate P_2 are "phenocopies" of each other.

To map this case to the Yampolsky–Stoltzfus model, we consider four different $\underline{B}$ loci at which mutation-and-altered-development results in phenotype P_2, as opposed to a single $\underline{A}$ locus for P_1. If all of the individual mutation rates are the same, the introduction of P_2 is favored 4-fold over P_1, and this will have the same effect as a mutation bias of magnitude $B = 4$ in the Yampolsky–Stoltzfus model.

This result is important. For decades, evo-devo enthusiasts have suggested that "integrating

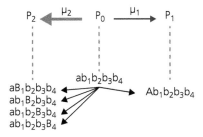

Figure 8.18 A bias induced by a GP map, *via* the phenocopy effect.

development into evolutionary theory" might require something more than *just adding new facts to a fixed set of pre-established principles*. In some cases, these claims involve biases in variation, e.g., Thomson (1985, p. 222) writes that

The whole thrust of the developmentalist approach to evolution is to explore the possibility that asymmetries in the introduction of variation at the focal level of individual phenotypes, arising from the inherent properties of developing systems, constitutes a powerful source of causation in evolutionary change.

The reformist concern for developmental constraints that emerged in the 1980s encountered stiff resistance, on two main grounds. One of the objections draws on opposing-pressures thinking to argue that developmental biases would be overcome by selection, as when Reeve and Sherman (1993) ask, in their rebuttal of evo-devo, "why couldn't selection suppress an 'easily generated physicochemical process' if the latter were disfavored?" That is, even though Maynard Smith et al. (1985) and others refer to the concept of a constraint or limit, which denotes *an absolute effect rather than a graduated one*, they and others (e.g., Thomson, 1985; Arthur, 2004) clearly mean to include the influence of *quantitative biases* due to some kinds of changes being quantitatively easier to achieve via alterations of development (e.g., Arthur, 2004).

Reeve and Sherman (1993) do not object to the causal efficacy of absolute constraints, but to the causal efficacy of quantitative biases in variation, and their objection is simply a form of the opposing pressures argument (Section 6.4.2).

The other objection (to this claim of causal novelty from evo-devo) referred to the Synthesis orthodoxy on what kinds of causes qualify as evolutionary causes. For instance, Mayr (1994) writes

I must have read in the last two years four or five papers and one book on development and evolution. Now development, the decoding of the genetic program, is clearly a matter of proximate causations. Evolution, equally clearly, is a matter of evolutionary causations. An yet, in all these papers and that book the two kinds of causations were hopelessly mixed up.

Similarly, having asked whether embryologists can contribute to understanding evolutionary mechanisms, Wallace (1986) answers negatively, arguing that "problems concerned with the orderly development of the individual are unrelated to those of the evolution of organisms through time." Likewise, Maynard Smith (1983) argues

If we are to understand evolution, we must remember that it is a process which occurs in populations, not in individuals. Individual animals may dig, swim, climb or gallop, and they also develop, but they do not evolve. To attempt an explanation of evolution in terms of the development of individuals is to commit precisely that error of misplaced reductionism of which geneticists are sometimes accused (p. 45).

Objections such as these were sufficiently influential that, for decades, advocates of evo-devo largely surrendered any place in discussions of fundamental evolutionary causes, and turned to (1) the "causal completeness" argument to the effect that development, because it produces the phenotypes that are the objects of evolutionary explanation, has to be included *somewhere* (see Section 7.5.4), and (2) the related "alternative explanatory narratives" position, to the effect that explanations for episodes of evolution must have two parts, the traditional dry population-genetics account of the properly *evolutionary* forces at work, and a wet biological narrative of changes in development that accompany evolutionary transitions (Wilkins, 1998).

Yet, the objections of Mayr, Reeve and Sherman, Maynard Smith, and Wallace lead to the wrong conclusion and are therefore mistaken. The mistake is either to ignore proximate causes in evolution (see West-Eberhard, 2003), or to fail to construe development as a true evolutionary cause. That is, if we accept the doctrine that evolutionary causes must emerge at the level of a population, then the mis-

take is in failing to recognize how the introduction process is actually *an emergent process at the population level*, in the precise sense that *whether a specific event of mutation is an introduction event cannot be determined from the event itself, but only by considering, in the higher context of the population, whether that event introduces something not already present* (see Section 8.8).

8.11 An interpretation of structuralism

The dominant strain of thought in evolutionary biology includes a preference for explanations that are externalist, functional, and mechanistic: the explanations have to do with external rather than internal causes; they explain form by relating it to function; and they are seen to rest on Aristotelean material and efficient causes rather than formal causes. Yet, there has always been a structuralist counter-culture, a minority tradition in which the focus is on explanations of the *forms* of things in terms of structural principles, e.g., relating the form of a mushroom to the form of a mushroom cloud. One would associate structuralism historically with D'Arcy Thompson (1917), and in more recent times, with authors such as Kauffman (1993), Goodwin (1994), and Fontana (2002).

Contemporary structuralists frequently present hypotheses and explanatory claims that relate the chances of evolving a particular form to its frequency or accessibility in state-space (e.g., Kauffman, 1993; Fontana, 2002). What do these formal arguments suggest about evolutionary causes? The distribution of a phenotype in genetic state-space sounds like some type of background condition, rather than a reference to causal forces. However, these claims can be interpreted as references to biases in the introduction of variation.

To understand this interpretation, consider an early argument from molecular evolution about the rough correspondence observed between the frequency with which an amino acid is found in proteins, and the number of codons assigned to it in the genetic code (which ranges from one to six). Originally, King and Jukes (1969) pointed to this correlation as an argument for neutral evolution, but King (1971) later recanted this position, arguing that the same correlation could arise under a stochastic model with selective allele fixations, and explained this position as follows:

Suppose that, at a given time, there are several possible amino acid substitutions that might improve a protein, and among these are changes to serine and to methionine; serine, with its six codons, has roughly six times the probability of becoming fixed in evolution. Once this has occurred, the mutation to methionine may no longer be advantageous.

Amino acids with more codons occupy a greater volume of sequence space. Because of this, they have more mutational arrows pointed at them from other parts of sequence space: they are more likely to be proposed, and therefore more likely to evolve, other things being equal. The resulting tendency for the frequency of an amino acid (phenotype) to correspond to its number of codons (genotypes) is not caused by natural selection, even if all changes are beneficial: instead, this pattern arises from the way in which the space of possibilities is explored by the mutation process.

This principle is implicated in a small but important body of theoretical studies on the evolution of RNA and protein folds, e.g., Cowperthwaite et al. (2008) define "abundance" as the number of genotypes assigned to a phenotype, and pursue hypotheses relating abundance to the chance that a phenotype will evolve. In "How development may direct evolution," Garson et al. (2003) consider evolution on an abstract model with a discrete map between genotypes and phenotypes, and find that the chance of evolving a phenotype is a function of the number of genotypes assigned to the phenotype. In *The Origins of Order* (1993), Kauffman refers to properties of systems (e.g., switching networks) that are widely distributed in state-space, and which (for this reason) tend to recur, as "ensemble properties."

A related idea concerns the role of the mutational accessibility of genotypes with a given phenotype. In the RNA-based models of Fontana (2002), sequences that have the same phenotype (fold) may be represented by a graph or network with one or more connected components. Figure 8.19 (after Fontana, 2002) shows the networks for three phenotypes, each of which has a single connected component. An RNA under selection to maintain function

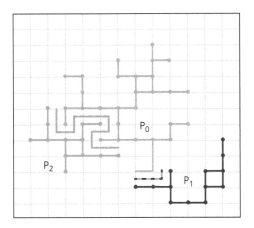

Figure 8.19 Networks of identical folds (phenotypes) in sequence space.

may (over long-term evolution) diffuse neutrally within the connected network for its current fold. Under such a scenario, the chances of evolving a different fold (phenotype) depend on the accessibility of the new fold over the entire network. For the networks shown in Fig. 8.19, a sequence with phenotype P_0 is more likely to evolve to P_2 than P_1, because a random step out of the P_0 network (grey) is more likely to yield P_2 (green) than P_1 (blue).

This effect is *not the same as* the effect of the number of genotypes assigned to a phenotype. To emphasize this, I have assigned the same number of genotypes to P_1 and P_2. The *local* accessibility of an alternative network is not a matter of its total size, but only the number of its genotypes within the mutational horizon (the set of one-mutant neighbors) of the network for the focal phenotype P_0.

Because evolutionary biology has lacked a theory of causation adequate to describe such effects of abundance and connectivity, their causes are often described in ways that are confusing. For instance, Kauffman (1993, p. 24) occasionally suggests that mutations provide the "back-pressure" to push the system toward ensemble properties, but he never makes this influence clear, prompting reviewers such as Fox (1993) to complain about his "almost magical" references to "self-organization." Cowperthwaite et al. (2008) describe the same kind of effects in terms of "constraints" on adaptation. After presenting several metrics of abundance and "mutational connectivity," they refer to them as aspects of the "fitness landscape," even though none of the metrics actually includes fitness.

The account of causation that makes sense of these structuralist arguments can be stated in generic terms as follows: evolutionary processes explore possibility-spaces via local processes of mutation that impose kinetic biases; this locally biased process of mutational exploration induces effects of adjacency and abundance (occupancy in possibility-space) that are the basis of structuralist claims. In the origin-fixation regime (or in a simulated proposal-acceptance process), these biases will be biases in the introduction (proposal) process. The biases in adjacency (local accessibility) are more relevant in the short-term, and the biases in total abundance are more relevant in long-term evolutionary diffusion in state-space.

To my way of thinking, this account of causation is the obvious explanation for the apparent magic of Kauffman's "self-organization." However, let us treat this interpretation as a conjecture subject to testing. To test it, one must implement an evolutionary simulation in which the mutation operator has some artificial options that allow it to escape certain effects of abundance or locality. The option of non-local mutation (long-jump mutation per Kauffman) will allow the current genotype to mutate to any other genotype (not just adjacent genotypes) with equal probability. Therefore, long-jump mutation will remove effects of adjacency. However, long-jump mutation will not remove the effect of the abundance of genotypes for an alternative phenotype. To counteract this effect, one must devise an artificial mutation process that samples alternative *phenotypes* uniformly, without regard to their number of genotypes.

If the conjecture is correct, then implementing these options will show that the effects of adjacency and abundance implicated in the structuralist literature are effects of the mutational operator and not effects of selection. In passing, note that other effects may complicate these expectations. For instance, in correlated fitness landscapes, higher peaks tend to have larger footprints (larger basins of attraction), which makes them more discoverable, even for the

case of long-jump mutation. Adjusting the mutation process to compensate for the footprint of a peak will not prevent valley-crossing algorithms from preferring higher peaks with larger basins of attraction (i.e., they will be preferred because they are higher, not because they are more findable).

Thus, if this conjecture is correct, a key concern of contemporary structuralism—understanding the relative chances for the evolution of various forms by considering their distribution and accessibility' in state-space—relies on the action of biases in the introduction of variation.

8.12 Parallel evolution

Shull (1935) said that "It strains one's faith in the laws of chance to imagine that identical changes should crop out again and again if the possibilities are endless and the probabilities equal" (p. 448). That is, it seems intuitively obvious that parallel changes are unlikely given a vast number of possibilities of equal probability. Instead, parallelism is more likely to the extent that (1) the number of possibilities is limited, and (2) they vary in probability.

We can give this intuitive idea a more precise meaning. Suppose that evolutionary change has some possible outcomes with probabilities $p_1, p_2, p_3,$ and so on. Then the chance of parallelism between two independent lineages is the sum of squared probabilities (Eqn. 8.6):

$$Pr_{para} = \sum_{i=1}^{n} p_i^2 \qquad (8.6)$$

This relation is illustrated with a graphical example of 13 unequal possibilities in Fig. 8.20, where the shaded area represents the chance of parallelism. For any such case of unequal probabilities, we can define an "effective number" of possibilities with equal probability, i.e., the number of equally likely possibilities that gives the same chance of parallelism. Thus, the effective number must be the inverse of the chance of parallelism, $n_e = 1/Pr_{para}$. For the example in Fig. 8.20, the chance of parallelism is 0.11, and so the effective number of possibilities is $1/0.11 = 9.2$.

Intuitively, it seems obvious that the greater the dispersion in elementary evolutionary probabilities,

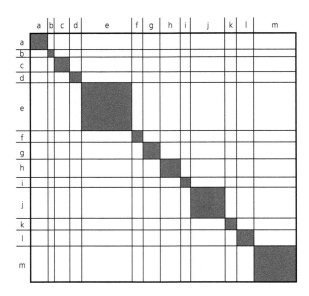

Figure 8.20 The chance of parallelism between two lineages is a sum of squares.

the greater the chance of parallelism. The relationship of the chances of parallelism to the variance in underlying probabilities becomes clear if we express the variance as

$$V_p = \frac{\sum_{i=1}^{n} (p_i - \bar{p})^2}{n} = \frac{\sum_{i=1}^{n} p_i^2}{n} - \bar{p}^2 \qquad (8.7)$$

This shows that one of the terms in the calculation of variance is identical to the probability of parallelism in Eqn. 8.6. Replacing this term with Pr_{para} and solving gives (Eqn. 8.8):

$$Pr_{para} = nV + \frac{1}{n} = \frac{C^2 + 1}{n} \qquad (8.8)$$

where C is the coefficient of variation, defined as the ratio of the standard deviation to the mean. When there is no variation in chances of occurrence ($V = 0$ or $C = 0$), the chance of parallelism reduces to $1/n$, as expected.

Our assumptions so far have been very minimal: change consists of steps with individual probabilities, and the two lineages are independent. The conclusions are that

- Pr_{para} depends on n and V (or C) alone
- Any factor that increases V (or C), also increases Pr_{para}

- For a given n, Pr_{para} is minimized by uniformity ($V = C = 0$)
- For a given C, increasing n decreases Pr_{para} linearly.

We considered parallelism for elementary events such as a specific nucleotide or amino acid change. However, there are many cases in which we might wish to aggregate over distinct events in the same equivalence class. One example would be the case of developmental bias (Section 8.10) implicating the set of alternative genotypes that encode the same alternative phenotype. Another example would be parallelism at the level of a gene or pathway, rather than an elementary event. That is, suppose that we wish to consider nonidentical changes in the same gene, or the same pathway, as a kind of parallelism (e.g., as in Tenaillon et al., 2012).

As an example of this kind of parallelism, Fig. 8.21 shows the same 13 possibilities we considered previously, assigned to three different equivalence classes (red, green, and blue). The classes may represent genes or pathways or phenotypes implicated by multiple elementary events.

Figure 8.21 shows geometrically why the solution to parallelism for equivalence classes is to sum the individual probabilities of events in a class, then

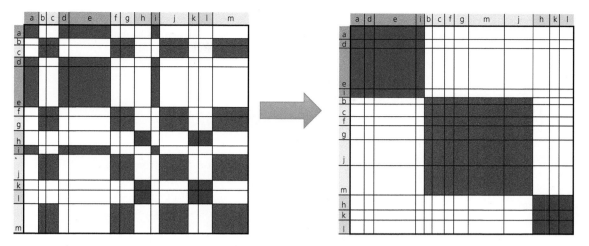

Figure 8.21 The chances of parallelism, when events are aggregated into equivalence classes, is simply a square of frequencies of equivalence classes.

sum the squares of those aggregate probabilities (Eqn. 8.9):

$$Pr_{para} = \sum_{j=1}^{k} \left(\sum_{i \in j} p_i \right)^2 \qquad (8.9)$$

Here, j is the index over the k equivalence classes, and i is the index over the elementary probabilities. We expect on this basis that, the greater the degree of aggregation, the greater the chance of parallelism (see Tenaillon et al., 2012). Note also that the effect of some kind of bias (e.g., green vs. red) is linear on the distribution of paths or sites, but has a squared effect when one counts events.

Finally, let us add origin-fixation dynamics, following Chevin et al. (2010). The result is simply an extension of Eqn. 8.8 to reference the variability in μ and in the probability of fixation π (Eqn. 8.10):

$$Pr_{para} = \frac{C^2 + 1}{n} = \frac{(C_\mu^2 + 1)(C_\pi^2 + 1)}{n} \qquad (8.10)$$

This relation holds if we can assume that μ and π are independent, and that μ^2 and π^2 are independent. Thus, whereas the conventional wisdom repeated in countless scientific papers is that parallel evolution is evidence of selection, under these conditions, parallel evolution is evidence of an enhanced joint probability of introduction and acceptance.

8.13 Conditioning on mutational effects

The possibility of mutation-biased adaptation induces a change in the forms of reasoning applicable to sampling (or otherwise interrogating) evolved changes.

When a sample from a population is conditioned on the sum of two random variables, this induces a negative covariance between the two variables in the sample, relative to the entire population. In the field of statistics, this is called Berkson's paradox. For instance, suppose that there is a population of individuals that differ in height, and that environmental and genetic variation in height are equal in magnitude and uncorrelated. If we select the top 1 % of individuals from this population, they will show a negative correlation between environmental and genetic contributions to height, because the ones that had a small contribution of genetics must have had a large contribution of environment, and vice versa.

Suppose that the chance of observing something in evolution is a joint probability of mutation and acceptance. For instance, consider the specific case in which the accumulation of beneficial changes is an origin-fixation process, with a rate proportional to μs. Our observation of beneficial changes is conditioned on the product of two variables. Because a product is simply a sum of logarithms, Berkson's paradox applies: a sample of evolved beneficial changes will show a negative correlation between μ and s if they are uncorrelated in the entire population of mutational possibilities, and more generally, the covariance of μ and s will be more negative in the sample than in the entire population.

For instance, Stoltzfus and McCandlish (2017) show that the transition bias observed in experimental cases of parallel adaptation persists when the number of repeated events is increased from two (the minimum required for a parallelism) up to eight, results that are repeated in Table 9.4. This was intended to show that the observed transition bias cannot be due to contamination (e.g., by neutral parallelisms or clerical errors), but what are the expected effects on the sample? When we restrict our attention to events with greater numbers of occurrences, we are biasing the sample toward higher values of μs. Thus, we expect higher values of μ, higher values of s, and a stronger negative correlation between the two. In fact, Table 9.4 shows that the transition bias tends to increase as the minimum number of occurrences is increased. This is expected, but it does not mean that the fitness effects are any less: again, we expect both higher μ and higher s, as the number of recurrences increases.

Due to Berkson's paradox, the discovery that a presumptively adaptive evolutionary change involves a CpG mutational hotspot immediately *depresses* our expectation of its fitness benefit and, indeed, depresses our expectation that it is genuinely adaptive. However, in the case of the Ile55Val change in Galen et al. (2015), such an effect is far outweighed by direct evidence (from altitude correlation and functional effects) of a fitness benefit. In cases where there is no direct evidence, one must be extremely careful. For instance, to estimate the mutation rate for a category of changes from its frequency among strains reported in the Loci of Repeated Evolution database (see Streisfeld and Rausher, 2011) would be unsafe, because this sample is conditioned on both mutation rates and fitness effects, in an unknown way.

More generally, this is a difficult technical problem in evolutionary analysis. We are constantly faced with samples that are conditioned in an unknown way on the variables that we want to estimate. We can develop some parts of a forward theory for parallelism, as in this chapter, or Chevin et al. (2010), but developing methods of inference from natural data is very difficult given that we do not know the underlying distributions.

8.14 Synopsis

What is the expected effect of a bias in mutational introduction on the outcome of evolution, particularly adaptive evolution?

In the latter half of the twentieth century, issues of causation were addressed under the dominant conception of evolution as a process that begins with abundant variation and does not depend on new mutations (which, in any case, were dismissed as "random"). From this perspective, the question about the effect of a bias in introduction was not relevant to evolution. The question apparently was not asked, much less answered.

Our exploration of this question reveals a new and simple idea, a kind of first-come, first-served principle by which a bias in the introduction of variants by mutation—or, in the case of phenotypes, by mutation-and-altered-development—is a cause of orientation or direction in evolution.

The basic idea is intuitively simple. We might even compare it to a completely different (and intuitive) argument made by the philosopher John Stuart Mill (1869, Ch. 2) about the triumph of truth:

the dictum that truth always triumphs over persecution, is one of those pleasant falsehoods which men repeat after one another till they pass into commonplaces, but which all experience refutes . . . The real advantage which truth has, consists in this, that when an opinion is true, it may be extinguished once, twice, or many times, but in the course of ages there will generally be found persons to rediscover it, until some one of its reappearances falls on a time when from favourable circumstances it escapes persecution until it has made such head as to withstand all subsequent attempts to suppress it.

Mill's depiction of the evolution of ideas as a dual process of *discovery* followed by *repression or triumph* makes it possible to have two different kinds of explanations for a differential pattern, analogous to biases in introduction and biases in acceptance. In this case, Mill argues that the advantage of truth resides in the influence of the discovery step. I do not mean to endorse Mill's conclusion (which seems to overlook the recurring introduction of facile falsehoods), but only to draw attention to the *form* of the argument.

Using models and equations to guide and verify our thinking, and adding some knowledge about mutation and genetics, we can extend this basic idea into a larger set of principles. Within the origin-fixation (sushi conveyor) regime, a bias in the introduction of variation is a possible source of orientation or direction in evolution. Any kind of bias can contribute to parallelism, but only certain kinds of biases can lead to directional trends.

The result is a quantitative theory of dual causation by internal and external causes, a welcome development in the history of evolutionary thought. Historic attempts to combine internal and external causes typically begin with a radical simplification, by which either (1) one kind of cause rules over the other (e.g., as in neo-Darwinism), or (2) one kind of cause can be reduced to binary terms (viable vs. inviable or possible vs. impossible), whereas the other kind of cause is allowed to have graduated effects. By contrast, this theory of dual causation does not require one kind of cause to be subordinated; it does not require absolute effects or "constraints;" it applies to mutational, developmental, and systemic biases.

The theory provides an alternative to the position on causation associated with the Modern Synthesis (and neo-Darwinism more generally). Fisher (1930b) famously argued that, once one accepts Mendelian genetics, Darwinism follows and all other views must be cast aside. As we saw in Chapter 6, the architects of the Modern Synthesis, and those influenced by their thinking, offered a bewildering array of arguments against attributing importance to mutation. These arguments were often presented with an air of finality, as if the irrelevance of mutation followed from unassailable facts or logic.

Indeed, the arguments in Chapter 6 were presented to previous generations of evolutionary biologists as though they were universal truths, but we can see now that they apply only under the narrow conditions of the shifting-gene-frequencies theory. That is, the harsh mathematical logic of population genetics does not force us to deny mutation any importance as a dispositional factor in evolution: the architects of the Modern Synthesis essentially *chose* this position, and then built an elaborate verbal and theoretical edifice to justify and explain it. The possibility of variation-induced trends that Simpson (1967, p. 159) ridicules as "the vagueness of inherent tendencies, vital urges, or cosmic goals, without known mechanism" cannot be dismissed, and the reason is not that we have discovered vital urges or cosmic goals, but that we have reconsidered population genetics without the tendentious assumptions needed to justify neo-Darwinism.

From this perspective, the key assumption of the shifting-gene-frequencies theory, part of its multi-faceted "gene pool" theory, is that *the alleles relevant to the outcome of evolution are present initially*. Stated differently, the key assumption is to redefine "evolution" as the sorting out of available variation in an abundant gene pool. Using a theoretical model, we can show that the power of mutation bias to influence the course of evolution disappears under this assumption.

That is, we can model both the shifting-gene-frequencies theory of the Modern Synthesis, and an alternative theory, and compare them. The difference is not merely a matter of words, nor is it an insurmountable divide. The two theories correspond to what we have called the "buffet" and the "sushi conveyor" regimes of population genetics. In the sushi conveyor regime, the power of mutation to influence the rate and direction of evolution is particularly strong; in the buffet regime, mutation has essentially no dispositional effect, but the effect of selection is enormous.

Now that we have articulated this difference in theoretical terms, we have a further theoretical issue, and an empirical issue. The theoretical issue, hinted at earlier, is to understand what the role of mutation is in intermediate regimes of population genetics, i.e., regimes that depart from the strict sushi-conveyor regime of origin-fixation dynamics, but lack some of the properties of the buffet or "gene pool" regime. McCandlish and Stoltzfus (2014) review this issue and suggest some directions for future theoretical work.

The empirical question is to understand the actual role of mutation in evolution. To what extent do biases in the introduction of variation influence the course of evolution? In Chapter 9, we consider some specific examples.

CHAPTER 9

The revolt of the clay

...the discovery that the same mutation happens over and over again, not only within the same species, but in different species, is, I think, one of the most interesting discoveries in recent genetic work. It means that certain kinds of changes in the germ material are more likely to occur than are others... the appearance of new variations in the hereditary material is something less a random process than we had hitherto supposed. **T. H. Morgan (1923)**

9.1 Introduction

A century ago, Poulton (1908) ridiculed non-Darwinian views of the role of variation as "the revolt of the clay against the power of the potter." Indeed, the neo-Darwinian theory holds that variation is sufficiently abundant and formless that selection can be construed as the agent that supplies initiative, creativity, and direction to evolution, whereas variation merely provides raw materials.

However, evolution does not necessarily work in this way. In Chapter 8, we explored modes of evolutionary change in which the role of variation is not merely to supply inert formless raw materials. Once we understand some of the ways that the process of mutation (or mutation-and-altered-development) may influence evolution, we suddenly have new questions, often very basic ones. How is it that some heterogeneities can result in a direction, whereas others can not? How are developmental biases like mutation biases? How important is the mutational path density (determined by the genetic code) connecting amino acids? How important are mutational effects of adjacency and abundance during evolution among genotypic networks?

In this chapter, we turn from theory to evidence. What is the evidence that biases in variation have influenced the course of evolution?

Hundreds of published studies from the field of molecular evolution contribute to the empirical case

that biases in mutation are influential in shaping genes and genomes. However, the vast majority of these studies are retrospective comparisons using models that simply postulate mutational effects without verification, as explained previously (see Section 7.5.5).

Although *some* studies, including one covered in this chapter, are more convincing, they do little to undermine the "mutation is random" doctrine, if we take this to be an explanatory claim about the importance of mutation in explanations of *things that evolutionary biologists care about*. Rightly or wrongly, most evolutionary biologists care neither about codon usage nor patterns that might represent neutral evolution, on the grounds that they represent the "boring parts" of evolution (see Box 3.1). Instead, they care about adaptation, and they care about visible phenotypes, especially morphological, behavioral, and life-history traits of large charismatic animals. As Dawkins (1987) says, "What I mainly want a theory of evolution to do is explain well-designed, complex mechanisms like hearts, hands, eyes and echolocation."

Thus, whereas this chapter is about evidence of mutational effects in evolution, it focuses on *a very particular kind* of evidence, based on what could undermine the explanatory irrelevance doctrine and establish that mutation imposes biases on evolution in important ways. The focus is on evidence for effects of mutation that are predictable, direct,

Mutation, Randomness, and Evolution. Arlin Stoltzfus, Oxford University Press (2021). © Arlin Stoltzfus. DOI: 10.1093/oso/9780198844457.003.0009

Figure 9.1 A potter at work shaping clay. Image reproduced under the Creative Commons CC0 1.0 Universal Public Domain Dedication.

large, and, as much as possible, related to adaptation rather than to neutral evolution.

9.2 A predictive model of protein sequence evolution

Comparative analyses of molecular sequence divergence often infer effects of mutation (or selection) using abstract models, without conducting further experiments to verify the postulated effects. However, one may construct a purely empirical model with no free parameters, and show that the model is predictive due to mutational effects. Stoltzfus and Yampolsky (2009) applied this type of model to predict patterns of amino acid replacement in human–chimp divergence. The bubble plot in Fig. 9.2 shows the relative frequencies of each type of replacement, considering the 150 "singlet" replacements, i.e., the ones that involve a single-nucleotide substitution. The symmetry across the diagonal reflects the strong correlation between the frequency of a replacement and the frequency of the reverse replacement.

Tendencies of amino acid replacement have been an object of analysis and interpretation for over 50 years, e.g., Zuckerkandl and Pauling (1965) remarked that "The inadequacy of *a priori* views on conservatism and nonconservatism [of amino acid replacements] is patent. Apparently chemists and protein molecules do not share the same opinions regarding the definition of the most prominent properties of a residue" (p. 33).

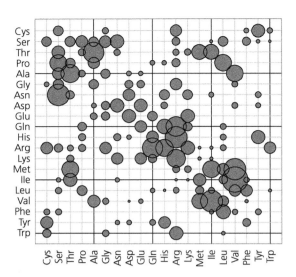

Figure 9.2 Relative frequency of evolution for 150 types of replacements in human–chimp divergence, using data from Stoltzfus and Yampolsky (2009), where row indicates the starting amino acid, and column indicates the ending amino acid.

To what extent are tendencies of amino acid replacement due to effects of mutation, and to what extent are they due to fitness effects? Stoltzfus and Yampolsky (2009) make a simple prediction model in which the frequency of occurrence is proportional to the product of two factors, one for the acceptability of a proposed replacement, and the other for the rate at which proposed replacements are introduced by mutation. The acceptance model was based on **EX**, a measure of exchangeability derived from results of mutation-scanning in proteins (Yampolsky and Stoltzfus, 2005). For the present purposes, I have repeated the analysis with an updated version of **EX** based on published fitness data from roughly 40,000 amino acid replacements (Stoltzfus and McCandlish, in progress), from the new technology of deep mutational scanning (Fowler and Fields, 2014).

The mutation model combines the effect of the genetic code discussed previously (Section 8.10) with nucleotide mutation biases estimated from the divergence of noncoding regions, shown in Fig. 9.3. The mutational effects include (in decreasing order of strength) the 9.1-fold effect of CpG context (C* and G*) on transitions, the 3.8-fold effect of transition-transversion bias, the 3.5-fold effect of

CpG context on transversions, and the 1.7-fold effect of $GC \rightarrow AT$ bias among transitions. These values are inferred from patterns of divergence of noncoding regions of mammal genomes by Hwang and Green (2004), as explained by Stoltzfus and Yampolsky (2009).

To generate a predictor for the frequency of replacements, we simply multiply these two factors. Note that this is an empirical prediction model in the sense that the information used to assign values to predictive factors are independent from the data to be predicted. The model of fitness effects uses an exchangeability measure from experimental mutation scans in various proteins (Yampolsky and Stoltzfus, 2005).

The resulting prediction of relative frequencies for replacements is illustrated with the bubble plot in Fig. 9.4 (left). Note that the visual similarity with Fig. 9.2 owes much to the pattern of singlets vs.

blank space (doublets and triplets), rather than to matching sizes of bubbles, but the prediction model is evaluated considering only the singlets (thus it is not misleading in this regard). In fact, the correlation of observed and predicted values for singlets is substantial, with $R^2 = 0.59$ (right panel, Fig. 9.4).

The results for the mutational predictor alone are shown in Fig. 9.5. The R^2 value for the correlation of observed frequencies with the exchangeability predictor is 0.24 (not shown), respectively. Thus, the success of the joint predictor is not primarily due to exchangeability, but mutation. Note that the two factors interact in some unknown way, which is why the correlation for the full model is not the sum of its parts; however, the two factors are not highly correlated with each other ($R^2 = 0.034$), that is, the success of the mutational predictor is not due to being strongly correlated with exchangeability.

This model is missing potentially important factors such as fitness effects at the level of codons or RNA, for which empirical fitness models are lacking. However, perfection is not required in order make a rough assessment of the contribution of mutation to an evolutionary pattern. The results show that the rate at which replacement mutations are introduced (which reflects both the genetic code and nucleotide mutation biases) is a powerful factor. This result was not a foregone conclusion: it was not guaranteed by any principle of evolution-

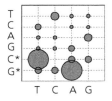

Figure 9.3 A simplified model of nucleotide mutation rates.

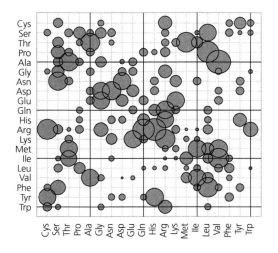

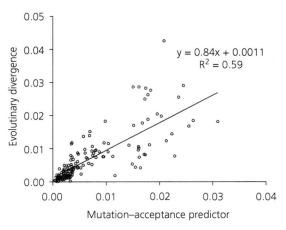

Figure 9.4 Prediction of observed tendencies of amino acid change. Left, predicted values of relative frequency of amino acid changes from an empirical model. Right, correlation of observed and predicted frequencies.

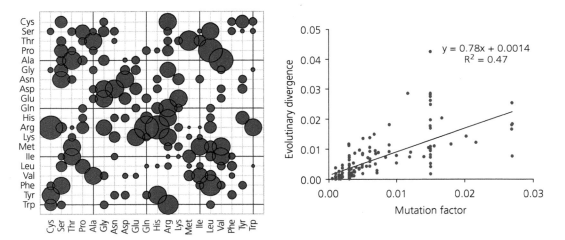

Figure 9.5 The mutational component of the model, considered alone. Left, predicted pattern of relative frequency from mutational effects alone. Right, observed frequency of replacements vs. the predicted frequency from mutation alone.

Box 9.1 The conservative transitions hypothesis

A common observation in molecular evolution is that nucleotide transitions occur more often than transversions, even though the number of possible transversions is always greater. When this pattern occurs among amino acid replacements, explanations often rely on a proposed effect of selection, on the grounds that transitions cause changes that (on average) tend to be "less severe with respect to the chemical properties of the original and mutant amino acids" (Rosenberg et al., 2003; see also Wakeley, 1996; Keller et al., 2007).

This idea, though present in the literature for decades, was not tested directly until Stoltzfus and Norris (2016) gathered data from studies that directly characterize the fitness effects of mutant proteins, including 544 transitions and 695 transversions. The results indicated that a transition has a roughly 53 % (CI, 50 to 56 %) chance of being more conservative than a transversion, compared to the null expectation of 50 %. Using the model of Tang et al. (2004), they showed that this slight preference is not large compared to that of most biochemical factors (e.g., polarity, volume), and is not large enough to explain the several-fold bias observed in evolution.

We can revisit this issue using the much larger set of data from deep mutational scanning introduced in Section 4.5. As shown in Fig. 9.6, the distributions of fitness ranks for 3,440 transitions (red) and 7,706 transversions (blue) look very much the same. Specifically, a transition has a 51 % chance of being more conservative, relative to the null expectation of 50 %, even smaller than the effect-size of 53 % reported by Stoltzfus and Norris (2016).

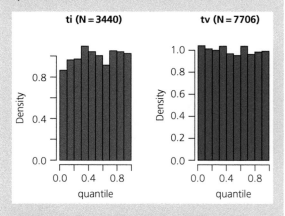

Figure 9.6 Fitness effects (quantiles) for replacements by transition or transversion mutation, from deep mutational scanning experiments.

This comparison depends on the entire fitness distribution, which allows the use of a large set of data but is otherwise not ideal. If selection filters out (for instance) the bottom 80 % or 90 % of missense mutations, then what we really want to know is whether the top 20 % or 10 % is biased toward transitions. Accordingly, Stoltzfus and Norris (2016) applied a successively higher threshold to filter out less conservative mutations, finding no significant pattern of conservativeness of transitions.

Finally, Stoltzfus and Norris (2016) compared fitness effects of transitions and transversions in studies that focus only on *beneficial mutations*. For the combined set of 111 beneficial mutants from the four studies listed in the table below, the AUC—representing the chance that a transition is more conservative than a transversion—is 0.40 (95 % CI, 0.28 to 0.51), lower than the null expectation of 0.5 (the P values are for a one-sided test where the alternative is that transitions are more fit). This suggests that beneficial transitions are slightly *less fit* than beneficial transversions.

Study	n	AUC	P
Schenk et al. (2012) TEM1	38	0.49	0.53
MacLean et al. (2010) RpoB	31	0.35	0.93
Miller et al. (2011) ID11	27	0.25	0.96
Ferris et al. (2007) phi6	15	0.39	0.76
Combined	111	0.40	0.95

Thus, out of multiple lines of evidence, none of them support the hypothesis that the evolutionary bias toward transitions is due to a fitness advantage. Stoltzfus and Norris (2016) note that the original 50-year-old claims of conservativeness were not based on direct measurements of mutant effects, but emerged from plausibility arguments that attempt to account for observed evolutionary patterns by referring to biochemical factors. That is, the observed evolutionary pattern (favoring transitions) was given an explanation in terms of fitness effects, and over time, this explanation achieved the status of lore.

ary biology. We might have discovered that the mutational factor accounts for only 1 % or 5 % of the pattern, and on the basis of such a finding, we might have dismissed mutation as being *quantitatively unimportant* in shaping the outcome of sequence evolution. Instead, we found that the mutational factor accounts for half of the pattern. Note that the mutational factor would be even more important if we had included doublets and triplets, the low frequency of which (in evolution) is much more an effect of lower mutation rates than of being particularly disruptive (see Fig. 4.3).

9.3 Mutation-biased adaptation in the lab

The previous section showed predictable effects of mutation bias among a large set of amino acid changes. However, such genome-wide patterns might reflect neutral evolution, thus they would fail the test of representing subjectively important, nonboring parts of evolution. What evidence do we have that modest biases in rates of mutation can influence the course of adaptation?

Relevant information is available from various experimental studies of laboratory evolution. Previously, we considered the results of MacLean et al. (2010), who measured fitness effects and mutation rates for 11 resistance-conferring single-nucleotide substitutions in the *rpoB* gene. They reported the mutation rates, and noted that the most frequent outcome of adaptation was the variant accessed by the highest mutation rate, but did not apparently explore the general relationship between frequency evolved and mutation rate, shown in the center panel of Fig. 9.7.

As mentioned earlier, this study is worth emphasizing because the experimenters measured (1) mutation rates, (2) fitnesses, and (3) frequency of evolving. The results present a paradox relative to the "forces" theory (Section 8.8). The frequency with which an outcome evolves is strongly correlated with the mutation rate (center panel of Fig. 9.7), but not with the effect on fitness (left panel), even though selection is supposed to be a stronger "force." This paradox is resolved by considering origin-fixation dynamics. A 50-fold range of mutation rates has an expected 50-fold effect on the chance of evolving. By contrast, the 11 rifampicin-resistant mutations studied by MacLean et al. (2010) have measured fitness effects ranging from $s = 0.3$ to $s = 0.9$. The probability of fixation, using Kimura's

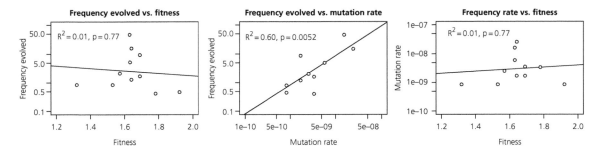

Figure 9.7 Inter-relations of mutation rate, fitness effect, and frequency of evolution for 11 rifampicin-resistant variants (data from MacLean et al., 2010).

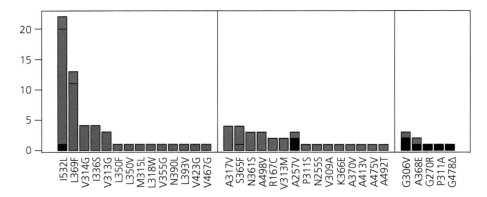

Figure 9.8 Frequency of evolution of cefotaxime resistance from parents with three different mutation spectra (after Fig. 3A of Couce et al. 2015). The resistant mutants from all three parents are distributed among three boxes depending on the type of mutation: $A : T \rightarrow C : G$ transversions (left), which are favored by *mutT*; $G : C \rightarrow A : T$ and $A : T \rightarrow G : C$ transitions (center), which are favored by *mutH*; and other types of mutations (right), including a deletion of glycine 478 (rightmost column; figure generated from data kindly provided by Alex Couce).

formula $\frac{1-e^{-2s}}{1-e^{-2Ns}}$, ranges from 0.45 to 0.83 (assuming $N = 10^6$, though the conclusion is not sensitive to increasing or decreasing N by several orders of magnitude). That is, in the regime that population geneticists under the influence of the "forces" paradigm call "weak mutation, strong selection," the difference-making power of mutation is strong and that of selection is weak.

A more elaborate experimental example is provided by Couce et al. (2015), showing a striking effect of the mutation spectrum on the outcome of evolutionary adaptation. Couce et al. (2015) subjected replicate lines of *E. coli* to increasing concentrations of the antibiotic cefotaxime. They carried out the experiment with replicate cultures from wild-type, <u>mutH</u>, and <u>mutT</u> parents, the latter two being "mutator" strains with elevated rates of mutation and different mutation spectra.

The results are illustrated in Figure 9.8. Like MacLean et al. (2010), the authors counted the number of times a particular mutant rose to prominence in a resistant culture. Most of the cultures that survived the highest levels of cefotaxime were from the mutator parents, shown in red (*mutH*) or blue (*mutT*), whereas a smaller number of resistant cultures emerged from wild-type parents (black bars). Fig. 9.8 shows the mutations affecting the PBP3 protein (other mutations affecting the TEM-1 protein are not shown). The striking difference in coloration between the three boxes results from the match between evolved changes and favored mutations: resistant cultures from the <u>mutT</u> (blue) parent tend to adapt by $A : T \rightarrow C : G$ transversions, whereas resistant cultures from the *mutH* (red) parent tend to adapt by $G : C \rightarrow A : T$ and $A : T \rightarrow G : C$ transitions (see figure legend).

In a study using bacteriophages rather than bacteria, Rokyta et al. (2005) carried out 20 one-step adaptive walks with a laboratory population of phage ID11, with results shown in Table 9.1. The nine rows represent the nine amino-acid variants that arose in laboratory adaptation, some of which occurred multiple times, as shown in the column labeled n. The column labeled with s represents the measured selection coefficient.

Rokyta et al. (2005) compared these results to a "mutational landscape" model (Orr, 2002), although

Table 9.1 Results of parallel adaptation of phage ID11 from Table 1 of Rokyta et al. (2005). Each row gives the genomic position, nucleotide and amino acid changes, the number of evolutionary events, the fitness, and the selection coefficient relative to the parental strain.

Site	Nt	Aa	*n*	*w*	*s*
2,534	G to T	V to L	1	20.31	0.39
3,665	C to T	P to S	5	20.05	0.37
3,850	G to AT	M to I	3	19.45	0.33
2,520	C to T	A to V	6	19.29	0.32
3,543	C to T	A to V	1	19.13	0.31
3,857	A to G	T to A	1	19.04	0.3
2,609	G to T	V to F	1	17.56	0.2
3,567	A to G	N to S	1	16.74	0.15
3,864	A to G	D to G	1	16.22	0.11

the model is not actually designed for this kind of experiment. Gillespie's original "mutational landscape" concept is about how the mutational neighborhood changes from one step to the next in an adaptive walk (Gillespie, 1984), but the study of Rokyta et al. (2005) is a study of one-step adaptation, thus there is no change in the mutational neighborhood—only repeated exploration of the initial wild-type neighborhood with its nine more-fit alternatives.

Nonetheless, Rokyta et al. (2005) used the nine ranked beneficial mutations in Table 9.1 to generate predictions from the Gillespie–Orr model, which are compared with observations in the left panel of Fig. 9.9. The fit between expectations (grey) and observations (black) is not strong, mainly because the top-ranked alternative is only seen once, yet the fourth-ranked is seen six times.

Rokyta et al. (2005) proposed that this deviation is an effect of mutation, noting that, in these bacteriophages, there is a strong mutational bias toward transitions. The top-ranked alternative is only accessible via a transversion mutation, whereas the seond-, third-, and fourth-ranked alternatives each can be reached by a transition mutation.

The last four columns of Table 9.2 illustrate how to compute an origin-fixation expectation using Haldane's $2s$ and relative rates of mutation based on path density (from the genetic code) and the mutation biases inferred by Rokyta et al. (2005, p. 433) from a separate set of data on the natural diver-

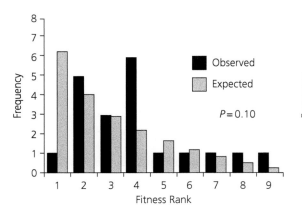
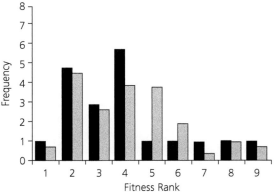

Figure 9.9 Expected (grey) vs. observed (black) numbers of genotypes resulting from experimental adaptation for Orr's mutational landscape model (left), which lacks mutational effects, and an origin-fixation model with mutational bias (right). Left figure kindly provided by Dr. Darin Rokyta.

Table 9.2 Origin-fixation expectations from the data of Rokyta et al. (2005). The first six columns are as in Table 9.1. The next four columns provide (respectively) the list of nucleotide mutations that result in the observed amino acid change, the expected combined rate for these mutations (u), $2su$, and the expected numbers of occurrences (out of 20). To evaluate this model, one compares the observed number n to the expected number $E(n)$.

Site	Nt	Aa	n	w	s	Paths	u	$2su$	$E(n)$
2,534	G to T	V to L	1	20.31	0.39	G to TC	1.23	0.96	0.71
3,665	C to T	P to S	5	20.05	0.37	C to T	8.5	6.33	4.71
3,850	G to AT	M to I	3	19.45	0.33	G to ATC	5.54	3.67	2.73
2,520	C to T	A to V	6	19.29	0.32	C to T	8.5	5.45	4.05
3,543	C to T	A to V	1	19.13	0.31	C to T	8.5	5.26	3.91
3,857	A to G	T to A	1	19.04	0.3	A to G	4.31	2.61	1.94
2,609	G to T	V to F	1	17.56	0.2	G to T	1	0.4	0.3
3,567	A to G	N to S	1	16.74	0.15	A to G	4.31	1.26	0.93
3,864	A to G	D to G	1	16.22	0.11	A to G	4.31	0.95	0.71

gence of phage genomes. That is, we can combine the rate of mutation u with $2s$ to assign an origin-fixation weight $2su$ to each alternative: normalizing the weights and multiplying by 20 (the number of replicates) gives the expected number of outcomes $E(n)$ for each alternative, shown in the last column of Table 9.2. Figure 9.9 (right) indicates that this model gives a better fit to the data than the mutational landscape model.

Rokyta et al. (2005) also calculated expectations of an origin-fixation model, using a slightly different formula for the probability of fixation (more suited to phage growth), finding that an origin-fixation model with mutational effects increased the likelihood of the observed results 21-fold relative to the mutational landscape model.

A concurrent commentary by Bull and Otto (2005) framed this study as "the first empirical test of an evolutionary theory," saying that it "provides support for a mutational landscape model underlying the process of adaptation." However, as mentioned, this study was not a critical test of the Orr–Gillespie model. The results of Rokyta et al. (2005) provided support for the relevance of origin-fixation dynamics, but in regard to the distinctive features of the mutational landscape model, (1) the mutational landscape was not tested at all, (2) the uniform mutation assumption was rejected, and (3) the implication of extreme value theory (an exponential distribution of fitness intervals between ranks) was sustained but not critically tested, because the data set of just

nine fitness ranks was too small to rule out other similar distributions.

Note again how far such results stray from the shifting-gene-frequencies theory of the Modern Synthesis. This theory holds that each species has a "gene pool" abundantly full of variation, and infinite heritable variation is generated by recombination each generation, and so on, with selection ultimately delivering an optimal multi-locus distribution of gene frequencies. Evolution is literally defined as a process that begins with population variation. Advocates of this theory believed that to imagine a unique ancestral "type" (as in Rokyta et al., 2005) was to commit an error of "typological thinking," showing a failure to understand "population thinking" (e.g., Mayr, 1980).

Sackman et al. (2017) followed up the study of Rokyta et al. (2005) by applying the same 20-replicate protocol to three closely related phages. The combined results (Fig. 9.10) show a strong bias favoring transitions: 29:5 for paths, and 74:6 for events.

Meyer et al. (2012) carried out replicate adaptation of λ (which represents a quite different family of bacteriophages), identifying 24 cases in which a new host specificity evolved. They compared J gene sequences of isolates from these 24 lines, and 24 other evolved lines. The complete set of 241 differences from the parental sequence is shown in Fig. 9.11 (after Fig. 1 of Meyer et al., 2012).

Of interest here is the possible influence of transition-transversion bias, which is known to

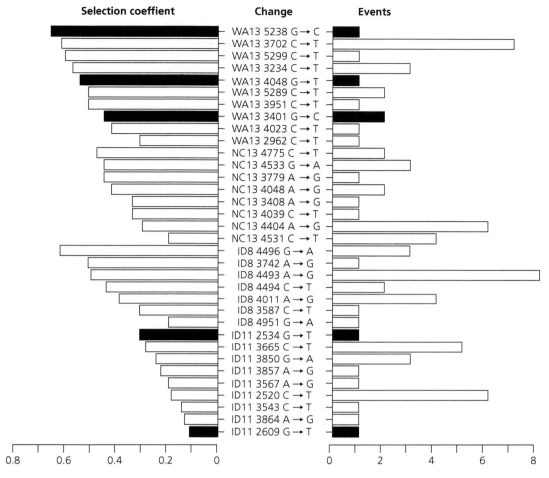

Figure 9.10 Results of one-step adaptation using data from Table 1 of Sackman et al. (2017), showing transitions in white and transversions in black. Left: measured selection coefficients; center, parental bacteriophage (four types) and mutational change; right: number of events in 20 replicates of one-step adaptation.

be present in the host organism, *E. coli* (see Katju and Bergthorsson, 2019). Among the 22 changes shown in Fig. 9.11 that have occurred at least twice (asterisks), 16 are transitions (black) that have occurred a total of 181 times, and six are transversions (grey) that have occurred a total of 42 times. Thus the transition:transversion ratio is $16/6 = 2.7$ if we count by paths, and $181/42 = 4.3$ if we count by events. The null expectation for a transition:transversion ratio, assuming the canonical genetic code, uniform codon usage, and no significant effect of selection, is 0.42 or about 1:2.5 (Stoltzfus and Norris, 2016). That is,

the data in Fig. 9.11 reveal a strong preference for transitions.

The role of each change has not been verified experimentally, and in such experiments, there is always a risk of "hitchhiking," i.e., hitchhiker mutations may be carried to high frequency with a "driver" mutation. How many hitchhikers are in the set of 241? The authors report that none of the changes in the J gene are synonymous, and this fact alone suggests that the frequency of hitchhikers is very low, by the following calculation. The chance of seeing zero events of some type, when three are expected, is $exp(-3) = 0.050$ (from the Poisson

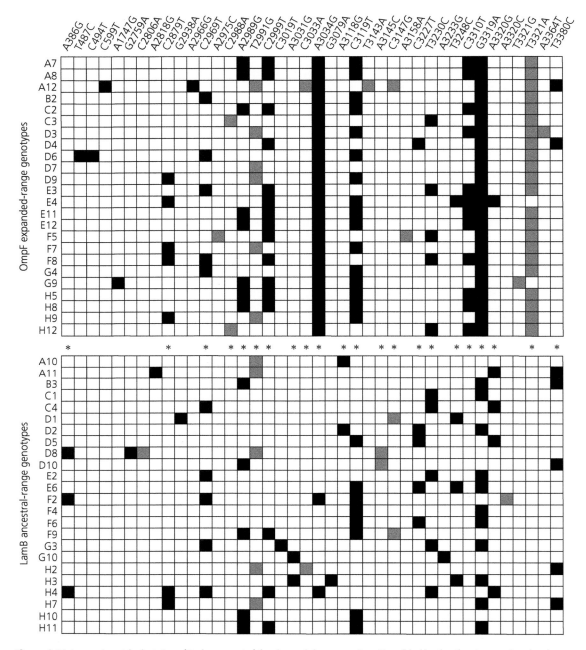

Figure 9.11 In experimental adaptation of λ phage, most of the observed changes are transitions (black) rather than transversions (grey).

distribution). So, if all hitchhikers are synomymous, and the expected frequency of contamination by hitchhikers is 3 / 241, it would be unlikely ($P = 0.05$) to find zero synomymous changes: this suggests that the frequency of contamination is less than 3/241. Or, we could suppose more conservatively that only half of hitchhikers are synomymous: then, by the same kind of calculation, the frequency of

Table 9.3 Events (frequency evolved) or paths of parallelism by transition (Ti) or transversion (Tv)

Phenotype	Taxon	Target	Paths		Ti Events		Tv Events	
			Ti	Tv	Counts	Sum	Counts	Sum
High-T host adaptation	ϕX174	genome	17	8	4, 3, 2, 2, 4, 2, 2, 3, 4, 4, 2, 2, 4, 2, 4, 3, 2	49	2, 3, 2, 4, 3, 3, 2, 4	23
Host adaptation	Lambda	Jprotein	16	6	3, 7, 10, 13, 16, 2, 26, 2, 24, 5, 10, 4, 12, 35, 5, 7	181	2, 11, 2, 2, 3, 22	42
Rifampicin resistance	P. aeruginosa	RNAPol	7	4	4, 35, 2, 5, 2, 4, 9	61	3, 3, 2, 3	11
Increased fitness	ϕX174	genome	3	0	5, 2, 6	13		0
Kanamycin resistance	E. coli	KNTase	0	2		0	7, 2	9
totals			43	20		304		85

contamination is likely to be less than 6 / 241. Using the latter calculation, the odds are better than 235 to 6 (39 to 1) that an individual change is a driver rather a hitchhiker.

Could there be a selective bias favoring transitions? Decades ago, protein sequence comparisons revealed a widespread evolutionary bias favoring transitions, and since that time, researchers have suggested that this pattern is due to transitions being more conservative (for review, see Stoltzfus and Norris, 2016). However, as explained in Box 9.1, multiple lines of direct evidence on fitness effects of mutations fail to support the premise of the conservative-transitions hypothesis, which is that transitions are much more conservative. Meanwhile, mutation-accumulation experiments have verified the premise of the mutational hypothesis, namely that transition bias in mutation is widespread (for review, see Katju and Bergthorsson, 2019). Thus, evidence of substantial transition bias among adaptive changes is evidence for an influence of mutation biases on the course of adaptation.

On this basis, Stoltzfus and McCandlish (2017) gathered data from published cases of repeated changes in experimental evolution, and compared the observed numbers of transitions and transversions to a null model, using a conservative null expectation of 0.5 for the transition-transversion ratio (Table 9.3). The cases included Meyer et al. (2012); the rifampicin resistance study of MacLean

et al. (2010); the ϕX174 study of Rokyta et al. (2005); the study by Liao et al. (1986) identifying seven temperature-resistant mutants of an E. coli kanamycin nucleotidyl-transferase expressed in Bacillus stearothermophilus, and finding two parallelisms; and the study of Crill et al. (2000), who extended earlier experiments of Bull et al. (1997), propagating lines of ϕX174 through successive host reversals (switching between E. coli and Salmonella typhimurium), observing numerous reversals and parallels.

All of the experiments involve bacteria or bacteriophages. The observed transition:transversion ratios are $43/20 = 2.2$ (95% binomial CI of 1.3 to 3.8) for paths and $304/85 = 3.6$ for events (95% bootstrap CI of 1.7 to 8.7). These ratios are 4-fold and 7-fold (respectively) higher than the null expectation of 0.5, and both results are highly significant ($p < 10^{-5}$; see Stoltzfus and McCandlish, 2017).

Table 9.4 shows the result of restricting the data to paths that have occurred in parallel k or more times for $k = 2$ through 8. This should decrease the frequency of hitchhikers and other neutral contaminants. However, the results remain qualitatively similar and highly significant.

In summary, studies of experimental adaptation show the effects of mutation biases: the heritable changes implicated in adaptation are enriched for changes favored by the process of mutation. What about adaptation in nature?

Table 9.4 The bias toward transitions persists even when the minimum number of parallel events is increased from 2 to 8 (column 1) in an effort to filter out spurious parallels (data from Stoltzfus and McCandlish, 2017).

		Paths					Events		
Cutoff	Ti	Tv	ratio	p		Ti	Tv	ratio	p
2	43	20	2.2	$<1 \cdot 10^{-5}$		304	85	3.6	$<1 \cdot 10^{-5}$
3	30	12	2.5	$<1 \cdot 10^{-5}$		278	69	4.0	$<1 \cdot 10^{-5}$
4	26	5	5.2	$<1 \cdot 10^{-5}$		266	48	5.5	$<1 \cdot 10^{-5}$
5	17	3	5.7	$<1 \cdot 10^{-5}$		230	40	5.8	$1.7 \cdot 10^{-5}$
6	13	3	4.3	$1.16 \cdot 10^{-4}$		210	40	5.3	$1.41 \cdot 10^{-4}$
7	12	3	4.0	$2.85 \cdot 10^{-4}$		204	40	5.1	$2.32 \cdot 10^{-4}$
8	10	2	5.0	$5.44 \cdot 10^{-4}$		190	33	5.8	$4.38 \cdot 10^{-4}$

9.4 CpG mutational hotspots and altitude adaptation

The thinner air at high altitudes poses a special challenge for birds: an aerial lifestyle requires more energy due to lower air density and stronger winds (Altshuler and Dudley, 2006), and yet there is less oxygen in a given volume of inspired air—about half as much at 5,500 m as at sea level (Peacock, 1998). The bar-headed goose, which migrates over the Himalayas, represents a famous case of altitude adaptation, with various morphological and physiological manifestations (Hawkes et al., 2011).

Galen et al. (2015) present a case for the role of CpG mutational hotspot in altitude adaptation of *Troglodytes aedon*. This case is of particular interest due both to its thoroughness, and to the generality provided by a recent follow-up study (Storz et al., 2019).

We encountered the CpG effect previously (Sections 2.3, 4.6, and 9.2) without any detailed explanation. The "p" in CpG disambiguates between (1) a C followed by a G on the same strand, linked by a **p**hosphodiester bond, and (2) a C:G base-pair, i.e., C and G on opposite strands (linked by hydrogen bonds). In mammals and birds, the C in CpG is often methylated, and this has a specific effect on mutation. Ordinary C and 5-methyl-C are subject to oxidative deamination, which converts C into uridine (U) and 5-methyl-C into thymine (T). If the T:G or U:G pre-mutation persists through the next round of replication, it will give rise to a

$CpG \rightarrow TpG$ transition (a $CpG \rightarrow CpA$ transition on the opposite strand). Whereas U is recognized by DNA repair machinery as non-DNA and efficiently removed, T is a natural component of DNA that is not as efficiently removed. This is the basis for the CpG hotspot effect. In mammals, the CpG context elevates $C \rightarrow T$ transitions about 10-fold higher, and also elevates $C \rightarrow G$ and $C \rightarrow A$ transversions (Mugal et al., 2015).

Galen et al. (2015) report on a CpG polymorphism that is linked to altitude in Andean house wren populations. Position 55 of the β^A chain of hemoglobin—the major β-hemoglobin chain in the blood of adult birds—has either the ancestral state of valine, or a derived isoleucine. The CpG straddles codon 54 and codon 55 for Val, so that a $CpG \rightarrow CpA$ mutation changes a GTH codon to an ATH codon for Ile ("H" is the single-letter code for "not G," i.e., C, A, or T). The frequency of the Val55Ile variant in separate populations is positively correlated with the altitude of the population. Introducing the Val55Ile change by genetic engineering results in a hemoglobin with 34 % higher oxygen affinity. In other words, this change appears to be adaptive: birds at higher altitudes are more likely to have the feature (a hemoglobin with higher oxygen affinity) that we expect *a priori* to be beneficial due to the reduced oxygen in high-altitude environments.

By examining hemoglobins of several dozen songbirds related to the wren, the authors found seven other lineages that had undergone the Val55Ile change, and a few more that had undergone a $CpG \rightarrow CpC$ transversion, i.e., Val55Leu. Once the site changes away from CpG, it ceases to be a hotspot. The authors do not measure the mutation rates directly, but infer a 5-fold elevated rate at the hotspot using a phylogenetic model.

What, precisely, is the proposed mutational effect? The suggested role of the hotspot is simply that, out of all the things that might happen in evolution, this change is probabilized by its higher rate of mutational appearance. We see the Val-to-Ile change, not because it represents the fittest alternative, but because this change is so strongly favored by the "arrival of the fitter." One does not know how many times the mutation occurred in wrens. Perhaps it was only once. Perhaps the polymorphism appeared prior to invading higher altitudes, and simply pre-adapted wrens carrying the mutation to survive better during a process of altitudinal diffusion, or (more interestingly) biased their upward diffusion by giving affected birds greater energy to explore higher altitudes.

Though this case involved a large amount of effort, the results pertain to a single pathway of change. Does altitude adaptation exhibit a more general pattern? The way to find out is to gather a systematic set of cases large enough to support statistical hypothesis-testing. Indeed, Storz et al. (2019) recently presented a systematic study of 35 pairs of low- and high-altitude bird species. For each pair in which the highland species shows a hemoglobin with an increase in oxygen affinity, the amino acid changes underlying this difference were identified by protein engineering.

The resulting set of changes implicates ten different paths of change that occur a total of 22 times. Storz et al. (2019) analyzed this set of data relative to a null model in which the CpG context is irrelevant. Given the observed background frequency of CpG sites, the chance of implicating a CpG site in adaptation, under the null hypothesis, is 0.1. On this basis, one expects one of the ten paths and two of the 22 events to involve mutations at CpG sites. However, six of the ten paths, and ten out of 22 events, involve mutations at CpG sites, both of which are statistically significant results. One might object that, to be more rigorous, the Val55Ile change from Galen et al. (2015) must be excluded, on the grounds that this observation stimulated the follow-up study and therefore is not independent (for pur-

poses of statistical hypothesis-testing). If Val55Ile is removed, then five out of nine paths, and nine out of 21 events, involve mutations at CpG sites, a significant excess over null expectations (Storz et al., 2019). Thus, changes associated with altitude adaptation in birds show an excess of changes involving mutations at CpG sites, consistent with an effect of mutation rate.

9.5 Transition bias in natural parallelisms

In principle, the analysis of transition bias presented previously for laboratory adaptation could be applied to natural cases. The reason for choosing amino acid replacements, again, is that a prior expectation for the ratio of transitions and transversions under a null model can be established empirically, as explained in Box 9.1. The reason for choosing transition-transversion bias is that this is a widespread bias with a modest magnitude (typically 2-fold to 4-fold above null expectations; see Katju and Bergthorsson, 2019). The barrier to such an approach, until recently, would have been the lack of a sufficient number of validated cases of adaptation (associated with specific mutations) to support statistical hypothesis-testing. However, this situation has changed in the past decade due to the work of many researchers.

An excellent example is the case of resistance to glycosides mediated by changes in the sodium pump ATPα1 (Zhen et al., 2012; Aardema et al., 2012; Ujvari et al., 2015). ATPα1 is targeted by some naturally occurring glycoside toxins, including both cardiac glycosides produced by amphibians, and the cardenolides produced by milkweed and other members of the dogbane family (*Apocynaceae*). Species such as monarch butterflies (*Danaus plexippus*) have not only evolved sufficient resistance to eat *Apocynaceae*, they sequester the toxin in their tissues, making them noxious to predators. Resistance is conferred typically by changes in ATPα1, and many specific mutations have been explored via genetic engineering followed by functional and structural analysis.

Figure 9.12 illustrates the data on ATPα1 parallelisms reported by Zhen et al. (2012). Species that consume and sequester plants producing cardenolides are shown in yellow, and species that merely consume cardenolides are shown in grey.

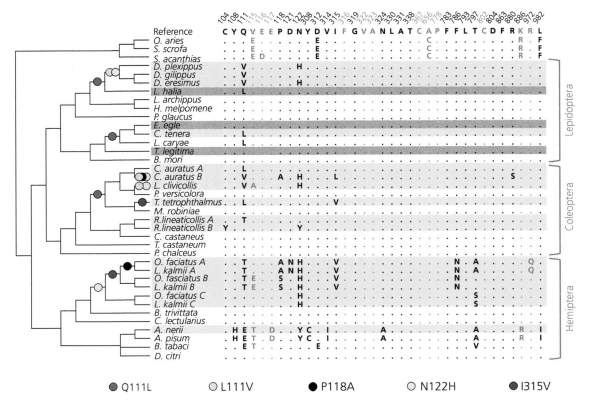

Figure 9.12 Parallel changes in ATPα1 implicated in cardiac glycoside resistance (after Fig. 1 of Zhen et al., 2012).

The columns shown in bold are the sites of experimentally verified effects, summarized in Fig. S1 of Zhen et al. (2012). A key source of information is the random mutagenesis and screening for ouabain resistance by Croyle et al. (1997), which implicates the T797A replacement as well as replacements at sites 111, 118, and 122.

In the first part of this chapter, we proceeded rapidly through various cases to focus on quantitative results. In this section, more attention will be devoted to (1) explaining the biological context, and (2) explaining the methodology. The reason to explain the context more fully is that we are now considering the prospect of identifying effects of mutation bias in regard to *nonboring aspects of natural adaptation*; thus we must consider the kinds of information that would establish biological significance.

The main reason to explain the methodology more fully, well known to aficionados of molecular adaptation, is that legitimate cases are very difficult to establish, such that the literature contains

multiple examples in which a specific mutation was alleged to confer a benefit that, upon closer examination, failed to materialize (Coombes et al., 2019). In the present context, an added complication is that we are repurposing retrospective studies of natural adaptation to test mutational hypotheses that were not considered originally. For each molecular path of change linked to adaptation, one must identify the possible underlying mutations, assess the probable history of events along the path, and evaluate the evidence that the path is actually adaptive, applying a consistent set of principles.

Below, the case of resistance to glycosides is presented in great detail, followed by a synopsis of other cases, so that the reader may gain a sense of the breadth of biological phenomena covered by the cases. The reader who trusts that these cases are carefully analyzed and that they represent a broad swath of interesting biology may simply skip ahead to Table 9.5 and the last few paragraphs of this section.

From Fig. 9.12, the replacements P118A (black dots), N122Y, I315V (magenta dots), and T797A clearly happen twice each. For site 122, the most parsimonious reconstruction calls for three changes to H (yellow dots) among the Hemiptera and Coleoptera, and either two changes to H, or one change plus a reversal among the Lepidoptera. This is counted conservatively as four changes. Site 111 is more complex. Q111L (blue dots) is a single transversion $CAR \rightarrow CTR$ that occurs several times. Q111V implicates two changes, which Aardema et al. (2012) argue is most likely $Q \rightarrow L \rightarrow V$ via $CAR \rightarrow CTR \rightarrow GTR$. In the minimal clade containing *Danaus plexippus* and *Lycorea halia*, the pattern of amino acids indicates $Q \rightarrow L$ (blue) in the ancestral lineage and then $L \rightarrow V$ (light blue) in the *Danaus* ancestor. Finally, site 111 allows for several equally parsimonious scenarios with five changes in the clade that includes *C. auratus* and *M. robiniae*; the most conservative scenario (for our purposes, the one with the fewest parallels) entails 2 $L \rightarrow Q$ reversals in *M. robiniae* and *P. versicolora*. No parallel can be inferred in the case of Q111T, because it requires a double-nucleotide change that cannot be inferred unambiguously.

Not all of the parallels are counted as adaptive parallels by Zhen et al. (2012). The occurrence of T797A and H122Y in the aphid clade has no strong correlation with cardenolide consumption, as there is one consumer (*A. nerii*) and one nonconsumer (*A. pisum*). Perhaps these parallel changes arose in an adaptive context, but their distribution is not consistent with an adaptive story in which derived resistant changes in ATPα1 are found only in cardenolide consumers and never in nonconsumers; therefore we must set them aside.

Additional parallel changes have been reported by Ujvari et al. (2015) among mammals and squamates (lizards and snakes) that consume glycoside-bearing plants, insects, or toads. Their Fig. 2 (see also their Fig. S1) includes the Q111L and N122H events seen in insects (noted earlier), four additional events of Q111L, and five other paths. One of the paths (Q111E) is seen only once; another path with two events (Q119D) is a double-nucleotide change, and thus is not counted here. The G120R change that occurs four times in squamates is ambiguously either $GGR \rightarrow AGR$ or $GGN \rightarrow CGN$:

inspecting columns 70 to 72 of the aligned sequences from Ujvari et al. (2015) (NCBI popset 928240786) indicates that the change in all four cases is $GGA \rightarrow AGA$. The remaining paths observed are Q111R $CAR \rightarrow CGR$ and N122D $AAY \rightarrow GAY$.

Using "ti" and "tv" to represent transition and transversion mutations, the complete set of changes is as follows:

- seven events of Q111L $CAR \rightarrow CTR$ (tv): implicated as a strong resistance-conferring mutation in the random mutagenesis of Croyle et al. (1997).
- two events of L111V $CTR \rightarrow GTR$ (tv): implicated by a strong resistance-conferring mutation in the random mutagenesis of Croyle et al. (1997).
- four events of Q111R $CAR \rightarrow CGR$ (ti): the combination of this and N122D confers 1,000-fold resistance to ouabain (Price and Lingrel, 1988).
- two events of P118A $CCN \rightarrow GCN$ (tv): implicated by a resistant mutant with only this replacement (Croyle et al., 1997; Schultheis et al., 1993).
- four events of G120R $GGY \rightarrow AGY$ (ti): in combination with Q111L and A119S, confers resistance in an engineered mutant (Ujvari et al., 2013).
- two events of N122D $AAY \rightarrow GAY$ (ti): the combination of this and Q111R confers 1,000-fold resistance to ouabain (Price and Lingrel, 1988).
- four events of N122H $AAY \rightarrow CAY$ (tv): this replacement confers resistance to ouabain when engineered in a *Drosophila* gene and expressed in human kidney cells (Holzinger and Wink, 1996).
- two events of I315V $ATH \rightarrow GTH$ (ti): implicated by analysis of engineered mutants (Qiu et al., 2005).

Using the case of resistance to glycosides presented in detail here, and other cases summarized later, Stoltzfus and McCandlish (2017) recently tested for an influence of transition-transversion bias on adaptive parallelisms. They identified cases until they had assembled a data set of at least 50 independent paths of exactly parallel amino acid changes that could not be attributed to an ancestral polymorphism, i.e., changes due to separate mutations. For each path, they required specific evidence that the change is a driver rather than a nondriver. A total of 55 paths and 231 events were identified. For 13 paths, the functional evidence was a genetic association with fitness (e.g., an allele that confers insecticide resistance), and for 42 paths, it was a separate experimental result that validated the proposed functional significance of the amino acid change (e.g., an assay for the effect of an engineered mutation). The complete set of cases is as follows.

ATPα1: resistance to glycosides (animals) This case consists of eight paths and 27 events, exactly as described above.

Various targets: insecticide resistance (insects) ffrench-Constant et al. (2004) review the literature of insecticide resistance in insects, and summarize parallel replacements implicated in three types of resistance targeting acetylcholinesterase, ligand-gated ion channels, and voltage-gated ion channels. In the first category, nearly all reported instances of resistance to cyclodiene pesticides involve site 302 (sometimes numbered 296) of the GABA receptor known as *rdl* (resistance to dieldrin) in *Drosophila*. Resistance was mapped genetically to the *rdl* locus, and 58 resistant isolates all showed a change at site 302, either *Ala → Ser* or *Ala → Gly* (ffrench-Constant et al., 1993). Ten different species of insects are implicated.

The second category involves DDT and the pyrethroid insecticides that bind to an insect sodium channel encoded by the *kdr* (knockdown resistance) gene. Parallel changes are listed in Soderlund (2005, Table 2), though verifying functional effects and identifying the underlying mutation (which is sometimes ambiguous when described at the amino acid level), requires reading various other papers and carrying out sequence database searches, as described in the supplement to Stoltzfus and McCandlish (2017). All together, 12 different species are implicated.

The third category cited by ffrench-Constant et al. (2004) implicates acetylcholinesterase mutations in the resistance of insects to organophosphates and carbamates. The key results are related by Weill et al. (2003), who established that the same G119S mutation is found in resistant strains of *Anopheles gambiae*, *Culex pipiens pipiens*, and *Culex pipiens quinquefasciatus*, and that the latter two are

independent mutations. The effect of G119S has been verified by genetic engineering (Weill et al., 2003).

Overall, these studies implicate eight paths and 29 events. ffrench-Constant et al. (2004) discusses other types of resistance that do not implicate parallel amino acid changes.

Sodium channels: resistance to tetrodotoxin (vertebrates) Tetrodotoxin (TTX) is a potent natural toxin used by pufferfish, many amphibians, and invertebrates from several different phyla; the interaction of TTX with voltage-gated sodium channels has been studied for decades, and much is known about the effects of replacements (Jost et al., 2008). Among snakes that prey on amphibians, resistance to TTX has arisen at least six times (Feldman et al., 2012). Jost et al. (2008) use published experimental information, as well as their own site-directed mutagenesis experiments, to identify functionally important changes in sodium channels in four species of pufferfish. This information, supplemented by the separate study by Feldman et al. (2012), implicates eight paths and 23 events.

ACCase: resistance to herbicides (grasses) Some grass herbicides (aryloxyphenoxypropionates and cyclohexanediones) target the plastid acetyl coenzyme A carboxylase (ACCase). Liu et al. (2007) summarize previous reports of resistance via ACCase replacements, and examine nine populations of Australian wild oats (*Avena sterilis* ssp. ludoviciana Durieu) finding I2041N, W2027C, and D2078G among herbicide-resistant strains (because introgression cannot be ruled out, each change is counted only once). They used genetic engineering of a wheat ACCase in yeast to verify functional effects. Additional information is found in Beckie et al. (2012). This case implicates six paths and 25 events in eight species of grasses (mostly *Avena* and *Lolium*).

Opsins: spectral tuning (vertebrates) The spectral sensitivity of opsins (visual pigments) has been the subject of much research reviewed by Yokoyama and Radlwimmer (2001). In multiple cases, the sensitivity of pigments has shifted from red (long-wavelength) to green (middle-wavelength) by the three changes S180A (tv), Y277F (tv), and T285A (ti). Other sites are also implicated. The functional effects of replacements are implicated partly by experimental verification by Asenjo et al. (1994), and more generally by a model that quantitatively accounts for the spectral sensivity of various natural and engineered opsins. Specifically, Yokoyama and Radlwimmer (2001) find that spectral sensitivity is explained by a "5-sites rule" to the effect that "S180A, H197Y, Y277F, T285A, and A308S shift the max of the LWS/MWS pigments toward green by 7, 28, 7, 15, and 16 nm, respectively, and the reverse changes toward red by the same amounts." The set of naturally evolved changes reported in Yokoyama and Radlwimmer (2001) and Shyue et al. (1995) implicates five paths and 19 events.

Prestin: convergence in echolocation (mammals)
Toothed whales and two different groups of bats have evolved sensitive systems for echolocation. The systems rely on an unusual ability to produce sound at specific frequencies, and an unusual sensitivity in detecting frequency modulation. Liu et al. (2014) carried out a genetic analysis of changes underlying convergence in biochemical properties of prestin, a motor protein that underlies the frequency sensitivity of mammalian hearing, focusing on two key parameters that differ systematically between echolocating and nonecholocating mammals. The set of changes reported by Liu et al. (2014) includes five paths and 11 events.

HIV Protease: resistance to ritonavir (HIV) Molla et al. (1996) analyzed HIV protease gene sequences from 41 patients who responded poorly to the protease inhibitor ritonavir, and used a cell-culture assay to show that viruses isolated from patients were indeed resistant to ritonavir. A specific set of nine sites in their Table 1 is implicated by the pattern that (1) patients who responded well to treatment (i.e., the patients with drug-sensitive virus) typically only had the wild-type sequence at these positions, and (2) the putatively resistance-conferring variant is found in post-treatment samples but not pre-treatment samples from the same

patients. Some of the reported results cannot be used because the authors do not provide sufficient information, e.g., separate counts are not given for sites with multiple changes. This study implicates four paths with eight events.

Hemoglobin: altitude adaptation (birds) Studies of altitude adaptation via changes in hemoglobin were discussed already (Section 9.4). Projecto-Garcia et al. (2013) studied hemoglobin properties in South American hummingbirds (*Trochilidae*) from high- and low-altitude populations. Functional effects were verified from some of the inferred changes by site-directed mutagenesis. In a separate study of low- and high-altitude populations of Andean waterfowl (*Anatidae*), McCracken et al. (2009) found levels of parallelism far in excess of the expectations of a simulated null model, including five kinds of recurrent changes in β-hemoglobin alleles. However, mutational effects were not verified. A subsequent study by Natarajan et al. (2015) supported some of the proposed parallels, while undercutting other claims that are not supported by functional evidence, or that reflect introgression rather than independent mutations. The data set combining adaptive parallels reported by Projecto-Garcia et al. (2013), McCracken et al. (2009), and Natarajan et al. (2015) includes five paths and 24 events.

Ribonucleases: convergence in foregut fermentation (monkeys) Foregut fermentation has evolved multiple times, in ruminants such as cows and goats, colobine monkeys such as langurs, and the hoatzin, a bird. These organisms, which lack genetically encoded enzymes to digest cellulose, have evolved an extra chamber (the foregut) in which plant matter is digested by microbes, which are then digested downstream in an unusually acidic environment. In a classic study, Stewart et al. (1987) argued for sequence convergence in lysozymes between cows and langurs. Foregut fermentation is also associated with elevated use of acid-tolerant ribonuclease (RNA-digesting enzyme) in the small intestine, to harvest the nitrogen contained in bacteria. African and

Asian leaf-eating colobine monkeys (*Colobinae*) separately evolved duplicate RNase loci that underwent changes that lower the pH optimum of the enzyme. Zhang (2006) used site-directed mutagenesis and enzyme assays to show that three changes observed in parallel are sufficient to cause the observed change in pH optimum. Yu et al. (2010) identified additional events of the same paths (R39W and K6E). The combined set of data implicates three paths and ten events.

β-tubulin: resistance to benzimidazole (fungi and some nematodes). Benzimidazole, which targets β-tubulin, is used in agricultural settings to control fungal infections of crops, and worm infections in livestock. Koenraadt et al. (1992) isolated resistant strains of ascomycete fungi, and sequenced their β-tubulin genes to identify changes. Elard et al. (1996) did the same with resistant worms. Together Koenraadt et al. (1992) and Elard et al. (1996) cite various references that isolate functional effects of individual changes. The combined data set from these two studies implicates three paths and 18 events.

The combined set of data from these ten cases is shown in Table 9.5. Cases 1, 4, 8, and 10 represent recent local adaptation of subpopulations, with a transition:transversion ratio of 11:10 for paths, and 69:48 for events, and the other cases represent species divergence, with ratios of 17:17 for paths, and 63:51 for events. Thus, the transition:transversion ratios are nearly the same for both categories. Cases of recent local adaptation have more events per path, but this difference is not significant ($P = 0.096$).

Stoltzfus and McCandlish (2017) compare these results to a null model in which mutation biases have no effect, so that the expected transitions: transversions is 0.5. On this basis, 18.3 transitions and 36.7 transversions are expected in the set of 55 paths. In fact, the observed ratio of 28 / 27 is 2-fold higher than the null expectation, a significant excess ($p = 5.3 \times 10^{-3}$ by a binomial test). Likewise, out of the 231 events, 132 of them are transitions, instead of the expected 77, a significant excess ($p = 3.0 \times 10^{-3}$, based on randomizations).

What are the alternative explanations for this pattern, and how likely are they? The available evidence from systematic studies of mutant fitnesses does not support the idea that transitions are substantially more conservative than transversions (Box 9.1).

Another possibility is that the data set is contaminated with changes that are not adaptive and which are biased toward transitions. Indeed, molecular changes in general are biased toward transitions (Wakeley, 1996). Natarajan et al. (2015) and Aardema and Andolfatto (2016) show that some

Table 9.5 Summary of changes identified in ten cases of natural parallel adaptation

Phenotype	Taxon	Target	Paths		Ti Events		Tv Events	
			Ti	Tv	Counts	Sum	Counts	Sum
Insecticide resistance	Insecta	Rdl, Kdr, Ace	5	3	2, 2, 5, 2, 3	14	9, 2, 4	15
Tetrodotoxin resistance	Vertebrata	Na channels	3	5	2, 6, 3	11	2, 2, 2, 3, 3	12
Glycoside resistance	Metazoa	Na$^+$/K$^+$-ATPase	4	4	4, 4, 2, 2	12	7, 2, 2, 4	15
Herbicide resistance	Poaceae	ACCase	2	4	5, 2	7	7, 2, 4, 5	18
Altitude adaptation	Aves	β-hemoglobin	2	3	4, 13	17	2, 3, 2	7
Trichromatic vision	Vertebrata	Opsins	2	3	2, 5	7	6, 4, 2	12
Echolocation	Mammalia	Prestin	3	2	2, 2, 2	6	3, 2	5
Growth in ritonavir	HIV1	Protease	3	1	25, 7, 9	41	4	4
Foregut fermentation	Vertebrata	Ribonucleases	3	0	2, 4, 4	10		0
Benzimidazole resistance	Ascomycota	β-tubulin	1	2	7	7	5, 6	11
totals			28	27		132		99

published reports of natural adaptive parallelisms in the literature are nonadaptive or nonindependent. If the null hypothesis (no effect of mutation bias) is true, and we assume (conservatively) that all misidentifications and other contaminants are transitions, then explaining the observed 28:27 ratio would require 25 % contamination—14.5 contaminants mixed with 40.5 genuine parallels that exhibit the null 13.5:27 ratio—, which seems unlikely, given that each path is accompanied by specific evidence. Stoltzfus and McCandlish (2017) also show that the pattern of an excess of transitions continues to hold when the minimum number of events for a path is increased from two up to eight, and this excess remains significant up to a cutoff of four. This result renders the contamination hypothesis less likely.

This methodology could be applied to many more data sets, e.g., the emergence of herbicide resistance in weeds (Baucom, 2019), or antibiotic resistance in microbial pathogens. Apropos, Payne et al. (2019) recently examined transition bias in two different curated databases of naturally occurring antibiotic resistance mutations in *M. tuberculosis*, finding an excess of transition mutations. For instance, they take advantage of the unusual case of Met-to-Ile replacements, which can take place by one transition (ATG to ATA) or two different transversions (ATG to ATT or ATC). Instead of this 1:2 ratio of possibilities, they see a ratio of 88:49 (Basel dataset) or 96:39 (Manson dataset), roughly 4-fold above null expectations. As all the replacements are the same type, the bias cannot be due to selection at the amino acid level.

9.6 Preferences for regulatory or structural changes

The issue of an evolutionary preference for regulatory vs structural changes—whether a preference exists, and if so, why—has been debated for decades. The issue has received renewed attention in evo-devo. Streisfeld and Rausher (2011) attempt to tackle the causes of preferences quantitatively. They argue that previous authors (Wray et al., 2003; Hoekstra and Coyne, 2007; Stern and Orgogozo, 2008) already recognized that mutation biases and fixation biases are two different types of causes, and they propose a previously missing mathemat-

ical framework. The missing framework that these authors supply is actually the origin-fixation framework explained in Chapter 8. The method of apportioning evolutionary preferences into biases in fixation and biases in mutation has the same formal basis that we explored earlier: in the origin-fixation regime, a bias of magnitude x will have an x-fold effect on evolution whether it represents a fixation bias or a mutation bias.

More specifically, Streisfeld and Rausher (2011) consider the relative contribution of some category of mutation i that represents a fraction θ_i of all mutations, so that the rate of introduction for that class of mutation is $2N\mu\theta_i$ where μ is the total mutation rate, and the origin-fixation rate is $K_i = 2N\mu\theta_i\pi_i$, where π_i is the chance of fixation. From this, one may compute the fraction of changes of this category (relative to all categories) that accrue over some period of time T (in generations), which is

$$r_i = \frac{K_i T}{\sum_j K_j T} = \frac{\theta_i \pi_i}{\sum_j \theta_j \pi_j} \qquad (9.1)$$

Either this equation (equivalent to Eqn. 1 of Streisfeld and Rausher, 2011), or Eqn. 8.3, could be used to compare two categories—they differ in that Eqn. 9.1 is expressed in terms of frequency before being converted to a ratio. Both formulas represent the ratio, for two equivalence classes, of changes expected under an origin-fixation process.

The approach used by Streisfeld and Rausher (2011) is worth examining in more detail. They carried out two separate analyses, one on floral traits (their specialty), and the other using all traits represented in the list of loci of repeated evolution from Stern and Orgogozo (2008). For floral traits, they considered three mutational effect categories: coding mutations in pigment pathway genes, cis-acting mutations in pigment pathway genes, and transcription factor mutations. The mutation spectrum, i.e., the relative rate at which a mutation in each category affects pigment traits, was estimated by gathering literature reports on floral pigmentation mutants, including a small number of segregating variations, supplemented with results of a large mutagenesis screen in petunia. For the same three categories, the spectrum of evolutionary changes, i.e., the "substitution spectrum" in the terminology of Streisfeld and

Rausher (2011), was estimated similarly by a literature screen designed to identify all cases in which floral pigmentation mutations underlying species differences were known.

For the analysis of other traits, Streisfeld and Rausher (2011) considered two categories of mutation (coding and cis-acting), and two categories of traits (physiology and morphology). They estimate mutation biases and evolutionary biases directly from the data of Stern and Orgogozo (2008), counting intra-specific variation and domesticated varieties as being indicative of the mutation spectrum, and inter-specific differences as being indicative of the substitution spectrum.

These simplifying assumptions are extreme. For instance, the "domesticated" category of Stern and Orgogozo (2008) includes cases in which resistance or tolerance has emerged repeatedly in domestic varieties, cases in which selection is obviously involved—indeed, earlier in this chapter, we treated cases of the emergence of resistance to anthropogenic substances such as pesticides and herbicides as cases of local adaptation, not cases of mere mutation. Even in a case of similar pigmentation arising repeatedly in domesticated chickens, we know that selection has affected the chances of this observation, because we know that the chickens survive well enough to be propagated readily. Likewise, Streisfeld and Rausher (2011) do not claim to have any direct information on the fixation spectrum for mutational categories. Instead, they simply define this, following origin-fixation dynamics, as the effect on the mutation spectrum that produces the observed substitution spectrum. That is, relative to Eqn. 9.1, they measure the substitution factor r_i, estimate the mutational factor θ_i from other data, and then infer the fixation probability π_i by division.

Thus, although the approach taken by Streisfeld and Rausher (2011) is a promising avenue for future research, it cannot lead to reliable conclusions without much further work.

Nevertheless, Streisfeld and Rausher (2011) have provided a more rigorous framework to interpret previous verbal arguments, particularly the claim of Carroll (2008) that cis-regulatory changes predominate in evolution because (1) morphological evolution is primarily about changing regulatory networks, and (2) cis-regulatory mutations are more likely to have this effect. In doing so, they have introduced a framework for causal attributions that is incompatible with the shifting-gene-frequencies theory of the Modern Synthesis. The equation of dual causation appeared only in 2001, and is based on origin-fixation dynamics, which emerged only in 1969 (McCandlish and Stoltzfus, 2014). As noted earlier (Section 8.10), this point is relevant to evaluating the argument of Lynch (2007a), *contra* Carroll and others, that evo-devo has not introduced anything new into evolutionary theory.

Box 9.2 The emergence of "molecular" evolution

The 1959 Darwin centennial was a milestone marking the establishment of the Modern Synthesis and its shifting-gene-frequencies theory as the dominant view of how evolutionary change occurs (see Tax and Callender, 1960; Stoltzfus, 2017).

Yet, at precisely the same time, a different vision of evolution was emerging. Biochemists had developed a way to sequence proteins, and began comparing proteins from different species. Each protein had a discrete sequence that differed from others at specific sites, obviously representing evolutionary changes. For a given protein compared between species, some parts differed, and others did not. From such patterns, Anfinsen (1959) developed a powerful new mode of inference—now a foundation of widely used methods of practical sequence analysis—by which a macromolecular subsequence (part of a gene, protein, or RNA) is inferred to be not "critical" or not "essential" in determining some functional property if the subsequence differs among species while the functional property does not (e.g., pp. 149 to 155 of Anfinsen, 1959).

In this new view, change is the default condition, and the lack of change indicates a conserving influence (in contrast to the Hardy–Weinberg paradigm). The comparison of proteins suggested evolution as a series of discrete amino acid substitutions, each determined by a mutation (e.g., Zuckerkandl and Pauling, 1962; Margoliash, 1963; Zuckerkandl and Pauling, 1965). The role of natural selection in "molecular evolution" was to "accept" or "reject" mutations, and it was depicted mainly as a filter preventing harmful changes (Eck and Dayhoff, 1966, pp. 161, 200). The changes appeared to accumulate over time with some regularity, thus one could compute rates of change and compare them between different proteins; the regular accumulation of changes could be used to probe relationships of species and even estimate their times of divergence.

These patterns of change were not predicted or explained by the shifting-gene-frequencies theory, in which change involves quantitative shifts in frequency at many loci simultaneously, the course of evolution is determined externally by the environment, and mutation rate is irrelevant (see Ch. 6). For instance, this theory did not suggest any way to interpret a pattern of discrete amino acid changes at a single protein site: examining a single site, or even a single locus, would be foolish, because selection does not act on individual loci, but on masses of variations recombining and interacting in the gene pool. Thus, to use a phylogeny from one protein, hemoglobin, to represent relationships of entire species was foolish— it "has nothing to tell us about affinities, or indeed tells us a lie," according to Simpson (1964).

The struggle to rethink evolution in the light of comparative molecular data was most obvious in the neutralist-selectionist controversy. In regard to the controversy, Maynard Smith (1975) recited King's argument about the correlation between codon number and amino acid frequency (see Section 8.11), with its two interpretations (adaptive and neutral), noting

> Hence the correlation does not enable us to decide between the two. However, it is worth remembering that *if* we accept the selectionist view that most substitutions are selective, we cannot at the same time assume that there is a unique deterministic course for evolution. Instead, we must assume that there are alternative ways in which a protein can evolve, the actual path taken depending on chance events. This seems to be the minimum concession the selectionists will have to make to the neutralists; they may have to concede much more.

This is important to understand: in 1975, an indeterminate view, even one based exclusively on selective fixations, was viewed as a "concession." For a surprisingly long time, our intellectual ancestors, and particularly theoretical population geneticists, were committed to analyzing evolution as though it were a deterministic process driven by selection to some optimum (e.g., see Grodwohl, 2016).

Recognizing the threat from molecular results, Mayr and Simpson went on a public campaign to circumscribe their relevance for evolution, claiming that they offered, at best, only a superficial window on the proximate causes of evolution (for review, see Dietrich, 1998). Many of the arguments about the role of mutation that we reviewed in Chapter 6 were stated with particular force and eloquence in the 1960s, when the challenge to neo-Darwinism came, *not from Bateson and Morgan, but from the pioneers of molecular evolution*.

If molecular sequence comparisons prompted an immediate rift and undermined the theory of evolutionary genetics underlying the Modern Synthesis (the shifting-gene-frequencies theory), why did Mayr and his cohort still claim to have unified biology? Where was the scientific accountability for failed predictions about the rate of evolution?

Jack L. King, one of the co-proposers of the Neutral Theory, saw a broad conflict with the pre-molecular orthodoxy. King (1972) described a classic view—no possibility of neutral changes; variation as raw materials, sufficient to respond to any challenge; no relation of mutation to the rate of evolution; directionality due entirely to selection—and insisted that none of these claims apply to molecular changes.

Yet, King was arguing, not to reject this orthodoxy, but merely to limit its scope: he claimed that the old doctrines remained true for morphology or physiology, i.e., he assumed there could be one theory for "molecular evolution," and another for everything else. Kimura eventually adopted this same position (Kimura, 1983). "Molecular" evolution seemed so different that, by 1971, it had emerged as a distinct field with its own journals, meetings, and key theories (the neutral theory, the molecular clock hypothesis). Meanwhile, Mayr and others claimed that the old view fully covers the visible features of evolution, and left the invisible molecular aspects to others.

Thus, the invention of a special "molecular" category of evolution preserved the classical orthodoxy by sacrificing the idea that there was one theory for all of evolution, e.g., Provine (2001, epilogue) lists the lack of a single unifying theory as one of the reasons that the evolutionary synthesis "came unraveled" in the 1980s. We still see the effect of this schism, e.g., when a contemporary volume on adaptationism defines the topic as "the claim that natural selection is the only important cause of the evolution of most nonmolecular traits and that these traits are locally optimal" (Orzack and Sober, 2001).

9.7 Developmental bias

The field of evo-devo is characterized by a central focus on how the internal details of development influence evolution, and it includes, in particular, claims to the effect that the course of evolution reflects developmental biases in phenotypic variation (Section 7.5.4). In general, the current state of the field is that developmental biases in variation are known to exist in many cases, and in some cases, the sources of these biases can be understood with mechanistic models of development (e.g., Lange et al., 2014; for review, see Uller et al., 2018).

However, it has proven difficult to establish conclusively that a particular evolutionary bias is due to developmental bias and not selection. Showing that a developmental model of variability predicts observed dimensions of evolutionary change (e.g., Richardson and Chipman, 2003; Kavanagh et al., 2013; Uller et al., 2018) is suggestive, but not conclusive. For instance, Houle et al. (2017) argue that, even though the experimentally measured mutational variability in fly wings "predicts 40 million years of fly wing evolution" (meaning that the major dimensions of variability predict those of evolution), the rate of evolution is very low compared to what the observed variability allows, so that selection alone may be determining the pattern of change. Furthermore, for the case of quantitative traits, theoretical models have revealed conditions under which a history of correlated selection may shape the spectrum of mutational effects to align with a beneficial correlation (see Houle et al., 2017).

Thus, we may insist that the morphotron is logically impossible, but it still remains to show that some evolutionary effect that correlates with patterns of variability is actually caused by generative biases rather than by selection. One must consider alternative hypotheses, and in particular, one must reject or place limits on alternative selective explanations.

The best available example may still be the oft-cited case of Alberch and Gale (1985), who characterized propensities of skeletal variation in the amphibian autopod (hand or foot) by treating developing autopods with colchicine, so as to produce aberrant autopods with missing digits (fingers or toes) or phalanges (the component bones of digits). In this approach, chemical perturbation of development is used as a surrogate for genetic perturbation via mutations, following Cheverud's conjecture of a correlation between genetic and environmental components of variation (Cheverud, 1984; see also Waitt and Levin, 1998). This methodology avoids the problems posed by the rareness of mutations, and though indirect, at least ensures that the measured biases reflect potentialities of development, and not effects of fitness (as the treated individuals all survive the treatment).

The patterns of skeletal deviation observed in the laboratory for a salamander species and a frog species were compared taxon-specifically to patterns of evolutionary divergence, in a manner reminiscent of the arguments of Vavilov (1922). The results showed a clear correspondence between taxon-specific developmental preferences for digit and phalange loss, and taxon-specific patterns of digit and phalange loss in evolution, supporting the hypothesis that developmental propensities of variation *cause* propensities of evolutionary change. For instance, when an initial phalange or digit is lost in frogs, it tends to affect the innermost (pre-axial) digit, whereas in salamanders, it tends to affect one of the outermost (post-axial) two digits.

The strength of this line of argument lies in the *a priori* implausibility of suggesting that selection favors one pattern of loss in frogs and a different pattern in salamanders. That is, Alberch and Gale (1985) have discovered a dimension of phenotypic bias—loss of inner vs. outer phalanges and digits—that appears quite arbitrary. This enables them to make a compelling intuitive argument that, since the observed bias in variation appears to be the only unequal factor, it must be the source of the evolutionary bias.

Of interest is a more recent study by Baldi et al. (2011). In multiple cases, *Caenorhabditis* species have evolved androdiecy, with hermophrodites that produce sperm as late-stage larvae, then go on to act like females, producing eggs as adults. In every case, the hermaphrodite sperm are substantially smaller than male sperm, an effect that has been attributed to selection favoring male over hermaphrodite sperm. Previously, Baldi et al. (2011) reported that RNAi directed against a sex-determining gene

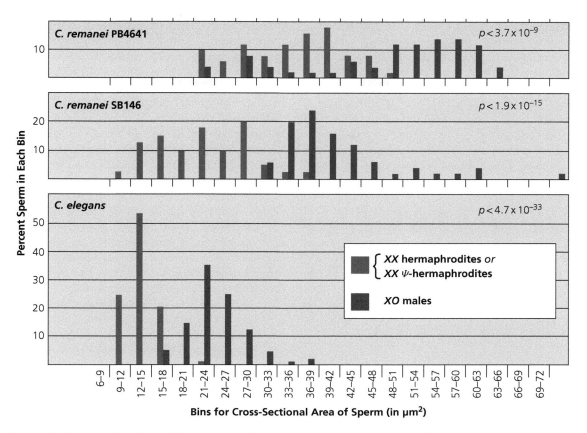

Figure 9.13 Hermaphrodite (red) vs. male (blue) sperm size distribution for natural hermaphrodites of *C. elegans*, and experimentally induced hermaphrodites of two other species. Figure kindly provided by Dr. Ron Ellis.

(*tra-2*) caused some females of nonhermaphroditic species to become pseudo-hermaphrodites, making sperm as juveniles before becoming female adults, just like natural hermaphrodites. The results in Fig. 9.13 indicate that, in the three cases tested, pseudo-hermaphrodites make smaller sperm than the corresponding males.

This asymmetry in size was not selected directly: these are novel phenotypes in species that do not have hermaphrodites naturally. That is, there appears to be a developmental bias toward smaller sperm when developmental pathways are deflected toward producing hermaphrodites, and the authors suggest that this bias has contributed to the observed evolutionary bias. Subsequent selection for smaller hermaphrodite (vs. male) sperm may also contribute to the bias seen in naturally evolved cases of facultative hermaphroditism.

As one attempts to make the case for developmental biases as distinct from hidden molecular biases such as transition-transversion bias, it becomes much less likely that the bias will have no fitness consequences. Again, Alberch and Gale (1985) began with a developmental bias that appears quite arbitrary. However, more generally, a developmental bias that affects interesting phenotypes is not likely to be perfectly orthogonal to the dimensions of fitness differences. Thus, there is a trade-off between the kinds of developmental biases that are most likely to be demonstrated convincingly, and the ones that are most likely to be considered important.

Furthermore, it seems as if the case for developmental bias affecting gross phenotypes will have to be built piecemeal, without taking advantage of the commonalities available in studies of molecular adaptation. That is, earlier we analyzed transition-transversion bias in data from bacteriophages, viruses, bacteria, fungi, plants, and animals. These organisms all share the same molecular GP map (the genetic code), they all are subject to transition bias (to varying degrees), and all (presumably) subject to the same assumption about the lack of a substantial fitness difference between transitions and transversions. These assumptions justify combining data from many cases, so as to achieve the statistical power to evaluate hypotheses of mutational effects.

However, once we go beyond the realm of molecular phenotypes, these diverse organisms do not have shared developmental pathways, shared GP maps, and shared variational biases. To build a broad case for developmental bias, it seems, will require many separate analyses, each with its own distinctive model of developmental effects on variation, and each with its own treatment of fitness effects.

9.8 Evaluating the argument

Let us reconsider the main evidence-based argument developed in this chapter. What, precisely, has been demonstrated? What are the possible limitations or weaknesses of the argument? How could it be established more generally?

The molecular results summarized in this chapter come from experimental evolution in the laboratory, and from nature. The analysis of experimental cases shows that biases in mutation influence the course of laboratory adaptation. This conclusion is strong but also narrow: the results implicate only biases in nucleotide mutations, only in bacteria and phages.

The case for mutation-biased adaptation *in nature* is broader, implicating diverse taxa including animals. The strongest cases available are ones in which (1) a phenotype-first approach is used to guide the search for changes in candidate genes, (2) implicated changes recur in a manner correlated with physiologically or ecologically relevant conditions, and (3) the changes are shown experimentally to have an effect consistent with the adaptive story.

In both the laboratory and in nature, the results show that a *modest* bias in variation, a several-fold

effect, can influence the course of evolutionary adaptation. Furthermore, this influence is not small, in the sense that, in evolutionary biology, an effect of 2-fold (for instance) is not a small effect.

If our rubric for importance is to account for the evolution of "hearts, hands, eyes and echolocation" (Dawkins, 1987, p. 265), then it is relevant that the natural cases implicate (1) echolocation, (2) eyes, via spectral tuning, and (3) hearts, via resistance to TTX, resistance to cardiac glycosides, and hemoglobin adaptation. The published studies of natural adaptation used in this meta-analysis represent the best work available today in the molecular dissection of adaptations. These studies were not filtered on any criterion other than their ability to supply evidence of the required type: amino acid changes that are robustly linked to adaptation and that can be ascribed to a known nucleotide mutation.

In short, the best available evidence shows that changes implicated in adaptation are enriched for mutationally likely changes, an effect predicted 20 years earlier via the theory described in Chapter 8.

Yet, the resemblance between observations and expectations may be only a superficial resemblance, or perhaps the resemblance is real but carries little significance for evolutionary biology. Therefore, let us consider some objections to this line of argument:

- the observed biases are actually due to selection
- the connection with theory is scant
- the influence of biases in mutation is trivial compared to the work of selection
- this only shows an influence of mutation on the boring parts of evolution.

9.8.1 Cryptic fitness biases actually explain the data

Could the evolutionary biases noted be due to cryptic fitness biases that happen to align with mutational preferences?

This alternative can be rejected for every study in which the authors actually *measure* fitness. The data from MacLean et al. (2010) reveal no correlation of mutation rate with fitness (see Figure 9.7). As explained, Rokyta et al. (2005) pursued a mutational

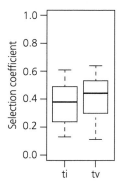

Figure 9.14 Selection coefficients for beneficial transitions and transversions from Sackman et al. (2017). The medians are 0.38 and 0.44 for transitions and transversions, and the means are 0.37 (CI 0.053) and 0.40 (CI 0.18), respectively.

explanation precisely because selection coefficients did not explain which mutant appeared most frequently. Figure 9.14 compares the fitnesses of transitions and transversions from Sackman et al. (2017), which includes the results of Rokyta et al. (2005). Transitions do not have higher selection coefficients: instead, transversions have a slight (insignificant) advantage.

Thus, for many experimental cases, we can rule out an alternative explanation based on fitness.

What about the analyses of natural cases, most of which involve transition bias? Here, fitnesses were not measured. Could there be a fitness advantage of transitions that accounts for the evolutionary bias?

As explained in Box 9.1, multiple lines of evidence bear on this question. First, we have the analysis of deep-mutational-scanning data, essentially an update of the earlier work of Stoltzfus and Norris (2016). Both studies show a minuscule benefit of transitions, considering the entire DFE. Second, because the key issue is not the entire DFE, but the right tail—whether *beneficial* transitions are better than *beneficial* transversions—we may consider data from the studies of laboratory adaptation collated by Stoltzfus and Norris (2016). The results show that transversions, not transitions, are slightly (insignificantly) better. Third, relative to this meta-analysis of four cases, the result just shown in Fig. 9.14 based on data from Sackman et al. (2017) represents an additional case showing the same thing. Finally, Payne

et al. (2019) report evolutionary biases toward transitions that cannot be due to protein-level selection, including transition bias in noncoding changes, and the excess (noted earlier) of Met-to-Ile transitions over Met-to-Ile transversions.

This evidence weighs strongly against the conservative transitions hypothesis. That is, transitions tend to predominate in evolution across the tree of life, and if the general explanation were selective, we would expect to see a general enrichment of transitions at the high end of the fitness distribution, but this pattern is not seen. Of course, a small sample of specific mutations might exhibit a bias, and for this reason, one cannot rule out a fitness-based explanation for *a specific narrow case*. For instance, the case of echolocation (see Table 9.5) implicates just three transition and two transversion paths, and perhaps these three transitions happen to be more advantageous than the two transversions. This is entirely possible. However, to propose this kind of explanation generally is to contradict available evidence.

9.8.2 The connection to theory is thin

The molecular changes involved in adaptation appear to be enriched for changes that are mutationally favored. This is a new and unexpected result, established on the basis of detailed analyses of adaptation. However, let us reconsider this result, not merely as a novel empirical pattern, but as evidence for a specific theory. That is, let us compare this chapter to the previous one.

Chapter 8 tells us that, when evolutionary change is dependent on the dynamics of the introduction process, it may exhibit effects that are not expected under the classic shifting-gene-frequencies theory. We can identify biases in the introduction of variation as a new kind of cause, and articulate a specific theory of its operation, based on the following distinctive characteristics (see Stoltzfus, 2019):

1. It poses a directional, quantitative dependence.
2. It is regime-dependent.
3. It depends on the rareness of mutations.
4. It establishes a condition of parity with selection.
5. It establishes a condition for composition and decomposition of causes.

6. It provides for a causal influence of developmental biases.
7. It provides for a causal influence of mutational accessibility in genotype-space.

Each of these statements could be given a more precise form (given some assumptions), following the development of theory in Chapter 8. For instance, in an origin-fixation model, introduction biases have a linear effect (point 1), mutations are assumed to be vanishingly rare (point 3), the effects of mutation bias and fixation bias are independent (point 5), and the two effects have the same (linear) form (point 4).

On this basis, we can ask how well the evidence reviewed in this chapter matches these diverse theoretical predictions. From this perspective, the connection between evidence and theory is narrow. The prediction that modest biases in variation may impose biases on evolution, without requiring neutrality or absolute constraints, is borne out, and this is a crucial departure from classical population genetics (Section 6.4.2). However, many other aspects have not been evaluated critically, mainly because this line of argument is new.

Therefore, let us consider some expected patterns, both to illustrate the many implications that have not been tested so far, and to point the way toward testing them in the future.

Graduated effects According to the arguments in Chapter 8, a larger bias will have a larger effect, and this relationship will be linear within the regime of origin-fixation dynamics. The range of possible biases in mutation is enormous, e.g., a 1.2-fold transition bias or a 2-fold AT bias in yeast (Zhu et al., 2014), a roughly 7-fold transition bias in HIV (Cromer et al., 2016), a roughly 10-fold effect of CpG bias in mammals (Mugal et al., 2015), and many other larger effects, when one considers the universe of mutations (Fig. 2.1, Appendix B). The best way to test for a graduated effect, perhaps, would be to compare episodes of evolution subject to different magnitudes of the same bias, e.g., different magnitudes of transition-transversion bias in different species. For instance, a robust analysis of this expectation might be done on the basis of a range of mutational transition bias in cases collated by Stoltzfus and McCandlish (2017), but this is

not possible currently, because robust estimates of transition bias are available only for a small set of model organisms (Katju and Bergthorsson, 2019).

Effects of mutational path density Chapter 8 presented several arguments that refer to mutational path density, regarding (1) the genetic code as GP map, (2) developmental bias, and (3) structuralist concerns about the size and accessibility of genotype networks. Some effects of the first type are already incorporated into results presented in this chapter. For instance, consider three possible adaptive changes from Rokyta et al. (2005), shown in Table 9.2: (1) from ACN (Thr) to GCN (Ala) by $A \rightarrow G$ at site 3857; (2) from GTR (Val) to TTR or CTR (Leu) by $G \rightarrow T$ or $G \rightarrow C$ at site 2534; and (3) from ATG (Met) to ATH (Ile) by $G \rightarrow T$, $G \rightarrow C$ or $G \rightarrow A$. The relative mutation rates (column 8 in Table 9.2) are 4.31, 1.23, and 5.54, respectively. The last number, 5.54, is the sum of the first two numbers, because it includes the same three types of nucleotide mutations implicated by the first two changes. If all individual mutation rates were the same, the relative rates for these amino acid changes would be 1, 2, and 3, respectively. This is the effect of path density.

In some of the analyses presented, such effects are not distinguished from biases in individual mutation rates, as in the analysis of Rokyta et al. (2005), and in other cases, such as Stoltzfus and McCandlish (2017). In order to evaluate the importance of this effect, it will be necessary to parameterize path density separately from the effect of biases in individual mutation rates.

Parity of biases in introduction and in fixation In the limiting case of origin-fixation dynamics, effects of mutation and selection achieve a kind of parity in which a bias in rates of mutational origination and a bias in chances of selective fixation have the same impact on evolution. This is a distinctive prediction relative to a neo-Darwinian theory that makes mutation a different kind of cause, a material cause subject to selection. This prediction is relevant in the analysis of Rokyta et al. (2005), because the match between observations and origin-fixation expectations tends to confirm parity. However, this is a relatively weak confirmation in the sense that, whereas origin-fixation dynamics predict the same pair of relationships *evolution* $\propto$ *origination* ($\propto$, proportional to) and *evolution* ∝ *fixation*, the set of data is so small that we would not be able to eliminate various other subtly different relationships.

Furthermore, in nearly all of the other cases described, the hypothesis-testing strategy begins by identifying a kind of variational bias that is orthogonal to fitness effects, so that they can be ignored, e.g., we choose to examine transition-transversion bias because transitions and transversions hardly differ at all in fitness effects. This makes establishing an effect of mutation easier, but makes establishing parity impossible.

Equivalence of diverse biases In the literature of evolutionary biology, the effect of mutational biases on genomic properties and the effect of developmental biases on phenotypes are treated as separate issues. To suggest conditions under which they will follow identical rules is a striking prediction. Other things being equal, mutational biases and certain developmental biases (the ones that are effects of path density induced by a GP map) are expected to have the same degree of effect. For instance, suppose that we have three completely distinct types of biases with the same arbitrarily chosen magnitude $B = 2$, one based on asymmetries imposed by a GP map, another based on a nucleotide bias, and a third based on micro-deletions vs. micro-insertions. Obviously, to make predictions, we need a quantitative model of fitness effects, or else we need to assume that these biases are orthogonal to fitness. However, in either case, estimating the magnitude of effect on evolution is possible, in principle, and this would represent an approach to evaluating the predicted equivalence of consequences.

Regime-dependent effects The theory developed in Chapter 8 links effects of the introduction process to a regime of population-genetics in which evolution depends on new mutations. The simulations of Yampolsky and Stoltzfus (2001) indicate that these effects are not limited to the strict origin-fixation regime, and the recent work of Gomez et al. (2020) provides analytical theory for the concurrent-mutations regime. Does the available evidence address population-genetic regime? In experimental studies of adaptation with microbes, episodes of evolution are initiated with cultures grown from purified stock, conditions that typically preclude inheritance of ancestral

polymorphisms, thus the mutations underlying adaptation must be new mutations. The relevance of origin-fixation dynamics is consistent with the fit of an origin-fixation model in Rokyta et al. (2005), though this is a weak inferential link for reasons noted earlier. Natural cases of adaptation or parallel adaptation typically provide little basis for drawing conclusions about population-genetic regime. One way to test for this kind of effect would be to seek out cases in which a reduced effect of variational biases is expected due to large μN, noting that (unlike for many other effects in population genetics) effects of the introduction process depend more on the census size N than the effective size N_e.

To summarize, whereas we can celebrate mutation-biased adaptation as a new and unexpected phenomenon with an empirical pattern that is easy to recognize, more work must be done before celebrating any particular theory of evolutionary genetics.

9.8.3 Selection did all the hard work

One may accept the evidence for an influence of mutational biases on adaptation, yet object that this influence is trivial, given that selection has done all the hard work of discriminating the beneficial mutations from out of the great mass of mutations, most of which are neutral or deleterious.

This objection represents whataboutery (the *tu quoque* fallacy), i.e., the fallacy of responding to a claim, not by challenging its validity, but by demanding attention for a second claim (here, a claim about selection). Was it ever suggested previously that predicting something about adaptation is trivial? The primary claim is significant because it contradicts the general expectation of neo-Darwinism as well as the more specific implications of the shifting-gene-frequencies theory, documented extensively in Chapter 6. Leading thinkers of the twentieth century argued at length for the irrelevance of mutation and mutation rates (see Section 6.4.3 and Appendix D), going back to the classic "opposing forces" argument of Fisher and Haldane (see Section 6.4.2). These arguments played an influential role in evolutionary discourse for 80 years. To the extent that science proceeds in

Popperian fashion, by falsification, showing that these arguments fail is important.

However, let us set aside these concerns and compare the discriminating power of selection with the discriminating power of mutation. In fact, selection frequently fails to discriminate in regard to effectively neutral or slightly deleterious alleles, and this is precisely why we experience so much difficulty when we attempt to study adaptation at the molecular level, i.e., it is difficult to achieve a high level of certainty that a change is adaptive, because so many changes are not (Coombes et al., 2019).

One way to understand the discriminating power of mutation is to consider how the biological processes operating in a cell impose a sampling distribution—the mutation spectrum **u**—on the universe of mutations. Familiar kinds of mutations, such as nucleotide substitutions, are outnumbered by an astronomical number of possible changes that occur at very low rates: the total universe of mutations is over 20 orders of magnitude larger than the set of point mutations (see Appendix B). This means that, in the molecular examples cited in this chapter, the process of mutation has elevated nucleotide substitutions from out of an enormously larger set of possibilities. Indeed, we calculated in Section B.10 that the recurrence of a mutation, i.e., any case in which we see the same mutation twice, is so statistically improbable under uniformity as to confirm the nonuniformity of mutation.

If we judge the power of selection by the probability of fixation, then it discriminates beneficial effects with probabilities of acceptance ranging from a maximum of nearly 1, down to values on the order of $\frac{1}{N_e}$, and it discriminates deleterious effects with probabilities of acceptance ranging from practically 0 (an allele so deleterious that fixation is practically impossible in billions of years of evolution) to values on the order of $\frac{1}{N_e}$. The entire range of discrimination combines these two scales. The range of mutation rates extends from practically zero (a mutation so rare it would be unlikely in billions of years of evolution) to 10^{-3} for STR mutations.

This kind of argument is interesting and deserves further attention. Certainly selection has an enormous discriminating power, but the process of mutation also imposes a drastically uneven sampling distribution on the universe of possibilities.

9.8.4 Mutation only affects the boring parts

A final objection would be that, although the influence of mutation bias on adaptation is real and surprising, this influence is limited to hidden details of evolution. The natural cases discussed relate to "hearts, hands, eyes and echolocation," but this relationship is distant and not very interesting, i.e., transition-transversion bias and the CpG hotspot effect are still "the boring parts" of evolution, even when they are involved in interesting phenotypic adaptations. Like studies of developmental changes in evolution, they reveal only "the hidden history of what has happened in evolution" (Charlesworth, 2005). They provide some of the details of the "how" of adaptation, but not the "why."

Therefore (to continue this objection), the traditional view in which selection rules evolution is still quite safe, and we have learned only that the course of adaptation is subject to internal biases that shape the underlying molecular details. Selection still determines what happens at the broad level of phenotypes, whereas the manner in which selection accomplishes its wonders depends on the specifics of each biological system.

This type of argument has been used for 50 years to argue against the relevance of molecular results to the major issues of evolutionary biology (see Box 9.2). Underlying the argument is the presumption that, for practical purposes, the major issues of evolution are *defined* at the level of visible phenotypes of charismatic megafauna, and therefore can be *resolved* only via visible phenotypes of charismatic megafauna. One might object that, as discussed later in Section 10.4, the *explananda* (things to be explained) in evolutionary biology have changed, and now include molecular and microbial evolution. That is, some evolutionary researchers study transposable elements (or repetitive DNA, codon usage, introns, etc.) not because they want to know why leopards have spots, but because they want to know why genomes have transposable elements (or repetitive DNA, codon usage, introns, etc.), i.e., they have an intrinsic interest in the phenomena of molecular evolution.

But let us set aside that concern, and accept the presumption that the major issues in evolution are defined in terms of visible morphology and behavior of animals, i.e., in terms of hearts, hands, eyes, and echolocation. Given this presumption, one aspect of the "unimportant molecular details" objection must be (partly) correct, and another aspect ultimately must be incorrect.

The *part that must be correct* relates to the indirectness between generic repetitive features of molecules shared throughout the tree of life, and token instances of phenotypes beloved by animal biologists. Hands, hearts, and eyes are complex features built up from repetitive parts. The amino acid methionine, for instance, in the context of any given species, literally plays different roles in thousands of different proteins, in hundreds of different pathways, such that methionine must be involved in some way in every major phenotype and behavior. When we conduct tests for effects of mutation on amino acid changes involving methionine and other amino acids, we do this precisely because the repeated use of the 20 amino acids in different contexts provides the basis for collecting large numbers of instances. This is how we achieve statistical power. But this repeated use in different contexts also means that generic mutational effects defined at the level of amino acids are unlikely to have strong differential effects on specific high-level phenotypes.

The *part that must be incorrect* is the implication that selection chooses the high-level course of adaptation, and internal factors merely determine the details. This hypothesis suggests a hierarchy of causes, or a division of labor, by which selection is the high-level cause that determines what outcome happens, and internal factors determine how the outcome emerges.

To understand why such a dichotomy is logically impossible, consider that, in any given situation, an organism might evolve adaptively in a variety of ways, some distinctly different from others (or it might fail to evolve adaptively). A mammal species, for instance, finding itself in increasingly desert-like conditions, will not necessarily evolve large ears that radiate heat, as for the jack-rabbit. Mammals have adapted to desert conditions by means of various morphological, physiological, and behavioral

traits. For the sake of this example, it does not matter if the outcome of adaptation is, in fact, highly repeatable for a given taxon. What matters is that, considering evolution generally, multiple outcomes are possible.

Let us begin by considering just one possible outcome for our case of a mammal facing desertification, e.g., the outcome of evolving large ears. Let us call this outcome A, and suppose that A1, A2, and A3 represent three detailed routes of developmental-genetic change to achieve outcome A. That is, A is the high-level outcome, and A1, A2, and A3 represent three different implementations.

Note that probing the details of some observed adaptive outcome A is a meaningful exercise for scientists, to the extent that this detailed implementation is difficult to predict. In quantitative genetics, this is called finding the "structure" or "architecture" of adaptation, and in some parts of the literature, e.g., Lind et al. (2019), "predicting" adaptation does not mean predicting a priori which direction adaptation will take, but predicting the hidden changes underlying an observed outcome. As the study by Lind et al. (2019) shows, differences in rates of mutation may render one route to an outcome more likely than others. That is, a mutational factor may enhance the chance of A2 (for instance) relative to A1 and A3.

Now, let us consider a second outcome, B. If outcome A represents the evolution of large ears that radiate heat, outcome B might represent a behavioral adaptation such as a nocturnal lifestyle. As with outcome A, we can imagine multiple routes to outcome B, such as B1, B2, and B3. And we can make the same argument that a mutational factor might render B3 (for instance) more likely relative to B1 or B2.

However, by introducing a second outcome, a second kind of conclusion follows logically. Any factor (including a mutational factor) that enhances the chance of A2 not only probabilizes A2 over A1 and A3, but also probabilizes A over B, relative to the absence of that factor. That is, it is not logically possible for factors to exist that merely influence the details of how a high-level adaptive outcome is encoded, without also influencing the chances of that high-level outcome relative to other high-level outcomes. To suppose otherwise is to assign to selection a top-down governing power incompatible with bottom-up mechanistic reasoning.

This conclusion is not evident when the study of adaptation follows the traditional paradigm in which a token case of adaptation is analyzed retrospectively (for discussion, see Section 10.4). For instance, the research program of dissecting the "genetic structure of adaptation" begins retrospectively with an adaptive trait, and the aim is to identify the loci involved, the sizes of contributions from each locus, and their genetic interactions. In the case of experimental adaptation, an experiment is conducted in which organisms are exposed to some conditions, and those that adapt are chosen for further study. The usual study systems are the ones in which the same kind of adaptation can be made to happen repeatedly. Adaptation has happened, and the challenge is to understand how. When the idiosyncratic details are discovered, they may be assigned to mutation, but considering how they probabilized one high-level outcome over another is not part of the paradigm.

Such approaches mask a hidden presumption of teleology, by which the direction of adaptation is predetermined by selection independent of tendencies of variation. This presumption remains hidden so long as we neglect to examine a larger context, with a more open-ended set of possibilities. For instance, going back to the example of Galen et al. (2015), the CpG hotspot not only probabilizes a specific amino acid change, it probabilizes an adaptive increase in the oxygen affinity of hemoglobin, distinct from a great variety of other modes of behavioral and physiological altitude adaptation; it also probabilizes higher-level outcomes such as altitude adaptation, as distinct from other adaptive or nonadaptive outcomes.

9.9 Synopsis

Punnett (1913) argued that color variations in mice and rabbits were the result of the same genetic factors, and thus that melanic or albino forms, which are found among many mammal species, might be the result of parallel mutations. Morgan (1923) highlighted "recurrent and parallel mutants" as "one of the most interesting discoveries in recent genetic

work," citing cases in which similar mutations happen in closely related species. The idea that parallel variations may lead to parallel patterns of species differences was taken to great lengths by Vavilov (1922).

In Chapter 8, we developed a formal basis to consider such ideas, i.e., we developed a conceptual and theoretical understanding of the generation of variation as, not merely a source of raw materials, but a source of initiative and direction in evolution. The novelty of this theory arises from the novelty of recognizing the introduction process formally as an evolutionary cause.

The resulting theory is not a truism. Evolution does not necessarily operate in a manner that depends on biases in the introduction process. Even if evolution shows mutational biases, this is not necessarily due to biased dynamics of introduction. Even if biased dynamics of introduction are sometimes influential, they are not necessarily an important cause of developmentally mediated effects, or of the kinds of effects implicated in structuralist arguments about accessibility in genetic spaces. A key prediction of this theory is that the efficacy of biases in the introduction process does not require neutrality, absolute constraints, or high mutation rates. Instead, the course of adaptation may reflect modest quantitative biases in ordinary mutations.

In this chapter, we considered the evidence. We began with a model of anonymous changes of amino acids at large numbers of sites. This demonstrated that empirical priors can be used to make a quantitative predictive model, with the result that mutational tendencies of amino acid replacement are a substantial part of the explanation for evolutionary tendencies of amino acid replacement. Yet, this kind of example is subject to the criticism that the pattern is a matter of "the boring parts" of evolution, not of much interest to the typical evolutionary biologist.

Therefore, we considered cases of phenotypic change, or cases in which molecular changes are linked to adaptation. Though there has been little direct attention to the role of mutation biases in molecular adaptation, available data can be leveraged to make a strong case for mutational effects. In regard to experimental studies of adaptation in the laboratory, quantitatively large effects have been shown for three different kinds of mutational biases: transition-transversion bias (Stoltzfus and McCandlish, 2017), the distinct nucleotide substitution preferences of two different mutator strains (Couce et al., 2015), and idiosyncratic context differences in rates of nucleotide mutations at specific sites (MacLean et al., 2010). In regard to retrospective studies of natural adaptation, the evidence for mutation-biased adaptation implicates substantial effect-sizes, a broad range of taxa, and two different kinds of mutation biases, transition-transversion bias (Payne et al., 2019; Stoltzfus and McCandlish, 2017) and CpG bias (Storz et al., 2019).

The natural cases, as a whole, represent a step in the direction of having a better predictive theory for the evolution of "hearts, hands, eyes and echolocation," if one is interested in *how sensitivity in echolocation is achieved, how eyes are tuned to different spectra, and how circulatory systems resist toxins and deliver oxygenated blood to tissues at high altitudes.*

Yet, much more could be done both to (1) validate aspects of the theory developed in Chapter 8, and (2) assess the role of biases in variation in evolution. In particular, we have not identified a similarly strong case for the impact of developmental bias in systematically biasing the course of evolution. The ideal case would be one involving a predictable morphological direction to biases in variation, shown to be effectual in cases that are understood to be adaptive, or at least, ecologically important. Studies patterned after Alberch and Gale (1985) would be helpful. A further interesting possibility is that there may be cases, like the case of the small hermaphrodite sperm, in which the hypothesis of a developmental bias is an alternative to a previously stated adaptive hypothesis.

CHAPTER 10

Moving on

In those days few people read such mathematical discussions as Haldane's and Fisher's. We got our theory in words from Dobzhansky, Mayr and Simpson.

George Williams, *Natural Selection: Domains, Levels and Challenges*

10.1 Introduction

In this book we have pursued a new understanding of the role of variation as a difference-maker in evolution. We have come a long way from the initial focus on mechanisms of mutation and the randomness doctrine. Probing the randomness doctrine leads away from the biology of mutation, toward what scientists mean by "randomness," and toward conceptualizations of the role of variation in evolutionary theories. We concluded that the randomness doctrine cannot be taken literally, but is best understood as part of a larger position contrasting the power and influence of selection with the weakness and irrelevance of mutation.

That is, the randomness doctrine did not emerge from mutation researchers attempting to find an empirically accurate characterization of mutation as a biological process, but instead arose among evolutionary biologists as a way to dismiss any dispositional role of mutation. Trying to capture the essence of a complex biological process with one word was never a good idea. Therefore, let us not attempt to amend the doctrine by saying that mutation is "nonrandom" or "indifferent." The remedy for trying and failing to put mutation into a black box is to stop trying to put mutation into a black box. What is mutation? Mutation is what mutation does, the sum of all processes that account for the mutation spectrum.

A general consideration of the role of variation led us to consider the morphotron and the larger problem of variation in evolutionary theory. Due to the influence of neo-Darwinism, the mainstream of evolutionary thought in the twentieth century developed without a substantive theory of the generation of variation. The inadequacy of this position, in turn, led to calls for reform beginning several decades ago. Thus the first two-thirds of this book might serve as the introduction and foundation for appreciating various departures from the morphotron in areas of evo-devo, genomics, and so on, briefly reviewed in Chapter 7.

The final chapters, however, focused on one specific line of argument, which is that biases in the introduction of variation are an important but previously unrecognized dispositional factor in evolution.

The aim of this concluding chapter is to summarize the main points, and to take advantage of the material already presented to raise some broader issues relevant to understanding evolutionary discourse.

10.2 Summary as historical narrative

Darwin faced an exhausting challenge from critics who characterized "natural selection" as a matter of luck or chance, e.g., he was disappointed when John Herschel, a philosopher whom he admired, called it "the law of higgledy-piggledy." Close colleagues were unsettled by his reliance on variations that arose by unknown means in an unguided manner (Beatty, 2010). How can evolution depend so

Mutation, Randomness, and Evolution. Arlin Stoltzfus, Oxford University Press (2021). © Arlin Stoltzfus. DOI: 10.1093/oso/9780198844457.003.0010

profoundly on a process of variation and not be shaped by it? The generation of variation, they assumed, must play a dispositional role, with a direct influence on the course taken by evolution. If so, *one cannot construct a working theory of evolution without a substantive theory describing which kinds of variations tend to occur, under which circumstances.*

Yet Darwin clearly disagreed with this assumption: he offered a theory of evolution *while claiming explicitly to be ignorant of the laws of variation* (Darwin, 1872, Ch. 5). In the theory he called "natural selection," the role of variation was not dispositional, but was a matter of "chance." To explain and justify this concept to his colleagues, Darwin developed (over many years) an analogy comparing variations to the numberless uncut stones used by a builder (Beatty, 2010). Imagine that the builder constructs a castle from these stones. Each stone is unique, and we can imagine that each fragment broke off and fell from a cliff following physical laws,

but there is no relation between these laws and the purpose for which each fragment is used by the builder. In the same manner the variations of each creature are determined by fixed and immutable laws; but these bear no relation to the living structure which is slowly built up through the power of selection (Darwin, 1868, Ch. 21)

Here chance is *not* depicted as an intrinsic property of what Darwin sees as a law-governed process of variation. Instead, chance is a way of denoting the explanatory irrelevance of these generative processes: because of the way evolution works (in Darwin's theory), the higher law of selection accounts for the outcome, to which the laws of variation "bear no relation." The structure built by selection is like a sandcastle: each grain of sand may be unique, with a unique history, but this hardly matters when explaining the shape of the sandcastle, which is due to the architect. Evolution results from a marriage of variation and selection, but this is an unequal marriage, in which selection makes all the important decisions and gets all the credit.

This dichotomy of roles persisted in the neo-Darwinian view that so heavily shaped the twentieth century, e.g. Fisher (1930b, p. 120) explains that the researcher who understands the weakness of mutation

...will direct his inquiries confidently toward a study of the selective agencies at work throughout the life history of the group in their native habitats, rather than to speculations on the possible causes which influence their mutations.

Ayala (1970) says that "the ultimate source of explanation in biology is the principle of natural selection." For Gould (2002), the key premise of neo-Darwinism is that "nothing about the provision of raw materials—that is, the sources of variation—imparts direction to evolutionary change" (p. 140). In the entry on "natural selection" in the contemporary *Oxford Encyclopedia of Evolution*, Ridley (2002, p. 800) writes:

In evolution by natural selection, the processes generating variation are not the same as the process that directs evolutionary change. Variation is undirected...What matters is that the mutations are undirected relative to the direction of evolution....Natural selection differs from most alternative theories of evolution in the independence between the processes that direct variation and that direct evolution...Darwin's theory is peculiar in that evolution is not an extension of the mutational process.

Though this is a contemporary source, it is atypical. For contemporary scientists, the distinction between selection and drift is paramount: the phrase "evolution by natural selection" is used to emphasize that allele replacements take place *by selection rather than drift*, not to emphasize that external factors shape outcomes via selection whereas internal causes of variation do not. Yet, the latter distinction is what Darwin, Fisher, Gould and Ridley mean to emphasize.

Guided by the selection-drift dichotomy, scientists today often assume that neutralism is the antithesis of neo-Darwinism and that the main axis of contention in evolutionary discourse is whether most fixations happen by selection or drift. Historically, however, neutralism is not the main contrast-case for neo-Darwinism. Instead, neo-Darwinism is the theory that selection (rather than a divine creator) is creative, and the centerpiece of historic criticism, from Mivart (1871) to the neo-Lamarckians to the early geneticists to evo-devo, is a belief in the importance of internal dispositional factors manifested in the generation of variation.

Darwin claimed ignorance of the laws of variation and yet, as the author of a two-volume work on variation in plants and animals under domestication, he was abundantly aware of peculiarities or regularities of variation. How did he conclude that the laws of variation "bear no relation" to the outcome of evolution in nature?

As pointed out by many other authors (Mayr, 1982; Gould, 2002; Beatty, 2016), Darwin developed his conceptions of the roles of selection and variation in a top-down manner, from the premise that organisms are exquisitely adapted, down to the finest detail, i.e., from empirical adaptationism (Box 3.1). Given this premise, evolution must work in a way that ensures exquisitely detailed adaptation, and prevents maladaptation. The role of variation, then, must be to provide an abundance of infinitesimal effects that selection shapes into adaptations: the exquisite adaptation of organisms is understandable only if variation is the clay and selection is the potter. This is why Darwin, in spite of knowing about many different kinds of variation, based his conception of "natural selection" on a process of "indefinite variability" that generates an abundance of infinitesimal differences in every trait, every generation (rather than on macromutations or other discrete peculiarities).

This is a strong theory, in the formal sense that a theory is strong when it promises to simplify the world, making it easier for us to construct explanations and to focus our research, telling us what to notice and what to ignore. If this theory is correct, then we can ignore peculiarities of variation and attribute traits to selection, in the same way that we attribute the shape of a piece of pottery to the potter and not to the clay, which merely provides substance. If this theory is correct, we can draw the same conclusions as Fisher and Ayala: we must focus our research on selection and not variation, because selection is the ultimate source of explanation in biology.

The mutation-is-random doctrine found in textbooks seems like a completely different kind of claim—a bottom-up claim invoking randomness as if it were a physical property of mutation. Yet, defining this property has proven difficult (Sarkar, 1991; Eble, 1999; Merlin, 2010; Razeto-Barry and Vecchi, 2016). Mutations emerge from a complex suite of processes. No ordinary definition of randomness fits, thus a special evolutionary definition of randomness is offered (Eble, 1999). Yet, the rhetorical purpose of the randomness doctrine (i.e., the manner in which it is employed) does not differ from Darwin's doctrine of chance as irrelevance: the doctrine is used to reject the explanatory potential of mutational factors as alternatives to selection. Because the only constraint on the "evolutionary" definition of randomness is that it must apply to mutation but not selection, the randomness doctrine has gravitated towards asserting, ever more plainly, that mutation is random because it is nonidentical with selection, e.g., it does not always favor higher fitness (Merlin, 2010).

This narrowing of the randomness doctrine is illustrated starkly by some recent responses to CRISPR-Cas defense systems. As we discovered (Section 5.3.3), such systems violate Luria–Delbrück randomness, and directly contradict historic statements denying the possibility of a mutation system that incorporates external cues into DNA (e.g., Mayr, 1959a; Section C.26). Yet, recent commentaries claim that CRISPR-Cas spacer acquisition is random because it does not guarantee beneficial changes (Weiss, 2015; Charlesworth et al., 2017).

Thus, although the random-mutation doctrine once seemed to support Darwin's theory of the irrelevance of variation, it appears to be evolving into a narrow claim of little practical importance.

Meanwhile, the challenge of rethinking the role of variation in evolution is complicated by the ongoing influence of neo-Darwinism on basic concepts used to negotiate questions of causation and explanation, including ubiquitous concepts such as "raw materials" and even population-genetic "forces." Mutation can convert one allele into another, an operation with two different implications. If we consider a model of evolution in which some relevant alleles are initially absent, an event of mutation may introduce a new allele, something that selection and drift cannot do. By contrast, if the alleles relevant to evolution are already present, then the occurrence of mutations does not introduce novelty, but merely shifts relative frequencies in a manner that is analogous to selection or drift.

The architects of the Modern Synthesis stated explicitly that individual events of mutation are not evolutionary causes, because evolution (in their shifting-gene-frequencies theory) uses the supply of variability built up in the gene pool (Section 6.4.3). This theory was offered as a contrast to the "lucky mutant" view, in which the timing and character of evolutionary change depends on the timing and character of mutations. The extrapolationist doctrine of the Modern Synthesis was that all of evolution follows from the short-term process of shifting allele frequencies (microevolution) observable in experimental populations of animals and plants, a process by which selection creates new types without mutation (Provine, 1971).

Generations of theoreticians built models under the influence of this conception of evolutionary genetics. When the alleles relevant to the outcome of evolution are present initially, mutation cannot introduce anything new, but may act only as a mass-action pressure or force that, like selection or drift, can shift the relative frequencies of alleles, possibly driving an allele to a high frequency. This force of mutation is exceedingly weak, because mutation rates are small, e.g., the force of a mutation rate of 10^{-9} is a million times weaker than a fitness difference of just 0.1 %. On the grounds that mutation is a weak force unable to oppose selection, Fisher, Haldane and Wright argued that tendencies of variation cannot cause evolutionary tendencies (Fisher, 1930b,a; Wright, 1931; Haldane, 1933), making the irrelevance of mutational tendencies seem like an inevitable mathematical result, e.g.,

If ever it could have been thought that mutation is important in the control of evolution, it is impossible to think so now, for not only do we observe it to be so rare that it cannot compete with the forces of selection but we know this must inevitably be so. (Ford, 1971, p. 361).

Since orthogenesis can only operate when mutation pressure becomes high enough to act as an agent of evolutionary change, empirical data on low mutation rates sound the death-knell of internalism. (Gould, 2002, p. 510).

Thus, the architects of the Modern Synthesis believed they had found a rigorous justification for neo-Darwinism, a magic bullet to kill internalism: theoretical population genetics proves that the generation of variation cannot play a dispositional role in evolution, because mutation rates are too small to overcome the opposing pressure of selection.

This classic argument presented an obstacle to the claims of Maynard Smith et al. (1985) in their seminal "developmental constraints" article:

[The issue] is whether biases on the production of variant phenotypes (i.e., developmental constraints) such as those just illustrated, cause evolutionary trends or patterns. Since the classic work of Fisher (1930) and Haldane (1932) established the weakness of directional mutation as compared to selection, it has been generally held that directional bias in variation will not produce evolutionary change in the face of opposing selection. This position deserves reexamination.

Though Maynard Smith et al. (1985) called for a re-examination, they remained trapped within the thinking of the shifting-gene-frequencies theory, and offered no resolution, other than a vague suggestion of neutral evolution.

The basis for a re-examination of this issue began to emerge in a completely different area. The lucky mutant view of the mutationists had already re-emerged in the 1960s, as soon as biochemists started comparing protein sequences. In 1969, theoreticians interested in "molecular evolution" developed origin-fixation models relating the rate of evolution directly to the rate of mutational introduction (McCandlish and Stoltzfus, 2014).

The emergence of a more sophisticated mutationist view following King (1972) was then derailed for a very long time while the selection-drift dichotomy became the axis around which, it seemed, all discussion of molecular evolution revolved, with no room for a second dichotomy. For a quarter-century after emerging, origin-fixation models were used almost exclusively to address neutral or deleterious mutations in treatments of molecular evolution (McCandlish and Stoltzfus, 2014).

Yet, eventually, theoreticians interested in adaptation began to explore models based on new mutations, resulting in a minor renaissance (Orr, 2005b). Various authors cited previously (Box 8.2) drew attention to the absence of an introduction process in classical theories, e.g., Yedid and Bell (2002) write that

In the short term, natural selection merely sorts the variation already present in a population, whereas in the longer term genotypes quite different from any that were initially present evolve through the accumulation of new mutations. The first process is described by the mathematical theory of population genetics. However, this theory begins by defining a fixed set of genotypes and cannot provide a satisfactory analysis of the second process because it does not permit any genuinely new type to arise.

Finally, in the 2000s, the influence of biases in the introduction process received direct attention as a novel cause with distinctive implications (Yampolsky and Stoltzfus, 2001; Stoltzfus, 2006a,b; Stoltzfus and Yampolsky, 2009; Stoltzfus, 2012). When evolutionary dynamics depend on mutational introduction, evolutionary change may be biased by mutation, in the sense that, if the rate of mutation for $A \rightarrow B$ is greater than for $A \rightarrow C$, this elevates the chances of evolving from A to B relative to A to C.

When the introduction process acts as an evolutionary cause, biases in the generation of variation become difference-makers in evolution, circumventing the opposing-pressures argument and leading to a novel prediction: mutation-biased adaptation. Recent studies confirm the predicted influence of simple mutation biases on molecular changes involved in adaptation (Rokyta et al., 2005; Couce et al., 2015; Stoltzfus and McCandlish, 2015, 2017; Sackman et al., 2017; Payne et al., 2019; Storz et al., 2019; Leighow et al., 2020).

The potential relevance of these discoveries is not limited solely to sequence evolution. Biases in the introduction of variation also may emerge from effects of development, including the asymmetric mapping of phenotypes to genotypes, providing a basis to assert an evo-devo claim to novelty by appealing to a newly recognized kind of causation (Stoltzfus, 2006b; Stoltzfus and Yampolsky, 2009; Stoltzfus, 2012). Likewise, certain claims about self-organization (e.g., Kauffman, 1993; Fontana, 2002) implicate biases in the introduction of novelty owing to differential relations of proximity in the architecture of genetic spaces (Stoltzfus, 2012, 2019).

Thus, the mutationist alternative to neo-Darwinism that emerged in the 1960s among biochemists—a new molecular version of "mutation proposes, selection disposes" (decides)—did not fade into irrelevance, though it has taken a very strange path. Today, origin-fixation models

represent a major branch of theory with many applications (McCandlish and Stoltzfus, 2014). The radical implications of recognizing the introduction process have been slow to emerge, and have not displaced traditional notions of causation and explanation that are still taught to students, and still applied by researchers when constructing hypotheses and explanations. The ultimate importance of recognizing the introduction process for addressing the major issues of evolution remains to be determined.

10.3 A synopsis of key points

The main points of the preceding chapters may be summarized briefly as follows.

Randomness as independence The proposition "mutation is random" sounds like a claim to the effect that the thing called mutation has a property of randomness, but claims of this sort typically do not stand up to scrutiny. The most reasonable claims refer to some kind of independence, but the independence of mutation from the rest of biology is not a valid biological principle, and is not an empirical finding based on systematic experiments. Mundane aspects of mutation make it nonindependent of fitness. Systematic data on the fitness effects of nucleotide mutations suggest that correlations of fitness effects with rates of occurrence are inevitable. Because mutations typically are completed before their fitness effects are felt, mutagenesis systems generally lack feedback loops by which incipient fitness effects of mutational intermediates influence the occurrence of mutations advantageously. This narrow sense of independence does not justify the historic randomness doctrine, and is not very useful. Rates of mutation can show correlations with fitness for idiosyncratic reasons and because mutation-generating systems can evolve to enhance the chances of useful mutations.

Randomness as irrelevance The persistence of the randomness doctrine makes more sense when we view it as an explanatory doctrine specific to neo-Darwinism, to the effect that selection is powerful, creative, and influential, whereas mutation is none of those things. To proclaim that mutation is random is to signal a commitment to a neo-Darwinian theory in which the details of mutation are irrelevant to anything of genuine biological interest, and selection is the ultimate source of meaning and explanation in biology. This position is somewhat subjective in outlook, but not entirely subjective. For instance, the architects of modern neo-Darwinism believed that theoretical and empirical results proved that mutation was only a source of raw materials and could not be a difference-maker in evolution, and this position is subject to refutation by showing *theoretically* that the generation of variation can be a difference-maker, and by showing *empirically* that it actually plays an important role as a difference-maker in evolution.

The indifference principle Nothing prevents us, as scientists, from dispensing with all versions of the randomness doctrine. The use of probability in scientific reasoning has a rigorous logical justification that does not rely on any ontological claims of "chance" or "randomness." The idea that randomness is something that exists in the world is, from this perspective, a mind-projection fallacy. In conducting research, one may assume that mutation maximizes disorder or entropy, except to the extent that one recognizes orderly properties of mutation on the basis of prior information. This is not a defensible interpretation of the historic mutation-is-random doctrine, but a generic approach to uncertainty that can be justified by appeal to the "objective Bayesian" position in statistics.

Specially evolved mutation systems At least six different types of mutation-generating systems in microbes have the hallmarks of specially evolved mutation-generating systems, i.e., systems that evolved by virtue of enhancing the chances of useful mutations: multiple inversion shufflons, phase variation (switching), multiple cassette donation systems, diversity-generating retro-elements, mating-type switching, and CRISPR-Cas spacer acquisition. CRISPR-Cas systems, which evolved in the context of the largest evolutionary arms race in the biosphere, do what was alleged to be impossible: they

systematically incorporate specific environmental information into a DNA mutation in such a way as to defend against a threat. This conceptual development recalls the case of physiological adaptation (discussed in 4.3) that, over a century ago, went from being seen as a mysteriously teleological property of living tissues, to a disparate set of topical evolutionary adaptations that each provide for a different kind of responsiveness. Likewise, specialized mutagenesis systems are not the manifestation of a mysterious internal property that makes mutation-generating systems inherently smart. Instead, entirely separate systems have emerged repeatedly, pieced together from ordinary kinds of mutational processes, often in the evolutionary pressure cooker of an arms race.

Biased mutational introduction as an evolutionary cause Mathematical and computational models show that bias in the introduction of variation is a theoretically possible cause of orientation in evolutionary change—a basic principle of population genetics not formally recognized in the twentieth century. This principle is *not* an implication of classic "mutation pressure:" it depends on the smallness, not the largeness, of mutation rates; it does not depend on neutrality, but can apply when fixations are selective. The effect does not require absolute "constraints" separating the possible from the impossible: any minor quantitative bias is potentially important—opening up enormous possibilities given the enormous range of mutation rates. Because of this causal principle, patterns of evolutionary change can have mutational causes, not simply in the sense that mutation is materially necessary for evolution, but in the sense that tendencies of mutation can be difference-makers, causing one kind of thing to happen more often than another.

Development and evolutionary causation The position on causation developed by the architects of the Modern Synthesis has far-reaching implications, one of which is that development, because it acts on the wrong "level," or is merely a "proximate cause," and so on, cannot be construed as an evolutionary cause. The same arguments apply to mutation, and we found that those arguments are insufficiently general because they depend on assuming the buffet regime of the shifting-gene-frequencies theory of evolution. By extending the concept of biases in the introduction of variation to developmental biases in mutation-and-altered-development, we can see that this classic position is also mistaken in regard to the possible role of development in evolution. Developmental biases in variation can be understood as theoretically valid causes of orientation. Whether they are actually important remains to be seen.

The case for mutation-biased adaptation The idea of mutation-biased evolution is familiar in the field of molecular evolution, typically in relation to features that a critic might dismiss as "the boring parts" of evolution. However, recent research has examined whether biases in mutation are difference-makers in cases of *adaptation*, both from laboratory evolution, and from studies of natural adaptation, including altitude adaptation in birds, trichromatic vision, cardiac glycoside resistance, and so on. In both the laboratory and in nature, the results show that *modest* biases in variation can influence the course of evolutionary adaptation. Furthermore, this influence is not small, in the sense that, in evolutionary biology, a factor that makes one outcome several times more likely than another is not a small factor. The results implicate several kinds of mutation bias (transition-transversion bias, CpG hotspots, and effects of sequence context) and a broad taxonomic range of organisms.

10.4 The objects and forms of explanations

To discuss the *explanandum* of evolutionary biology—Latin for "that which is to be explained" (plural: *explananda*)—is to be explicit about what, precisely, we are trying to explain. For instance, evolutionary research is often divided into two main thrusts (e.g., Futuyma, 2001): understanding the general causes of change, and reconstructing historical paths of descent. This book is squarely in the first category, focusing on the general causes of change.

The past century has brought major changes in what evolutionary biologists are trying to explain. Two major changes in the *objects of study* have occurred, and also a major change in the *form of explanation* has emerged in some areas of evolutionary biology. These are not the only changes in *explananda*, but merely the changes most relevant to the topics addressed in this book.

First, the entire field has been transformed by the molecular biology revolution in ways that transcend historical disputes.

The foundations of the Modern Synthesis were laid prior to the elucidation of the structure of DNA (1953), the operon (1960), mRNA (1961), and the genetic code (1966). For those who have shifted their attention from looking at the outsides of organisms, to understanding their detailed insides, the theory that the internal details do not matter for understanding the evolution of gross phenotypic traits has become irrelevant. For those of us who have struggled to understand the origin and evolution of discretized molecular structures, e.g., introns, proteins, or transposons, a neo-Darwinian theory that reduces evolutionary creativity to the accumulation of infinitesimal shifts in continuous quantities is typically useless, other than to provide a formal alternative to realistic models (e.g., consider the neo-Darwinian model of lateral transfer in Box 7.3). Much more could be said about this.

Second, the unveiling of the prokaryotic world is of enormous importance (as also pointed out by Koonin, 2011). Prokaryotes have dominated the biosphere for most of its existence. During the first three billion years of evolution, they ruled the planet and even changed its atmosphere, oxygenating it in a way that affected all further evolution. Subsequently, prokaryotes continued to dominate marine environments; initially, they dominated the land as well, but some time in the past 10^8 years, vascular plants came to dominate the planet as a whole, by greatly increasing the biomass of terrestrial environments (see Bar-On et al., 2018; Whitman et al., 1998). The large charismatic animals that have occupied the careers of so many evolutionary biologists are minor players in evolution.

That is, the objects of evolutionary research have changed, affecting research both on general causes and on reconstructing history. In regard to history, most of evolutionary history is prokaryotic history. Prokaryotes have dominated evolution, and they are secretly at work inside animals and plants, in the form of mitochondria and chloroplasts (endosymbiotic organelles that, like prokaryotes, reproduce in an asexual, haploid, low-recombination regime of population genetics). Interestingly, there is another world of diversity to be explored: viruses, which outnumber cellular organisms in most environments (see references in Koonin, 2017).

The historic Modern Synthesis was based on results from animals and plants. Stebbins (1959) commented that the theory might not apply to microbes, which were poorly known at the time. Indeed, the shifting-gene-frequencies theory of evolution relies on the maintenance of variation in the "gene pool" (see Box 6.2), yet the processes that give the gene pool its variation-maintaining power—recessivity, sexual mixis, chromosome assortment, chromosome recombination, heterosis, and frequency-dependent selection—were based on populations of diploid, sexual, obligately out-crossing animals. Of these factors, most are absent or severely diminished in prokaryotes: only frequency-dependent selection could play an analogous role.

A third point about explanations relates, not to the scale or taxonomic identity of what researchers aim to explain, but to the form of what is considered an acceptable evolutionary explanation. I refer to the more recent pattern of explanation as the *Dayhoff paradigm*, after computational chemist Margaret O. Dayhoff, inventor of the PAM matrix of amino acid replaceabilities (although this paradigm also may have roots in numerical taxonomy). In the Dayhoff paradigm, evolution is understood as a process of character-state change with a spectrum of propensities represented by quantities, e.g., a matrix of amino acid replaceability. The quantities may be instantaneous rates or expected differences (sometimes extrapolated out to great distances).

Within this paradigm, the targets of evolutionary explanations are patterns in the propensities of evolutionary change. That is, studies of molecular evolution often pose questions or hypotheses comparing two types of change A and B, e.g., transitions vs. transversions. We have addressed many questions of this type already, sometimes implicitly. Why do singlet amino-acid changes happen more often than doublets and triplets? Why do transitions happen more often than transversions? A great variety of evolutionary issues can be transformed into a version of the question "why does A happen more often than B?"

The importance of this paradigm can be understood by considering what has happened to the classic adaptationist paradigm over the past 35 years.

In their defense of an adaptationist research program, Reeve and Sherman (1993) suggest that when evolutionists are not trying to reconstruct history they are addressing questions of "phenotype existence." To choose phenotype existence as the *explanandum* has a particular set of prerequisites and implications. It makes the most sense if evolution is a process that reaches some kind of end point or equilibrium state, e.g., if evolution is a process in which the environment changes, and then a new adaptive equilibrium is established. If evolution only moves toward such end points, as empirical adaptationists (Box 3.1) believe—if "evolution is progressive adaptation and consists in nothing else" (Fisher, 1936, p. 58)—then the job of the evolutionist is to explain why the observed states of organisms are attractors, i.e., why they are more adaptive than accessible neighboring states, "irrespective of the precise historical pathways leading to their predominance" (Reeve and Sherman, 1993). As Amundson (2001, p. 361) puts it,

Like thermodynamics, adaptationist biology can be seen as an equilibrium study, dealing with systems in which a very large number of different causal histories could be expected to converge on a stable state. This often relieves adaptationist biology from the need to reconstruct detailed phylogenetic or populational histories (Sober 1983; Reeve and Sherman 1993).

Sober (1984, Ch. 5) calls this "equilibrium explanation" and suggests that it spares the evolutionary biologist from the need to know various historical details.

Why are the details irrelevant in an equilibrium explanation? Consider the example (from chemistry) of a solution of NaCl. To make the solution, we measure out some amount of solid NaCl, then add water to a desired volume. Whether the mixture is shaken or stirred does not matter; nor does it matter if we add the salt to the water or the water to the salt; nor does it matter if we add the salt all at once, or one grain at a time. In every case, the system reaches an equilibrium state with properties (e.g., specific gravity, boiling point) that we can predict precisely from the list of constituent parts.

Each instance of such a system, in fact, *has a detailed history*, one whose description would require a book far longer than this one. This history can be understood as an account of Aristotelean material and efficient causes (molecules subject to chemical forces). Nevertheless, *such a history need not be written*: fast kinetics ensure that an equilibrium is reached, and the properties of the equilibrium ensemble can be predicted solely from (1) a list of starting components and (2) an account of the energies of the various configurations that these components may adopt. A detailed understanding of kinetics simply is not required.

A corollary of the irrelevance of history is its inaccessibility: *the traces of history are erased by fast kinetics*. The current state of the NaCl solution does not allow us to infer whether it was shaken or stirred, and does not allow us to infer whether the salt was added in a single portion or many portions.

Classical adaptationist thinking fits this paradigm of explaining what is observed by arguments about what is favorable. Ayala's (1970) claim for an "ultimate source of explanation in biology" (like Dobzhansky's similar claim for selection as the ultimate source of meaning) rests on the idea that biology can be characterized as a set of endpoints with ultimate explanations. In Darwin's original theory, the process of heredity was a gusher of "indefinite variability" (fluctuation), so that the constancy of a species from one generation to the next, against this incessantly dispersive flow, was due to selection holding it in place. Mayr's view was similar, with its "infinitely variable natural populations" (see Appendix D, or Mayr, 1959a), which is why Mayr (1963, p. 609) famously predicted that it would be futile to search for homologous genes or mutations underlying shared traits (at least, beyond very closely related species).

This equilibrium approach, therefore, is not inherently unreasonable. Nevertheless, its demise was inevitable. As noted earlier, the two main thrusts of evolutionary research are to assess general causes and to discern history. In fact, scientists always have been able to trace paths of descent by examining the current states of organisms, which would have been impossible if each lineage had sampled all possible states and settled on the optimal ones. Thus, *the kinetics of evolution are slow, organisms are not endpoints, and their features do not have final causes.*

The collapse of the view of evolved things as adaptive end-points leads to a predictable set of challenges and accommodations recognizable in the conceptual innovations of the 1980s. If fast kinetics do not ensure that an equilibrium is reached, the current state is *contingent* upon the initial state. History matters, and accounting for the current state of a lineage depends on reconstructing phylogenies and understanding the detailed basis of evolutionary transitions. Then, because this history relies on events that are physically indeterminate or merely practically inaccessible to inference, we must construe the outcome of evolution to depend on *chance*. Though we may be able to list all the factors at work, and even to make estimates of many of them, we can rarely work out the detailed kinetics of evolution, and thus we are forced to consider, as a crude approximation, the notion that some changes are possible and others impossible—*constraints*.

That is, the collapse of the end-directed adaptationist explanatory paradigm leads predictably to an interest in contingency (historicity), chance, and constraints (e.g., Futuyma, 2010).

However, these attempts to patch up the old paradigm are futile. Chance, constraints, and contingency do not restore the paradigm of perfection, and they do not represent a new paradigm so much as an admission of failure. Chance is not a cause, but the lack of one. Contingency is not a cause, but a verbal marker of the inadequacy of the assumption of equilibration. Constraint is an explanatory principle, not a cause: a constraint offered to explain a nonoutcome does not cause the nonoutcome, because nonoutcomes are not real and do not have causes. The attempt to link "constraint" with causal mechanisms is futile (Antonovics and van Tienderen, 1991). These concepts represent a generic toolbox of excuses for departures from an ideal, e.g., why the bridge collapsed, why the dog got loose, and so on. Any outcome in evolution, or in life, can be explained as either the achievement of an ideal, or as the failure to achieve an ideal due to chance, constraints, and contingency.

Let us consider how to apply both paradigms to the results of Rokyta et al. (2005), who adapted 20 replicate phage populations, finding that the 4th-most-fit variant was found the most often (six times).

Within the Dayhoff paradigm, we begin by noting that the experimental results implicate nine different alternative genotypes. We can assign a frequency of evolution to each one of these, and this observed frequency is an estimate of an evolutionary propensity. We may also note a large number of other possible unobserved outcomes, i.e., outcomes with frequencies of 0. This set of numbers is an estimate of the propensities of evolution over an entire list of possible single-step changes. Then, we may consider questions about which kinds of changes are more likely, and why.

How would the equilibrium paradigm apply? Because this paradigm applies to token cases, we must apply it 20 separate times. In the single case in which the most-fit alternative emerged, adaptive perfection is achieved. In this case, we might apply the shifting-gene-frequencies theory to hypothesize that selection chose the best variant from an abundant gene pool. In the other 19 cases, there was a constraint, because the best variant was absent; by chance, some other variant was present; the outcome was contingent on which variant was present initially.

Thus, the classic paradigm can be applied to Rokyta et al. (2005). The result of applying the equilibrium paradigm is an explanation in which chance, constraints and contingency prevent perfection. Within the Dayhoff paradigm, the same results reveal short-term evolutionary propensities, and lead to questions about why some types of changes happen more often than others.

Note again that these are explanatory paradigms. Of course we have causal theories in evolution, but we typically apply them within some story that satisfies our sense that a thing has been explained. To apply a causal model within the Dayhoff paradigm, we might consider the hypothesis that the 4th-most-

fit variant happened more often due to a higher rate of mutational introduction, applying an origin-fixation model of kinetics. When we know why the 4th-most-fit variant emerges more than the most-fit variant (why X happens more than Y), we have a satisfying explanation within the Dayhoff paradigm.

The same kind of paradigm can be applied to propensities in the evolution of discrete morphological characters, as in Alberch and Gale (1985).

However, the old paradigm was never meant to convey the richness of a world of propensities. Indeed, the old paradigm predates evolutionary biology: creationists also focused on explaining existence in token cases, given their theory that each species emerged by a special process of creation resulting in a state of perfection. The Dayhoff paradigm is more appropriate for science because it focuses on general causes rather than token causes, and it is more appropriate for evolution because it focuses on change instead of end-states.

To summarize, a genuinely different approach is possible: focus on evolution as a tendency to change in certain ways. To the extent that evolution is change, the job of the evolutionist is to understand change, to explain it, and, where possible, to predict it. The fact that our knowledge is subject to uncertainty has two implications. One of them is to avoid focusing on instances, on token cases, and the other is to express our understanding of change in terms of propensities. That is, evolution is a process with certain tendencies, and our job is to understand those. The best we can do, with the best possible understanding of change, would be to predict the tendencies of change in a given system, under a given set of conditions.

A narrower but more manageable approach, perhaps, is to ask the simpler question of why one kind of change happens more often than another. Various complex problems in evolution can be broken down into the challenge of comparing one kind of change to another. To the extent that we can understand why one kind of evolutionary change happens more often than another, we understand evolution.

10.5 The importance of verbal theories of causation

S.J. Gould's understanding of evolutionary biology was so highly verbal that he literally wrote a 1,500-

page book about the "structure of evolutionary theory" without including a single equation describing evolutionary change. The opposite kind of book, with equations but no words or pictures, does not exist, and cannot exist. Questions of causation and explanation in science rely crucially on words, metaphors, and analogies. Effective scientific discourse requires one to recognize these verbal theories, and to develop, evaluate, and apply them intelligently and with caution. When Leigh (1987) argues that

The reason why I am disinclined to describe developmental constraints as directing evolution is the reason why I would not describe an artist's material as directing his handiwork. (p. 229),

this is an example of a theoretician reasoning using a verbal theory, by drawing on its metaphorical or analogical content. The theory—a version of the neo-Darwinian dichotomy of the potter and the clay—is dubious, but Leigh is, in fact, applying it correctly: if variation represents an artist's raw materials, and if developmental constraints act merely via effects on variation, then it follows that developmental constraints must not be construed as directing evolution.

As we have seen, verbal arguments about the nature of causation have been used repeatedly to address disputed claims, e.g., from evo-devo (Section 8.10). That is, evolutionary biologists themselves clearly accept that key issues can be resolved using verbal theories of causation. Thus, correct verbal theories are vital to effective discourse about evolution.

The introduction of mathematics into evolutionary biology did not change the need for these verbal theories. Mathematics is not a language of causation: it is a notation for relating quantities, along with a set of rules for manipulating these relations. The "equals" sign has no direction: it does not represent an arrow of causation. For instance, the equation of force $f = ma$ does not say what causes what, and if we compose a formulaic sentence in which the equals sign causally links the content on the left with the content on the right, this formula will fail under mathematically allowable transformations such as $m = f/a$ or $f/m = a$.

Thus, the rules of mathematics, by themselves, do not compel any particular interpretation of the equations of population genetics. Little in the breeder's equation, $\Delta z = h^2 S$, differentiates variation and selection: they are two factors multiplied together to yield a product. If we choose to call the product "the response to selection," this is because our application of this equation is situated in some context that extends beyond mathematics. In fact, referring to the left-hand side as "the response to selection" makes sense because, in practice, selection has nearly always been the experimentally manipulated variable. In principle, we could manipulate only the heritable component of available variation in a population, and choose to call the left-hand side "the response to variation."

As explained in Section 7.5.1, after Lande and Arnold (1983) derived the multivariate version of the breeder's equation ($\Delta \bar{z} = \mathbf{G}\boldsymbol{\beta}$), quantitative geneticists began to understand that, in the words of Steppan et al. (2002), "Together with natural selection (the adaptive landscape), [**G**] determines the direction and rate of evolution." This verbal theory, in which the rate and direction of evolution are *jointly* attributed to selection and standing variation, was novel: it does not correspond to the previous verbal theories in which the rate and direction of evolution are *governed* by selection (with variation merely supplying raw materials), even though quantitative genetics is the branch of mathematical theory most closely aligned with the neo-Darwinian theory.

That is, Darwin's verbal theory led to Fisher's mathematical theory; further mathematical developments along with empirical results (Schluter, 1996; McGuigan, 2006) led to conflicts with the original verbal theory; subsequently, a new verbal theory has emerged that aligns better with mathematical models and empirical results (though the old verbal theory is still in use). This process continues to occur in the area of quantitative genetics.

The arguments in the preceding chapters draw attention to the need for a theory of causation that implicates the introduction process as an evolutionary cause, so that biases in the introduction process are recognized as dispositional causes. Let us summarize this argument and make it more explicit.

To begin, we may document the absence of this theory. The three founders of population genetics, along with most of the architects of the Modern Synthesis, used the opposing-pressures argument to rule out an internal variational basis for evolutionary tendencies (Section 6.4.2). This argument is incompatible with a theoretical knowledge of the implications of the introduction process. That is, to argue that mutation cannot be a dispositional factor in evolution, on the grounds that mutation rates are too small, indicates a lack of awareness of the implications of the introduction process. The architects of the Modern Synthesis also argued that rates of evolutionary change would not reflect mutation rates (Section 6.4.3).

Perhaps this blind spot is not surprising among scientists dedicated to a neo-Darwinian theory.

Yet, we also have seen cases in which scientists failed to implicate biases in the introduction process, even when this would have served their interests. As noted above in Section 10.2, Maynard Smith et al. (1985) lacked a theory to rebut the opposing-pressures argument and explain to their readers how developmental biases in variation might operate in a manner consistent with population genetics. Due to the lack of such a theory, evo-devo has lacked legitimacy in the eyes of theoreticians such as Lynch (2007a), who insists that evo-devo has added no evolutionary principles. Likewise, Kauffman (1993) presented a range of provocative results on what he called "self-organization," but did not specify biases in the introduction of variation (due to differential mutational accessibility of genotypic networks in state-space) as the population-genetic cause. Due to the lack of an appropriate causal theory, reviewers (e.g., Fox, 1993) were disturbed by his references to self-organization.

Of course, all of these authors, some of them eminent theoreticians, were exposed to familiar concepts relating to mutation and selection. However, an awareness of these parts, evidently, does not lead to an awareness of wholes. When we do not understand a theory, i.e., when the theory does not yet exist for us, we are like the blind men in the parable of the elephant, able to identify some parts of the theory, but unable to assemble the parts into a whole. Instead, we see the parts from the perspective of *other* theories that are already familiar to us.

Which theories? As we discovered in Chapter 6, the architects of the Modern Synthesis had an explicit position on whether events of mutation are evolutionary causes: they are not. Mutation acts in individuals, thus it acts at the wrong level to be an evolutionary cause. In the shifting-gene-frequencies theory, the causes of evolution act at the population level: they are like the laws of statistical physics. Within the forces theory, mutation is a mass-action pressure that shifts allele frequencies. The theory does not say that novelty-introducing events of mutation never occur, only that this aspect of evolution happens in the background, in such a way that it does not affect the main dynamics of change, which are based on shifting gene frequencies in an abundant gene pool.

For instance, in "Evolution, Science and Society," a white paper written by representatives of key professional organizations, Futuyma et al. (2001) state:

Evolutionary change within a population consists of a change in the proportions (frequencies) of alleles in the population. . . Changes in the proportions of alleles can be due to either of two processes whereby some individuals leave more descendants than others.

And they go on to list the two causes as selection and drift. Within this theory, the introduction of novel variants is not a cause, and mutation pressure may be ignored because it is weak, not an effective cause of fixation.

Of course, many scientists have recognized intuitively that mutation appears in two guises: as a force and as a source of variation, e.g., Haldane (1932) in quotation D.35. Freeman and Herron (1998, p. 145), in their excellent textbook, write:

In summary, mutation creates the genetic diversity that is the raw material for evolution. In principle, mutation can also produce changes in allele frequencies across generations. In practice, mutation's role as the source of genetic variation is usually more important than its role as a mechanism of evolution.

However, our focus here is on genuine theories, of the kind that can be used in reasoning. The forces theory is used in evolutionary reasoning, but is inadequate to describe the consequences of the introduction process, which operates by a different currency of causation (Section 8.8). Let us consider an example that illustrates this point, a ratio of different types of neutral evolution under a mutation bias $\mu_1 \neq \mu_2$:

$$r_1/r_2 = \frac{\mu_1 N(1/N)}{\mu_2 N(1/N)} = \mu_1/\mu_2 \qquad (10.1)$$

The correct reading of Eqn. 10.1, offering a causal explanation for an evolutionary bias, is that (1) random genetic drift is the cause of fixation, affecting both categories of changes equally, (2) the differential effect $\mu_1 N/\mu_2 N$ represents a bias in the origination process, and (3) the overall process represents neutral evolution biased by a mutational bias in the origination process. This kind of interpretation is not embedded in the equation. It does not fall out of the sky. It must be constructed by means of a trained human intelligence.

By contrast, the forces theory says that, in order for mutation to prevail, it must overcome the opposing pressure of selection, so that mutation-biased evolution would require unusually high mutation rates, or an absence of selection (see Section 6.4.2). Accordingly, for decades, molecular evolutionists habitually treated mutational explanations as references to neutral models (e.g., Sueoka, 1988; Gillespie, 1991; Gu et al., 1998; Knight et al., 2001; Lafay et al., 1999; Wolfe, 1991). In one sense, this is an error, and in another sense, it is simply the correct application of the forces theory. The forces theory suggests that, when mutation-biased evolution happens, this must be because the pressure of recurrent mutation is driving alleles to fixation in the absence of opposing selection.

Because the introduction process is not the same thing as mutation pressure, i.e., it is formally *not a pressure on allele frequencies*, it is not subject to the logic of opposing pressures: it can not be opposed by selection. For instance, suppose that some new alleles are introduced at a rate $k \ll 1$, so that the waiting time for the next event is $1/k$. If the next allele is deleterious, this does not cause its introduction to be delayed by the opposing pressure of selection. Likewise, the term μN in an origin-fixation model is not a pressure on allele frequencies. An origin-fixation model lacks allele frequencies (recall the explanation for Fig. 8.11) and

typically represents a model of aggregate rates summing over many loci (McCandlish and Stoltzfus, 2014). That is, there is a differential effect in Eqn. 10.1, and it must somehow reflect the operation of mutation, but the differential effect does not arise from a mass-action force shifting an allele frequency.

The exact meaning of the μN term differs depending on whether the origin-fixation model is (1) a model of sequential fixations, characterizing the succession of alleles fixed at a single locus, or (2) an aggregate-rate model characterizing the behavior of an infinite set of loci (McCandlish and Stoltzfus, 2014). The two models often have the same mathematical form, but they are derived differently and the terms differ subtly in meaning. In the context of a sequential fixations model, μN acts like a point process or event-generator that, at random intervals, shifts the system into a potential transition state. In an aggregate-rate model, μN can be understood as a mass-action pressure, i.e., the mass effect of infinitely many particles, but the particles are infinitely many loci that are shifted into a potential transition state (for the next origin-fixation event), rather than infinitely many copies of an allele that may undergo mutation to a different allele, shifting its aggregate frequency.

Going back to Eqn. 10.1, we could look at the fully reduced equation $r_1/r_2 = \mu_1/\mu_2$ using the forces theory, and interpret this as two pressures pushing against each other to determine an equilibrium allele frequency, e.g., forward and backward mutation rates between two alternative alleles. However, this is not what the ratio means in the context of Eqn. 10.1.

As a final example, relevant to the theme of this section, note the difference just described between sequential-fixations models and aggregate-rate models. Before this distinction was made clear a few years ago (McCandlish and Stoltzfus, 2014), authors routinely confused the two, e.g., Stoltzfus (2006a) used a sequential-fixations model while citing Kimura (aggregate-rate models) instead of Gillespie (sequential fixations). Again, the models have the same form but a different meaning, emphasizing the importance of verbal theories that represent causal relations.

10.6 Discerning theories and traditions

Theories are at work in our thinking, whether we like it or not. The theory called "neo-Darwinism" is demonstrably at work in evolutionary discourse, and it (or something very much like it) would be at work today even if Charles Darwin had never lived. Theoreticians adore the kind of theory that is universal or substrate-independent, with behavior that converges on some important global outcome based on an optimizing principle (see Grodwohl, 2016). Neo-Darwinism was a valuable theory when little was known about the genetic and developmental details underlying evolution. Even today, the theory persists.

Perhaps some pragmatic readers will find this situation unfathomable. Why would anyone continue to entertain a theory known to be inadequate? One way to respond to this question is to note that scientists value theories for several distinct reasons, e.g., Levins (1966) characterizes models in terms of realism, precision, and generality. Because the three different scales of valuation have unknown relative weights, no global metric of value can be defined, and no method can guarantee to resolve disagreements between individuals who place different weights on (for instance) realism vs. generality. We can say that neo-Darwinism is a bad theory on the grounds that it sacrifices so much realism to achieve generality, or we can say that neo-Darwinism is a good theory on the grounds of achieving so much generality, albeit at the expense of realism.

This is why neo-Darwinism will always be a part of evolutionary biology. By being a facile and general theory, it has achieved a place of perpetual importance, regardless of a lack of realism.

The forces theory is also at work in evolutionary thinking. Concrete examples of the operation of this theory were given in Section 10.5, Section 6.4.2, and Appendix D. To understand evolutionary thinking, and particularly to understand disputes about internalist ideas and about the nature of causation, one must learn to recognize the forces theory.

Likewise, the shifting-gene-frequencies theory is evident in statements quoted previously from Mayr, Stebbins, Dobzhansky, Simpson, and others, and shaped approaches to mathematical modeling

for decades. To understand why Castle's experiments with the hooded rat are the touchstone of the Modern Synthesis for Provine (1971), but Johannsen's earlier selection experiments (demonstrating selection of pre-existing types) are dismissed as meaningless, one must understand the importance of demonstrating that selection can *create new types without mutation*, merely by shifting gene frequencies in the gene pool. Likewise, the accomplishments traditionally claimed for the Modern Synthesis (Futuyma, 1988) follow logically from accepting the shifting-gene-frequencies theory: neo-Darwinism is restored on a Mendelian basis; selection is established as the ultimate source of explanation in biology; Lamarckism, mutationism, and orthogenesis are excluded; and high-level rules—evolution as smooth adaptation to changing conditions based on abundant standing variation—emerge to provide a unified basis for paleontologists, systematists, and geneticists to understand evolution, even without knowing the underlying details.

Yet, the writings of the Synthesis era are heterogeneous, inconsistent and often ambiguous. They are not simply an explication of the shifting-gene-frequencies theory as the master theory covering all of evolution. For instance, Simpson and Mayr entertained theories of genetic revolutions, i.e., catastrophic theories, distinct from the macromutational catastrophism of Goldschmidt, which was treated as a heresy. Like de Vries, Stebbins was an expert on plant speciation by karyotypic macromutations (e.g., hybridization), although he dismissed the importance of this process on the grounds that it resulted in evolutionary dead ends. The idea that different taxa have inherited predispositions due to internal tendencies of variation, i.e., orthogenesis per Eimer (1898), was ridiculed as an appeal to mystical inner urges, yet there is a particular passage from Mayr (1963) that appears to welcome this effect, relating "parallelism and polyphyletic evolution" to "developmental and evolutionary limitation set by the organisms' genotype and its epigenetic system" (p. 608). Futuyma (2017), responding to recent claims about the importance of developmental bias, cites precisely this passage as evidence that the "Synthesis" includes developmental bias.

Does this mean that the Modern Synthesis includes catastrophism, one-step speciation, internal biases, and any other idea mentioned in passing by leading thinkers in the mid-century period? Does neo-Darwinism need to be extended to include a role for internal variational bias as a cause of orientation or direction? How do we account for the fact that Mayr ridiculed internalist theories and was disdainful of evo-devo, yet wrote a passage that seems to allow taxon-specific potentialities based on internal organization, a key theme of internalist critics and of evo-devo?

To respond to such questions requires some theory about scientific theories—what they do, how to discern theories in written works, and how to distinguish theories from persons or traditions (and from the rhetoric used to advocate for or against a theory). A common-sense position on theories is that they serve as provisional domain-specific extensions of logic, providing generalized yet precise guides to thought that lead us from inputs or conditions to provisional hypotheses or explanations, which are then subject to evaluation. When attempting to discern theories in textual matter, one must treat them typically as an unseen genotype whose phenotypic manifestations reflect the influence of the author's environment and of other genotypic factors (other theories).

Why does Mayr depart from standard orthodoxy to present an internalist theory? Perhaps an explanation exists, but this is a biographical or psychological issue, not an issue of science. For instance, a biographer might offer the opinion that Mayr held contradictory views, or suggest that Mayr was not as rigid and dogmatic as his reputation suggests. In fact, scientists frequently call on alternative theories, sometimes unconsciously, and other times deliberately. Ideas—even ideas that we dislike—can be captivating. Scientists frequently hedge their bets to reduce the chance of being wrong, so that a scientist who believes strongly in the sufficiency of theory A may also mention the possibility of an alternative theory B, just to be safe.

The person of Ernst Mayr—the author of diverse passages, a person whom one might characterize as right, wrong, dogmatic, confused, etc.—is irrelevant to the tasks of discerning theories and assessing what may be reliably established. Once we remove

the person of Mayr and his socio-political network of influence, we simply have *two different theories*. In theory A, selection in "infinitely variable natural populations" (Mayr, 1959a) can meet any challenge and therefore, if the same outcome happens repeatedly in evolution, this must be because it represents the uniquely apt solution. In theory B, taxon-specific limitations on developmental variability are important for understanding parallelism and convergence.

Are these two theories part of some larger theory, e.g., could we say that A and B combine to form "Mayr's theory of evolution"? If theories are to serve as reliable guides to thought, we have no basis to combine two disparate theories into one theory, merely because they are both stated by the same person. To do so would be to negotiate the content of theories biographically, rather than scientifically. Of course, if an author specifies a larger framework C, such that A and B are special cases of C, then we have a genuine scientific basis for a composite theory.

This brings us, finally, to a different approach to theories that pervades high-level discussions of the state of evolutionary thinking (see Laland et al., 2014), in which thought-collectives or cultural traditions—sociological entities made of people and their ideas—are treated as scientific theories. This position began to emerge when the Modern Synthesis came under attack, and various historians and philosophers devoted serious effort to understanding what defines it. As Smocovitis (1996) writes:

Looking for theories and trying to fix [i.e., define] the synthetic theory at the core, they were frustrated by discovering that the synthesis was a "moving target." Spilling gallons of ink on the subject and engaging in heated disputes for nearly a decade, the growing numbers of commentators on what became the "synthesis" would only agree in making this count as a historical "event."

That is, whereas Provine (1971) examined classic texts and identified a synthetic theory *before* it "came unraveled" in the 1980s (Provine, 2001), Smocovitis and others arrived *after* the original theory began to unravel, and found that working scientists disputed the substantive content of the Modern Synthesis, with some choosing to treat it as a moving target. The only agreement was that a sociological event resulted in an organized discipline of evolutionary biology that was present in the 1960s but not in the 1930s. This is why, in the historical literature, the "Evolutionary Synthesis" or "Synthesis" (often capitalized) is understood to refer to a genuine historical event, but the idea of a master theory is considered problematic.

Meanwhile, the evolutionary literature continues to feature a view in which all of contemporary thinking is alleged to be grounded in a foundational theory from the Synthesis era. Oddly, this notional theory is no longer called "neo-Darwinism" or the "Modern Synthesis," but the "Evolutionary Synthesis" (ES) or "Standard Evolutionary Theory" (SET).

Even more oddly, advocates of this position (e.g., Futuyma, 2017; Svensson and Berger, 2019) claim that the neutral theory is now part of the Modern Synthesis *qua* ES or SET, though what this means, precisely, is unclear. The shifting-gene-frequencies theory was understood to include minor quantitative shifts in allele frequencies by drift, in small populations, but not a voluminous stream of neutral mutation-fixation events, as in Kimura's theory. The neutralism that we know today entered evolutionary discourse in the 1960s, and this idea came, not from *within* the newly minted culture of evolutionary biology represented by the architects of the Modern Synthesis, but via biochemists, outsiders whose heterodox ideas prompted an orchestrated counter-offensive (Dietrich, 1998). Simpson (1964) lectured the upstarts, telling them that the consensus of experts is that neutral alleles do not occur. Mayr lived until 2005, long past the widespread acceptance of neutral evolution, without ever accepting the idea, preferring to define "evolution" in such a way as to exclude invisible molecular changes (Mayr, 2001, p. 199). One wonders, on what basis do contemporary authors assert that the Modern Synthesis *qua* SET now includes the neutral theory?

These defenses of SET often draw on arguments in which the meaning of SET is negotiated biographically. For instance, Svensson (2018) argues that reciprocal causation is part of SET, citing Lewontin (1985). Indeed, Lewontin clearly and explicitly advocates for the importance of reciprocal causation. Yet, the essay was published 26 years after the architects of the Modern Synthesis declared victory in 1959, in a book of essays focused specifically on *how to subvert conventional thinking*. Lewontin says *explicitly* that reciprocal causation is *not part of the received neo-Darwinian view*, in which the organism is merely the object of evolution, rather than (reciprocally) both the subject and object.

Thus, we would be justified in citing Lewontin (1985) to argue that reciprocal causation is *not part of the neo-Darwinian tradition*, because this is what Lewontin actually says. Instead, the main goal of Svensson (2018) is to defeat calls for an Extended Evolutionary Synthesis (EES) that features reciprocal causation; therefore, he uses Lewontin's advocacy of reciprocal causation to claim that reciprocal causation is already part of SET. The argument makes absolutely no sense scientifically: it exemplifies the way that the content of SET is negotiated *biographically* (see Stoltzfus, 2017).

This flexibility of SET suggests the possibility that it is merely a synonym for "mainstream thinking." That is, when a leading thinker writes something in a book, it must be mainstream and therefore part of SET. Clearly mainstream thinking has shifted over time, e.g., it has welcomed niche construction and absorbed the idea of widespread neutral evolution (though not Kimura's neutral theory, which holds that *most* of molecular evolution is neutral).

Yet, the advocates of SET do not claim to be shifting its content to match changes in thinking. Instead, they assert that evolutionary theory today is (except for the addition of neutral evolution) nothing more than an elaboration of an original master theory from the 1940s that remains perfectly intact, with "no sign that any of its components will have to be discarded" (Futuyma, 2017). Changes in evolutionary thinking over the past 70 years represent merely "shifts in emphasis and appreciation of the significance of previously acknowledged phenomena and processes" (Futuyma, 2010).

In fact, the original Modern Synthesis came "unraveled" for reasons given by Provine (2001), including the failure of the shifting-gene-frequencies theory and the gene pool ("one of the most artificial concepts of population genetics"), the lack of a unified theory for molecules and morphology, and so on. The shifting-gene-frequencies theory of evolution is not the foundation of contemporary evolutionary thought, and certainly is not the foundation of theoretical population genetics, but represents a special case (i.e., for highly polygenic quantitative traits in large outcrossed populations of sexual organisms). An example of a separate branch of formal theory$_A$ would be origin-fixation models, which emerged in 1969 in response to the discoveries of molecular evolution (McCandlish and Stoltzfus, 2014). Many other examples could be given, e.g., various models in phylogenetics, discussed briefly in Section 3.2, treat evolution as a stochastic Markov process of instantaneous transitions in discrete states. This kind of model, in which an entire species may flip from state A to B in an instant, is not based on the shifting-gene-frequencies theory, and indeed, contradicts the theory, which is why Simpson (1964) objected to it, as explained in Box 9.2.

The architects of the Modern Synthesis endorsed the creativity of natural selection, the gene pool theory, the opposing-pressures argument, and so on, all of which are vital for specifying a neo-Darwinian theory distinct from the views of the early geneticists. The creativity of selection is almost never defended today: the idea is so foreign that contemporary researchers may treat it as a popular "misrepresentation" of evolutionary thinking (Padian, 2013), instead of as a classic orthodoxy. The gene-pool theory led to claims about the rate of evolution being unrelated to the mutation rate (Section 6.4.3)—claims that are well known to be wrong. Defenses of SET do not include these discarded positions, but *only the parts of tradition needed to establish continuity with contemporary thinking*. Thus, the statement of Futuyma (2017) that no part of SET has been discarded is a truism: discarded ideas are excluded.

In short, the concept of SET is incoherent: it is neither a substantive falsifiable theory nor the historically accurate description of an evolving tradition.

Nevertheless, the rhetoric of the traditionalists is perfectly understandable under the assumption that nothing true is new and nothing new is true. Defenders of SET establish retroactive continuity, not merely by overlooking discarded ideas of the

past, but by projecting currently valuable ideas backwards in time, attempting to find historical antecedents of ideas that feature in contemporary disputes, including niche construction (Futuyma, 2017; Svensson, 2018; the "all is well" side in Laland et al., 2014), macromutation (Futuyma, 2017), developmental bias (Futuyma, 2017), and mutation-biased adaptation (Svensson and Berger, 2019). When faced with a claim that seems both reasonable and novel, the response is to attack its credibility, or to go back into the canon to anchor the new findings in tradition—or, quite often, *to do both at the same time* (e.g., Svensson and Berger, 2019; the "all is well" side in Laland et al., 2014). This behavior would be a genuine paradox unless, as suggested, the underlying belief is that nothing true is new and nothing new is true, in which case, attacking claims of novelty from both angles simultaneously is simply a bet-hedging strategy.

Those of us who are bystanders to this debate must bear in mind that it is more about politics than science. Conservative traditionalists construct narratives in which contemporary truths are anchored in a tradition, so that the author will benefit from the deference of readers who identify with tradition. By contrast, iconoclastic authors may align themselves with a minority tradition in order to evoke notoriety or conflict, e.g., Jablonka and Lamb (2005, p. 454) insist on using the term "neo-Lamarckian"—while literally admitting that the term provokes hostility and that no one today accepts Darwin's Darwinism or Lamarck's Lamarckism—on the grounds that avoiding this term would disengage them from a "tradition."

The primary issue at stake in these debates is who will have the authority to speak for the discipline, i.e., who is going to write the narrative of evolutionary biology, with attendant benefits regarding who will receive funding, recognition, and book sales. In the debate over "evolutionary theory" by Laland et al. (2014), the reformers and the traditionalists do not express major scientific disagreements about which forms of causation are possible or likely. Instead, the disagreement is primarily socio-political, and could be resolved with mutually agreed adjustments to (1) the allocation of funding, symposia, and other academic rewards, so that each topic receives attention according to its importance, and (2) the allocation of intellectual credit, so that traditional and contemporary sources each receive their fair share of credit for scientific developments.

The pressure to cast such a debate as a contest of "theories"—when it clearly is not a contest of theories—arises from the false premise that evolutionary biology has, or must have, a Grand Unified Theory. As noted earlier, 1959 was both the year that the architects of the Modern Synthesis declared victory, and the year that *The Molecular Basis of Evolution* (Anfinsen, 1959) inaugurated a new field of molecular evolution that quickly undermined the Modern Synthesis. Evolutionary biology clearly had no unified master theory in the 1970s and 1980s, when the schism between evolution and "molecular evolution" was paramount. In the absence of a genuine theory, traditionalists have constructed narratives that create an appearance of continuity, at the expense of torturing history, distorting the concept of a theory, and abandoning scientific norms regarding the attribution of credit (Stoltzfus, 2017).

In fact, leading thinkers today are perfectly capable of discussing the field and its future without the Synthesis myth, e.g., *Evolutionary Biology for the 21st Century* (Losos et al., 2013), written by a group including eminent theoreticians and experimentalists, does not refer anywhere to Darwin, neo-Darwinism, or the Modern Synthesis. Likewise, Lynch (2007b) largely side-steps the issue of whether there is a foundational mid-century consensus, and refers obliquely to the "next phase" of evolutionary biology in terms of "the post-Darwinian world" (p. 363–366).

However, to avoid taking a position on historically important claims, merely to avoid unresolvable disputes with traditionalists intent on shifting the goal posts, is an enormous concession. When a beautiful theory is killed by ugly facts, scientists have an obligation to set up a memorial inscribed with an epitaph, so that the death and its significance are not forgotten.

10.7 Synopsis

A grand challenge for twenty-first-century evolutionary biology is to develop a clearer and more potent understanding of the role of generative processes in evolutionary change. An important aspect of this challenge is to untangle the conceptual mess at the intersection of mutation, randomness, and evolution.

Running through this conceptual mess are the two main versions of the randomness doctrine: the independence claim and the irrelevance claim. The former can be preserved, in a sense, but only by diluting it down to the point of losing both its power to simplify the world, and its connection to the past. If we shift the goal posts so far that CRISPR-Cas mutations are now said to be "random mutations," then it is possible to (1) accept that evolutionary change is built from "random mutations" and (2) reject a long tradition of claims to the effect that evolutionary change is built from "random mutations." This is precisely why goal-post-shifting is contrary to the aims of science.

To repeat: the remedy for trying, and failing, to put mutation into a black box, is to stop trying to put mutation into a black box. The process of mutation, mutagenesis, is the sum of all processes that contribute to the emergence of mutations. One might study these processes for a lifetime and barely scratch the surface.

The irrelevance claim, which goes back to Darwin and is ultimately more important to understand, holds that variation is merely a material cause and not a dispositional factor. The influence of this claim, which coincides with the influence of neo-Darwinism, is most apparent in verbal theories of causation. If the reader understands how the words "mutation supplies the raw materials that selection shapes into adaptations" represent a provocative and falsifiable theory, not an empty truism, this book has been successful in focusing a spotlight on neo-Darwinism and the irrelevance doctrine.

With the benefit of hindsight, we can see why simplifying externalist theories have played such a large role in the development of evolutionary thought. A century ago, an enormous gap separated the promise of empirical reductionism within a genetical framework (i.e., the promise of applying techniques of genetic analysis to individuals and populations) and the actual power of this approach to address what evolutionary biologists wanted to understand. Evolutionists, having debated high-level theories for decades—having defined all of the major issues of evolution in terms of these theories—demanded some way to get from low-level principles of inheritance to high-level ideas about directionality, gradualism, adaptation, convergence, and so on. The early geneticists recognized this gap and simply asked for more time, e.g., Bateson (1914) said that "The student of genetics knows that the time for the development of theory is not yet. He would rather stick to the seed-pan and the incubator."

But the gap was too great, and theories rushed in to fill it—theories like the randomness doctrine, Fisher's geometric model, the infinitesimal model, the gene pool, and so on. These theories simplified the world and allowed scientists to proceed with studying the evolution of external phenotypes and behavior without having to understand the internal details. The Modern Synthesis, guided by neo-Darwinism, declared that the details do not matter, supplying simplified rules of evolutionary reasoning that the early geneticists were unwilling or unable to supply.

Over time, the substantive basis of the Modern Synthesis has been forgotten, obscured by the efforts of traditionalists to rationalize the past by misremembering it, so that the Modern Synthesis *qua* SET is now regularly treated as a generic framework that merely assumes a few basic principles about selection and genetics, i.e., the pre-Synthesis framework of Punnett and Morgan.

However, this shifting of the goal posts was superficial—merely an attempt to deflect criticism away from traditional authorities and historically important ideas that remained influential, nonetheless. Asserting that the Modern Synthesis is (or transformed into) a generic content-free theory—one that simply follows the implications of genetics—did not magically erase the patterns of thought and argument that were established

by decades of thinking in terms of the gene pool, shifting gene frequencies, population-genetic forces, and the neo-Darwinian dichotomy of the potter and the clay. This influence is evident in the delayed development of causal theories for the influence of biases in the introduction process.

This is why, today, we have ample opportunities to reconsider the role of variation in evolution and to develop new ways of thinking. Evolutionary change can be understood as the result of a compound process of the introduction, transmission, and sorting of variants in a hierarchy of reproducing entities, where sorting may be biased (selection) or unbiased (drift), and where biases in the introduction of variants may have mutational or developmental causes (note that this concise statement captures much, but certainly not all, of what is important in evolution).

An obvious possibility within this framework is that a bias in the outcome of evolution may be due to a bias in the introduction of variation. The search for such an effect was considered in Chapter 9.

As noted previously, this theory is not a truism. It is not necessarily true that evolution actually operates in a way that depends on biases in the introduction process. Even if evolution shows the impact of mutational biases, this is not necessarily due to biased dynamics of introduction. Even if biased dynamics of introduction are sometimes influential, they are not necessarily a primary cause of developmentally mediated effects, or of effects implicated in the self-organization literature. A key prediction of this theory is that the course of adaptation may reflect modest quantitative biases in ordinary mutations. The initial success of this prediction is promising, though much work remains to be done.

APPENDIX A

Mutation exemplars

This appendix provides four different accounts of a $C \rightarrow T$ nucleotide substitution mutation that differ in whether the mutation occurs by replication error (A.1), by damage and repair (A.2), symbolically via computer simulation (A.3), or as the result of human engineering (A.4).

A.1 A replication error

Most of the information in this section can be found in Maki (2002), which covers nucleotide replication errors in a larger review that includes other types of mutations. Stamos (2010) provides a more detailed review of proposed mechanisms of mis-pairing.

When a DNA polymerase adds to the end (terminus) of a growing strand, the added unit typically is the "complement" of the unit on the template strand, following the canonical pairing rules of A with T and G with C. However, a noncomplementary unit is added rarely. In the literature of mutation research, this is often called a "misinsertion" error (whereas elsewhere, "insertion" always implies a length-increasing change). The resulting "terminal mispair" is typically corrected by the proofreading capacity of the polymerase, and if not, it becomes an internal mispair (Fig. A.1).

For the case of *E. coli* polymerase III, misinsertions occur at a rate of 10^{-4} or 10^{-5} per site per replication for transition or transversion mispairs, respectively, and the effectiveness of proofreading is such that the rate of formation of internal mispairs is 50-fold to 100-fold lower than for terminal mispairs (Maki, 2002). The mispaired state is sometimes referred to as a "pre-mutation," i.e., an uncorrected misinsertion results in a pre-mutation, which a subsequent round of replication converts to a mutation. However, mismatch repair corrects most pre-mutations. These stages are illustrated in Fig. A.1. The overall rate of mutation from replication errors is small because (1) the misinsertion errors are rare, (2) proofreading removes most errors, and (3) mismatch repair removes most mispairs.

How does a misinsertion error originate? When Watson and Crick proposed the structure of DNA in 1953, they suggested that errors may reflect a specific type of "tautomeric shift" to an alternative form that favors noncanonical pairings: if the polymerase inserts a nucleotide while that nucleotide, or the template nucleotide, is in such an altered state, a noncanonical pairing is favored. This model was taught in textbooks for many years, but subsequent work has been unable to verify it (the tautomeric states are too difficult to detect), and various other models have

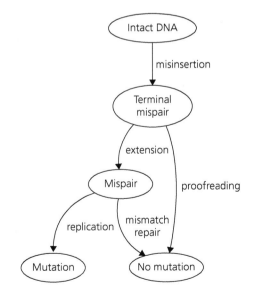

Figure A.1 Replication errors that arise by misinsertion face the possibility of proofreading and mismatch repair before becoming a mutation.

Mutation, Randomness, and Evolution. Arlin Stoltzfus, Oxford University Press (2021). © Arlin Stoltzfus. DOI: 10.1093/oso/9780198844457.001.0001

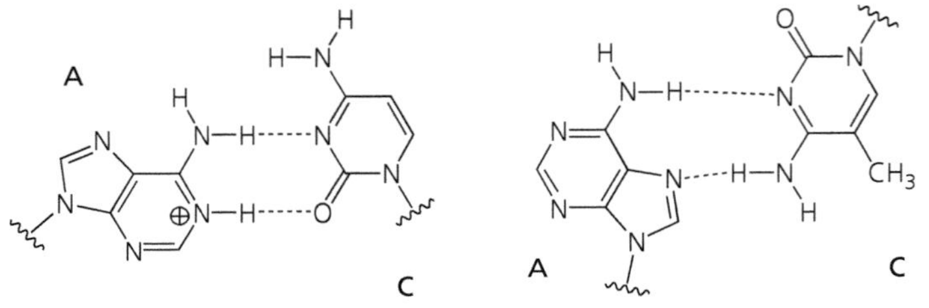

Figure A.2 Forward (left) and reverse (right) wobble pairs for A and C. Images by Yikrazuul reproduced under the Creative Commons CC0 1.0 Universal Public Domain Dedication.

emerged, some of which are supported by indirect evidence. For instance, many contemporary models rely on noncanonical "wobble" pairings of canonical bases that depend on the flexibility of the DNA backbone.

Stamos (2001) notes a proposed mechanism of misinsertion, based on conformational shifts due to quantum tunneling of an electron. As noted by Stamos, this means that ordinary replication errors may be subject to quantum indeterminacy (see Section 2.5).

Now let us consider the genesis of a specific mutation, e.g., a $C \rightarrow T$ mutation of a C:G pair, which may result from misinsertion of a T when G is on the template strand, or misinsertion of an A when C is on the template strand. The forward and reverse versions of an A:C wobble pair are shown in Figure A.2. At left, the figure shows an A:C wobble pair with two hydrogen bonds, one involving a protonated ring nitrogen of adenosine (positive charge indicated by $\oplus$). At right, Fig. A.2 shows a reverse A:C wobble pair with two hydrogen bonds. The reverse orientation means that each base is flipped around the bond to deoxyribose (wavy line), relative to the Watson–Crick orientation (to visualize how flipping allows alternative topologies, place your hands on a flat surface and make two bonds by joining them at the thumbs and index fingers, then flip both hands, and make two bonds by touching at fingers four and five).

If the resulting T:G or C:A mispair is not corrected by mismatch repair, then in the next round of DNA replication, one strand will have the unmutated C:G pair, and the other will have a mutant T:A pair, i.e., a $C \rightarrow T$ mutation will have happened.

The entire process is made complex by a number of factors. For instance, the tendency of the helix to allow wobble pairs is a local property affected by the local nucleotide sequence context. Proofreading and mismatch repair are likewise context-dependent.

A further complication, considered at length in the main text (see Section 2.9), is the influence of precursor pools on the mutation spectrum. A basic principle of biochemistry

is that enzymes change the rates of reactions but not the equilibrium constants. Therefore, even enzyme-catalyzed reactions such as replication are subject to the law of mass action, which means that the outcome of the reaction depends quantitatively on the concentrations of precursors, and in particular, the concentrations of dCTP, dATP, dGTP, and dTTP (Mathews, 2014; Kumar et al., 2011; Waisertreiger et al., 2012; Watt et al., 2015). Increasing the amount of dCTP, for instance, will increase the tendency for mutations resulting from misinsertion of C. Precursor pools change regularly during the cell cycle, and in organisms with large genomes, different parts of the genome may be replicated at slightly different times, resulting in different mutation spectra that may affect genome composition over the long term, as argued by Kenigsberg et al. (2016).

A.2 Error-prone repair of DNA damage

Mutations often arise from damage, and DNA can be damaged in a great variety of ways (Rastogi et al., 2010). Here we consider two common sources of damage: hydroxyl radical and UVB (ultraviolet B) radiation (i.e., electromagnetic radiation in the range of 280 to 315 nm). The treatment of damage and repair in this section draws heavily on *DNA Repair and Mutagenesis* by Friedberg et al. (2006), which addresses various kinds of DNA damage in Chapter 2, photo-reactivation in Chapter 4, and base excision repair in Chapters 7 to 10.

Hydroxyl radical, a kind of reactive oxygen species (ROS), arises regularly in cells when the energy of ionizing radiation is absorbed by water. It reacts quickly with nearby molecules, causing damage to proteins as well as to DNA, attacking either the backbone or nucleobases. A single particle of ionizing radiation can produce multiple radicals, resulting in a spatial cluster of damage (Friedberg et al., 2006, pp. 27–28).

The double bonds in the pyrimidine rings of Thymine (T) and Cytosine (C) are prone to absorb UV photons,

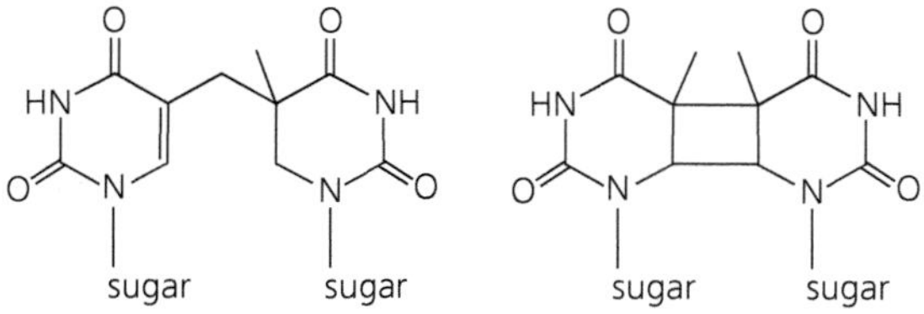

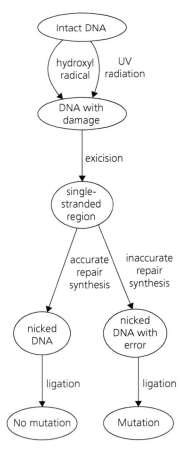

Figure A.3 Two kinds of pyrimidine photo-dimers, a 6,4-pyrimidine-pyrimidone dimer (left) and a cyclobutane dimer (right). Image by Smokefoot reproduced under the Creative Commons CC0 1.0 Universal Public Domain Dedication.

break, and become reactive. The activated molecules can bond to a neighboring pyrimidine, as shown in Fig. A.3.

TT dinucleotides are the most susceptible, and CC are the least. When our skin cells are exposed to sunlight, this kind of reaction occurs 50 to 100 times per second per cell (Goodsell, 2001). The effect of this kind of damage on DNA biology is like the effect of fusing two adjacent teeth on one side of a zipper, which can jam the zipper. Though DNA polymerases may sometimes bypass a pyrimidine dimer on the template strand, the presence of a dimer represents damage rather than mutation: replication by itself cannot create either (1) exact copies with pyrimidine dimers in both daughters or (2) a reverted copy. For instance, the human disease xeroderma pigmentosum, which arises from a defect in DNA repair, manifests most obviously as a skin disease (though it also has neurological symptoms) due to the extraordinarily high rate of UV damage to DNA in skin cells, leading (in the absence of repair) to skin lesions and melanomas. In severe cases, victims of xeroderma pigmentosum must avoid exposure to sunlight entirely.

For both of these sources (UV and hydroxyl radical), the distribution of damage to DNA is not uniform. For instance, for certain UV wavelengths, damage is more likely in nucleosomal linker regions (Niggli and Cerutti, 1982). In the case of hydroxyl-mediated damage to naked DNA oligomers, the susceptibility to damage is context-dependent, with most of the statistical variance in mutation rates being captured by a triplet model of context, i.e., a model with the focal nucleotide, plus one nucleotide upstream and downstream (Greenbaum et al., 2007). This variance seems to reflect local variations in solvent accessibility due to differences in local DNA geometry (Balasubramanian et al., 1998). Presumably such variations also would cause differences in the susceptibility to other kinds of damage.

Damage from radiation or chemicals may be repaired precisely or imprecisely, the latter leading to mutation. One pathway of DNA repair, called photo-reactivation, utilizes photons of a different wavelength to reverse

the photo-dimerization reaction. Photo-reactivation was the first mechanism of DNA repair to be discovered, based on work of Kelner and Dulbecco in the 1940s (see Friedberg, 1999). Both of these discoverers were trying to use UV mutagenesis, but were stymied by unaccountable variation in the effect of a dose of UV. Eventually this variability was traced to how UV-irradiated cultures were stored in places exposed to sunlight, which triggered photo-reactivation.

Two more general pathways of DNA repair, relevant to diverse sources of damage, are "base excision repair" (BER) and "nucleotide excision repair" (NER). The two differ in whether repair enzymes remove only the damaged nucleobase, or cut open the DNA strand and remove entire nucleotides (i.e., nucleobase-sugar-phosphate moieties). In the present context, we are more interested in NER (e.g., xeroderma pigmentosum is specifically a defect in NER). The process begins with proteins that recognize damaged

Figure A.4 Pathway from damage to mutation or no mutation, via nucleotide excision repair (NER).

or mispaired sequences. A segment of the damaged strand is excised and then re-synthesized using the other strand as a template. Repair synthesis is error-prone in the sense of having a higher rate of error than normal replication. DNA polymerases can extend a strand of DNA but cannot join it with another strand: closure of a nick is accomplished by a DNA ligase.

Now, let us combine this information to visualize the pathway by which a $C \rightarrow T$ mutation arises following damage (Fig. A.4). First, damage occurs at or near a particular C:G base-pair that is the focus of our attention. The damage might be a pyrimidine photo-dimer caused by UVB radiation, or hydroxyl-mediated damage to a nucleobase or to the DNA backbone. Second, the damage is recognized, and a piece of one strand is excised, including the focal site. Then, a repair polymerase fills the gap, and a ligase closes the nick. Whether this pathway results in a mutation depends on whether repair synthesis results in an error.

A.3 A symbolic mutation process in a computer program

Computer simulations of stochastic processes (Monte Carlo simulations) depend on the use of a pseudo-random number generator or PRNG (pseudo-random numbers also have other applications, such as cryptography). The concept is easiest to explain by reference to the simple form of the Linear Congruential PRNG, sometimes called the "Lehmer" or "Park and Miller" PRNG. As explained by Park and Miller (1988), a new number is derived from the product of the previous number and a multiplier, by taking the remainder of division by a divisor, i.e., $i + 1 = (x_i \times a) \bmod m$. Given constant values of a and m, the same x_i always leads to the same $x_{i+1}, x_{i+2}, x_{i+3}$, and so on. For instance, for $a = 6$ and $m = 13$, if we begin with $x = 1$, the resulting series is 1, 6, 10, 8, 9, 2, 12, 7, 3, 5, 4, 11, 1, . . . When x returns to 1, the pattern repeats.

The length of the repeating pattern is the "period" or "cycle" of the PRNG. With carefully chosen constants, the PRNG will proceed through many integers before repeating, i.e., it will have a long cycle. Assiduous users get their seeds from a secondary source (e.g., the table of random digits in an old statistics book), recording the seed so that each result can be replicated. Many PRNG routines obtain an initial seed from the computer's internal clock time—the number of seconds elapsed since 1 January 1970 00:00:00 UT—which produces unique, albeit highly correlated, seeds (seconds modulo L_{max} has a cycle of 136 years when $L_{max} = 2^{32} - 1$).

The Lehmer PRNG is valued for its simplicity and portability, but is not a good choice for cryptography, nor for serious scientific or engineering applications, due to serial correlations (L'Ecuyer, 2015). Many other types of PRNGs have been developed: Klimasauskas (2002) provides a helpful taxonomy. Testing of PRNGs is an ongoing area of research (e.g., Wang and Nicol, 2015).

With this background, let us imagine simulating mutation of a gene for purposes of simulating evolution. For the sake of illustration, let us limit our consideration to the eight types of single-nucleotide mutations possible at each site: three substitutions, one deletion, four insertions (A, T, C, G). Let us assume that the chance of each type of mutation at a particular site depends on the triplet consisting of the focal nucleotide and its two nearest neighbors, represented in an 8 × 64 matrix (given 64 possible triplet contexts). We can designate a particular value in the matrix as m_{ij}, where i represents the mutation type and j the triplet context.

In general, the way to use a PRNG to make a choice is to represent outcomes as the segments of a number line, each with a length proportional to the probability of the outcome, and then use the PRNG to pick a point on the number line, which identifies a segment and thus an outcome. For instance, suppose that we wish to represent nucleotide mutations in a coding sequence beginning with ATG. The first three segments on the number-line would represent mutations at the first sequence position, from A to T, C, or G. The next segment would represent a $T \rightarrow C$ mutation at the second sequence position. If this segment extends from p_L to p_R, then the $T \rightarrow C$ mutation will be chosen whenever a randomly generated value falls between p_L and p_R

Typically the mutation rate is so low that most alleles are not mutated in a given generation. In this case, for the sake of computational efficiency, one typically draws (each generation) a Poisson variate n with a mean equal to the expected number of mutations, and then one assigns n mutations given the model. The expected number would be based on applying the model for the mutation spectrum to the sequence of the allele. To be precise, this expectation must be recalculated each time the allele changes by mutation.

We can imagine several types of evolutionary models using such mutational changes, and it may be helpful to describe them briefly. When computer scientists implement "genetic algorithms," they typically impose arbitrary rules, e.g., give each gene three mutations, recombine it with one other randomly chosen gene, and then choose the top 10 % of the resulting genes for the next generation. Evolutionary biologists favor a more naturalistic approach. In the extreme case, one represents an explicit population of N individuals, each of whom contains genetic material subject to mutation. After mutation, an individual may mate and reproduce in a probabilistic way that takes into account the fitness effects of any mutations.

For long-term evolution, a useful simplification is to assume origin-fixation dynamics (see Section 8.6), such that the entire population acts as a single particle: in each generation, there is an opportunity for mutation, and any resulting mutation is either accepted or rejected probabilistically based on its fitness effects, using a standard formula for the probability of fixation.

Many models in population genetics do not depict N explicit individuals, but follow changes in the frequencies of genotypic classes in an implicit population, where the frequencies are subject to mutation, drift, selection, and so on. In a deterministic model, if the mutation rate from genotype i to j is μ_{ij}, then each generation, the frequency of genotype i decreases by $f_i\mu_{ij}$, and the frequency of genotype j is augmented by this same amount. A stochastic model would incorporate mutation from i to j by generating a stochastic variate from a distribution with a mean of $f_i\mu_{ij}$. Ideally this would be a realistic distribution incorporating an implicit population size of N. When the correct sampling theory is applied, this type of model is simply a more computationally efficient version of the individual-based model, though in practice it usually incorporates simplifications.

A.4 Human-engineered mutations

Most strategies to engineer mutations rely on automated chemical synthesis of oligonucleotides or "oligos" that can be ordered online and purchased cheaply in a quantity that typically represents about 10^{16} molecules. A single strand of DNA is synthesized using a stepwise chemical process that allows the synthesis of a specific sequence, or more generally, a sequence of length n specified by the $4 \times n$ matrix $\mathbf{S}$, where S_{ij} is the frequency of nucleotide $i = \{A, C, G, T\}$ at position j of the sequence. Synthesis typically begins with an initial nucleotide that is anchored to a substrate, and has a chemical blocking group that makes it nonreactive. After unblocking, the next nucleotide is added, then the substrate is washed. Synthesis proceeds by this unblock-react-wash cycle.

As part of a strategy for creating mutations, we might order the synthesis of an oligonucleotide that has an exact sequence differing from the parent sequence at a particular site, e.g., changing an ACA (Threonine) codon to a ATA (Isoleucine) codon, i.e., a specific $C \rightarrow T$ mutation. Alternatively, we could use an oligonucleotide with uniform nucleotide composition at the second position of a particular ACA codon, i.e., an ANA codon, resulting in 25 % ACA (Threonine), 25 % ATA (Isoleucine), 25 % AGA (Arginine) and 25 % AAA (Lysine). The oligonucleotide mixture can be used to synthesize a full-length gene, which is then introduced (by transformation or transfection) into a host, resulting in a mutant organism. Typically transformation is a scattershot process in which a mass of cells is subjected to a mass of mutant genes: cell lines in which a mutant gene has become established are isolated by virtue of some selectable (or screenable) marker linked to the mutant gene.

The acme of the scattershot approach is the deep mutational scanning experiment (Fowler and Fields, 2014), in which the method is applied (sometimes using robotics) to generate every single-amino-acid mutation in an entire protein-coding gene, e.g., the 4,997 variants generated in TEM-1 β-Lactamase by Stiffler et al. (2015).

In this complex chain of events, let us consider what determines the sequence of an isolated mutant. An isolated mutant strain reflects a particular parental cell that was transformed with a particular mutant construct; that mutant construct, in turn, reflects one or more rounds of library amplification from an originating event that combined one specific synthetic primer molecule with one specific biological template; and the particular primer molecule is one out of an astronomical number synthesized in bulk by a machine.

That is, the effect of this process of mutant isolation is to implicate a single primer molecule from out of a chemical mixture of an astronomical number of such molecules. Thus, if the mutant has a T at a site i of interest, where the synthesized oligonucleotide had an equal mixture of A, C, T, and G, the question of what determined the presence of T at site i reflects a series of events by which a particular oligo was synthesized, was incorporated into a mutagenized construct, found its way into a particular cell by transformation, and ultimately was identified via selection or screening. If we focus on the initial oligonucleotide synthesis, the presence of a T at site i reflects the events by which a T precursor (rather than some other precursor) diffused to the site of chemical synthesis and reacted with the growing chain. This series of events is subjectively unpredictable in that we cannot conceivably know which one of a mixture of nucleotides will be added to a particular oligonucleotide.

Counting the universe of mutations

> It strains one's faith in the laws of chance to imagine that identical changes should crop out again and again if the possibilities are endless and the probabilities equal.
>
> **A. Franklin Shull (1935)**

For a particular genome, the universe of mutations is the set of possible mutations, including all types or classes of mutations, such as nucleotide substitutions, insertions, deletions, and so on. The approach used in this appendix is to characterize this set of mutations in a formal but usually approximate way. As will become apparent, the size of a class of mutations is typically a very large number that differs from the sizes of other classes by multiple orders of magnitude. In this context, there is little reason to pursue highly accurate estimates when approximate estimates are within 50 % of the true value. However, often the approximations will be much closer than this to the anticipated true value.

B.1 Preliminaries

A genome is assumed to be a DNA sequence of n bp, composed of the usual four bases represented by T, C, A, and G. A sequence that follows the canonical base-pairing rules is fully specified by a single strand, and so the treatment below will refer to a single strand, leaving the complementary strand implicit. To avoid edge effects, we will assume that a genome is a single circle. A circular genome of length n has n nucleotide sites, and also n internucleotide bonds that might be broken and rejoined by some rearrangement. Arbitrarily, we will designate the link between nucleotide i and $i+1$ as internucleotide site i. Internucleotide site n connects the last base in the genome to the first, forming a circle.

Even though the genome is circularized to avoid edge effects, we will sometimes employ the assumption that the genome has an invisible point of origin (between $i = n$ and $i = 1$), for the purpose of identifying effects of redundancy. When the genome has an invisible point of origin, we will count five distinct C to T mutations in the 5-bp genome CCCCC, instead of just one, because each C has a different position in the genome. This is a useful

simplification because it allows us to ignore some trivial cases that would exhibit odd behavior, such as a genome consisting entirely of C. Any realistic genome will have a unique (i.e., nonrecurring) sequence somewhere, and the presence of a unique nonpalindromic sequence anywhere in the genome makes every other position in the genome identifiable, even if there is not a point of origin.

One must distinguish the biological or informational uniqueness of mutations from physical uniqueness. A change, if not inheritable, is not a mutation; and what is inherited (over the long term) is the pattern, not the physical substance. For instance, let us consider inserting a radioactive T nucleotide into each of the four inter-nucleotide sites in a circular ATCG sequence, resulting in four products distinguishable with a physical method such as mass spectroscopy. However, inserting T in the first or second internucleotide site gives the same sequence, either A**T**TCG or AT**T**CG, where **T** is the inserted nucleotide. As a result of this kind of effect, the number of distinct T insertions is not four, but only three.

As a further example, consider 3-bp deletions in a circle of 10^6 bp. If we designate a particular site as the point of origin of the chromosome, so that we can number each site, then 10^6 chemically different 3-bp deletions are possible. However, these are not all biologically different. If the sequence consists entirely of 10^6 Cs, then there is only one biologically distinct 3-bp deletion. More realistically, if the first ten nucleotides are AACTACTT, then the internal repeat of ACT means that the 3-bp deletions starting with nucleotides 2, 3, 4, or 5 all result in the same sequence AACTT.

If we consider longer deletions, the chances of redundancy are reduced, because a longer sequence is less likely to be present in multiple exact copies. More generally, the problems of redundancy are only quantitatively significant on the scale of single nucleotides or short segments.

In the following calculations, we sometimes begin by computing the physically distinct mutations, then proceed by making corrections for informational redundancy. When we count events, we enumerate them in the arbitrary order specified by genomic position, addressing redundancy by skipping over events whose informationally distinct outcomes have been counted already. Thus in the above example, we identify the 3-bp deletion at position 2 of AACTACTT, and the redundant events at 3, 4, and 5 are not counted.

B.2 A necessary simplification

Let us begin by making no concessions to biology: a genome is a circular sequence of n nucleotides that can mutate into any other sequence, either of the same length, or of a different length. In the former case, the universe of 4^n possible genomes includes $4^n - 1$ genomes other than the parental genome, thus $4^n - 1$ mutations, which is about $10^{0.6n}$, an inconceivably large number, even when we are talking about prokaryotic genomes with a typical size of $n = 2 \times 10^6$ bp.

If we consider the possibility of mutant genomes that are different in length from the parent genome, the number of mutations is the total number of possible genomes minus 1, which is infinite if there is no limit on length, or $\sum(4^n, 4^{n-1}, 4^{n-2}, \ldots) - 1$ if the limiting size is n. Note that each number in this series is $1/4$ of the previous number, and that the sum of the geometric series $1, 1/4, 1/16, 1/64, \ldots$ is $4/3$. Thus the universe of possible mutations is $\frac{4^{n+1}}{3} - 1$ where the limiting genome size is n, which is again on the order of $10^{0.6n}$, an inconceivably large number for natural values of n.

If we take a genomic rate of mutation of (for instance) $U = 0.01$ mutations per genome per generation, and divide it by this enormous number, the average rate for an individual mutation becomes vanishingly small. Thus, if we assume that a mutation can change one genome into any other genome, then the universe of mutations is easy to calculate, and its size is so enormous that, given uniform mutation, we would never see the same mutation twice.

Now, having considered this unrealistic scenario of mutation, let us move into more familiar ground and enumerate the universe of mutations as a sum over familiar topological categories of mutation, including:

1. point mutations
2. *de novo* insertions
3. deletions, inversions, and tandem duplications
4. transpositions
5. lateral gene transfers
6. compound events.

B.3 Point mutations

Let us define point mutations as those mutations involving only a single nucleotide. Then each genomic nucleotide is subject to one deletion and four physically distinct substitutions, and each internucleotide site is subject to four physically distinct insertions. The total number of such events is $9n$. Each of these events is physically distinct, e.g., if we substitute a radioactive T at each site, we will see exactly five chemically different T substitutions in any five-nucleotide sequence, whether the sequence is TTTTT or TACGC.

However, physically distinct mutations are not always informationally or biologically distinct. For instance, each site has only three (not four) distinct nucleotide substitution mutations, because removing nucleotide X_i and reintroducing the same nucleotide X_i (e.g., substituting a T with a T) does not create a different sequence and is therefore not considered a mutation, even though this is a physically (and biologically) possible event detectable with radioactive labels. Thus, the number of point mutations is no more than $8n$.

Similarly, the sequence TACGC is subject to five informationally distinct one-nucleotide deletions, but TTTTT is subject to only one.

Now, consider insertions of one nucleotide. For purposes of counting distinct outcomes, we will treat an insertion at site i as an insertion between i and $i + 1$, and in the case of redundant events, we count only the first one encountered (proceeding through the genome from start to end). Suppose that C occurs at position 4. The insertion of C after this, at internucleotide site 4, is redundant to the previously counted insertion of C at internucleotide site 3, i.e., $N_1 N_2 N_3 C_4 C N_5 N_6 N_7$ is always the same as the previously counted $N_1 N_2 N_3 C C_4 N_5 N_6 N_7$. In general, each site i has a nucleotide $X_i \in (C, A, T, G)$, and the insertion of the same nucleotide at i is always redundant to its insertion at $i - 1$, whereas the insertion of a different nucleotide is always nonredundant to any insertion at $i - 1$.

This is a logical necessity, not a statistical expectation: it has nothing to do with probabilities or sequence composition. Regardless of base frequencies or repeats, each site has three informationally distinct insertions. In the extreme case of unequal base composition, we have a genome that consists entirely of Cs (for instance), and every insertion of C looks the same, but every insertion of T, A, or G is unique (given our assumption that the chromosome has a known point of origin).

Note that this also will apply when we count longer (> 1-nucleotide) insertions: for any length of insertion x, there is always exactly 1 x-nucleotide insertion at site i that matches the sequence from $i - x + 1$ to i, and is therefore redundant to a previously counted insertion at $i - x$, e.g., the insertion of ATG *after* the ATG in a sequence,

NNNATG**ATG**NNN is redundant to the previously counted insertion *before* the ATG, NNN**ATG**ATGNNN.

Again, the number of informationally distinct one-nucleotide insertions per site is always three, not four. Thus, to summarize results so far, out of nine physically distinct point mutations per site, at most seven are informationally distinct, because one substitution fails to change the sequence, and one insertion is redundant to a previously counted insertion.

Now consider deletions of one nucleotide. The chance that the deletion at site i is redundant to that at site $i-1$ is just the chance that the nucleotide at i matches that at $i-1$. If sites are independent, this is the sum of the squared frequencies of nucleotides, $\sum_{X \in (A,T,C,G)} f_X^2$, which is $4 \times (1/4)^2 = 1/4$ in the case of uniform composition.

Unlike the case for insertions, this is not a logical necessity but a statistical expectation. We can imagine a sequence in which the same nucleotide never repeats (e.g., TCAG TCAG TCAG ...), in which case, every deletion is unique, resulting in n 1-nt deletions.

Some fraction of the expected 2-bp repeats of the same nucleotide are 3-bp repeats, therefore we must decrement the number of nonredundant 1-bp deletions by 1 for each 2-bp run that is also a 3-bp run, because the deletion at i is not only redundant to $i-1$, but to $i-2$ which was counted already. For the uniform case, the chances are obtained by multiplying once more by $1/4$, i.e., $1/4 * 1/4 = 1/16$. Considering longer and longer runs of the same nucleotide yields the geometric series that we encountered previously, $1/4 + 1/16 + 1/64...$, which equals $1/3$. Therefore, the expected number of deletions for a random sequence of uniform composition is $(1 - 1/3)n = (2/3)n$.

If the composition is *nonuniform* and independent, or uniform and positively autocorrelated (i.e., similar nucleotides cluster together), the expected number of distinct 1-bp deletions is smaller than $(2/3)n$. In the extreme case in which the genome is composed entirely of the same nucleotide, each of n physically distinct deletions gives the same result, thus the number of deletions is one. In the extreme case of positive autocorrelation, where the four nucleotides are all grouped together, as in AAATTTCCCGGG, then regardless of individual nucleotide frequencies, only four distinct deletions are possible. In such extreme cases, the total number of distinct 1-bp deletions is negligible compared to the $6n$ other point mutations. That is, the expected number of deletions in these cases is between 0 and $(2/3)n$.

In the case of negative autocorrelation (similar nucleotides repel each other), the expected number of unique single-nucleotide deletions can be larger. As noted earlier, in the extreme case in which no nucleotide ever occurs twice in a row, the number of unique single-nucleotide deletions is n.

Now we can return to tally up the total number of point mutations. Regardless of composition or noninde-pendence, this is at least $6n$, due to $3n$ substitutions and $3n$ insertions, and is never more than $7n$. For a uniform random sequence, the expected number of deletions is $(2/3)n$. Actual biological sequences typically are positively auto-correlated and have expected numbers of deletions less than $(2/3)n$. Thus we can propose $(6 + 2/3)n \approx 6.7n$ as an order-of-magnitude approximation for point mutations.

B.4 *De novo* insertions

Let us consider inserting a sequence of length d, where d is a small number. Presumably, long insertions actually are not *de novo* synthesis of a strand by aberrant polymerization, but duplications, transpositions, or lateral gene transfers of a pre-existing sequence. The reason that the calculation for *de novo* insertions is completely different than the one for lateral transfers (or duplications or translocations) is that, even for relatively short sequences, the number of distinct source sequences from the actual universe of genomes is limited relative to the number of possible sequences.

This can be explained briefly as follows. Imagine 10^7 species, each with a genome of 10^7 bp. Think of this as one large genome of 10^{14} bp that we can break down into 10^{14} overlapping 10-, 20-, 30-, or 40-bp sequences. This number of 10^{14} is an over-estimate of actual diversity, because genomes are related to each other, and internally redundant. Now, compare this to the number of possible sequences of length d, which is 4^d. For instance, the number of possible *de novo* sequences for $d = 10$ is $4^d \approx 10^6$ (a useful approximation when dealing with powers of 2 is that $2^{10} = 1024 \approx 10^3$, thus $4^{10} = 2^{20} \approx 10^6$). Likewise, for d of 20, 30, or 40, the numbers of possible sequences are 10^{12}, 10^{18}, or 10^{24}, respectively. This is why it is important to count *de novo* insertions separately from mutations that insert segments from the same or other genomes.

Now, let us return to the issue of counting *de novo* insertions, considering again that for $d = 10$, this number is $4^d \approx 10^6$. Shorter insertions of length $d - i$ will be 4^i-fold less numerous, e.g., insertions of lengths 9 and 8 will be 4-fold and 16-fold (respectively) less numerous than those of length 10. Therefore, to calculate the number of possible insertions for lengths up to r (range), we can make use of the series $1 + 1/4 + 1/16 + 1/64$, and so on. This sum is $4/3$ for an infinite series, and slightly less than $4/3$ for a finite series, e.g., for ten terms, the sum is 1.333332 (in general, the sum is off by a factor of $\approx 4^{-x}$ for x terms). So, given that we are considering a value of r in the range of 10, a very close approximation to the total number of insertions up to length r is $4/3$ times the number of insertions of length r, or $4^{r+1}/3$. For $r = 10$, that is 1.4×10^6.

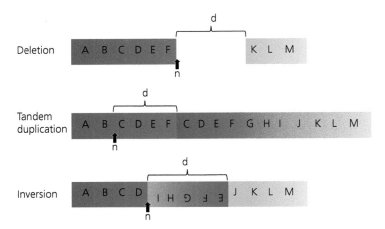

Figure B.1 Model for deletions, duplications, and inversions. Each type of mutation can be characterized by a point of origin n (arrows) and an extent d, and the number of possibilities follows from n and d. As a visual aid, the segment is marked with a color gradient and with letters A to M.

Previously, we found that for one-nucleotide insertions, the number of distinct insertions is always three, not four. The same kind of redundancy applies to longer insertions, e.g., insertions of length 2. Each site i in the genome, regardless of composition, precisely specifies a single dinucleotide sequence $X_{i-1}X_i$, and the insertion of this same dinucleotide at i (which is the internucleotide site after nucleotide i) has the same affect as inserting it at $i-2$. Thus, of the 16 possible dinucleotides that might be inserted at any given site, only 15/16 are nonredundant. Likewise, only 63/64 trinucleotide insertions are nonredundant at a given site i and $i-3$, and more generally, only $4^x - 1$ insertions of length x are nonredundant. The corrected formula would be, not $4^{r+1}/3$, but $\sum_{x=1}^{r} 4^x - 1 \approx 4^{r+1}/3 - r$.

Finally, note that we already counted one-nucleotide insertions, therefore a less redundant computation of insertions would be $\sum_{x=2}^{r} 4^x - 1 \approx 4^{r+1}/3 - r - 3$. Here we will use the simpler formula of $4^{r+1}/3$, which is accurate to within one part per million for $r = 10$. This formula gives the number of nonredundant insertions possible at each genomic site. To get the number for the whole genome, we simply multiply by n, thus the formula is $\sum_{x=2}^{r}(4^x - 1)n \approx (4^{r+1}/3)n$. This approximation gives 2.8×10^{12} for a prokaryotic genome of 2×10^6 bp, and 1.4×10^{15} for a eukaryotic genome of 10^9 bp.

B.5 Inversions, deletions, and tandem duplications

Inversions, deletions, and tandem duplications are similar in that each is defined by two points that are some distance apart in the genome. For each type of rearrangement, we can enumerate the possibilities by considering a point of

origin and a distance d beyond this point of origin. A circular genome of length n has n different segments of length d, e.g., n different segments of length 163.

Let us consider events up to length $d = r$ (range), where r might be something like 10^4 bp. The number of distinct events in a genome is computed from the number of source segments, which given a fixed length d is the same as the number of unique starting points (using our "marked origin" assumption). If a segment can be any length from $d = 1$ to $d = r$, the number of possible lengths is r, thus the number of source segments for rearrangements up to length r is $\sum_{d=1}^{r} n = rn$.

To be more precise, we should use $r - 1$, because we already counted insertions or deletions of length $d = 1$ as point mutations—but the scale of this error is only $1/r$, e.g., 1 part in 10^4 when $r = 10^4$.

Therefore, for inversions, deletions, or tandem duplications, the number of possible events is the number of possible source segments that may be deleted, inverted, or duplicated. The number of source segments is simply the number of origination sites multiplied by the range of possible segment lengths, thus rn for each type of rearrangement.

B.6 Transpositions (translocations)

Transpositions (also known as translocations) are more complex. In addition to a point of initiation and a length that together define rn possible source segments, a transposition also has a (1) a point of insertion, and (2) two possible orientations: direct or inverted. Furthermore, a transposition may be either duplicative or conservative, as shown in Fig. B.2. For the sake of illustration, the

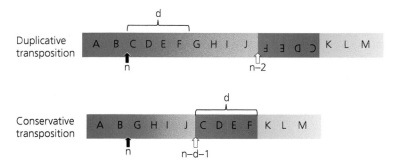

Figure B.2 Model for counting possible transpositions. A transposition is defined by a point of origin (black arrow), a length d up to $d = r$, and a destination (white arrow).

duplicative transposition is shown arbitrarily with an inverted orientation.

Of course, the biological reality is that transposition of segments, rather than applying uniformly across the genome, is overwhelmingly a phenomenon associated with the lifestyle of highly evolved *transposable elements*, as genomic parasites. Counting the universe of conceivable transpositions provides a formal basis to quantify the specificity of the transposition of mobile elements (insertion sequences and transposons), which generally shows great discrimination in which source segments are mobilized, and only a small amount of discrimination in the choice of destination.

A simple formula for the number of transpositions would assume that any of rn possible segments is inserted in any one of n sites, in either of two orientations, giving $2rn^2$ possibilities. This applies to both conservative and duplicative transpositions, thus the total for both types would be $2 \times 2rn^2 = 4rn^2$.

However, this simple formula implicates several kinds of outcomes that we do not want to count, because they are either impossible or redundant:

1. A set of rn null events of conservative transposition in which a segment returns to its origin in the same orientation, leaving the genome unchanged;
2. All inversions, because an inversion is simply a conservative transposition of a segment back to its origin, in opposite orientation;
3. All tandem duplications, because a tandem duplication is a duplicative transposition of a segment to the adjacent site before or after it;
4. A conservative transposition of a segment to a site within itself, which is not physically possible.

These redundancies are easily removed, starting with conservative transpositions. From the number of destination sites, we must subtract 1 to remove the chance of a

null reinsertion or an inversion. To exclude topologically impossible events of insertion of a segment into itself, we must further subtract $r - 1$ internal sites from the possible genomic insertion sites. That is, the number of possible destination sites is not n, but $n - r - 2$, so that total number of conservative transpositions is $2rn(n - r - 2)$.

For duplicative transpositions, we must treat the inverted and direct (noninverted) cases separately. For inverted duplicative transpositions, every topologically possible event is unique, and the number is $rn * n = rn^2$. For direct duplicative transpositions, we must subtract the two destination sites at the beginning and end of the segment where transposition would result in a tandem duplication. Thus, the number is $rn(n - 2)$.

The grand total for transpositions is then $2rn(n - r - 2) + rn^2 + rn(n - 2) = 2rn(2n - r - 3)$. Given that typically $2n >> r + 3$, $2n - (r + 3) \approx 2n$, thus $2rn(2n - r - 3) \approx 4rn^2$, i.e., the total number is close to the naive tally computed earlier. For instance, given $r = 10^4$ bp and considering a prokaryotic genome size of 2×10^6 bp, $4rn^2 = 1.6 \times 10^{17}$, whereas the more exact formula gives 1.5959988×10^{17}, i.e., the exact and approximate formulas differ by a factor of 0.0025. For a eukaryotic genome of 10^9 bp, the same approximation gives 4×10^{22}.

B.7 Lateral gene transfers

Lateral gene transfers are like transpositions in that a source segment, defined by a point of origin and a length, is inserted into a particular destination site, in one of two orientations. The main difference from transposition is that the source segment comes from one of many possible donor genomes that are not the same as the recipient genome. Because we are concerned only with the number of mutations in the recipient genome, we have no need to distinguish conservative vs. duplicative events, and furthermore, the concerns expressed in Section B.6 about null events, double-counting inversions, and so on, do not apply—every event of taking a defined segment from a donor genome and inserting it into a recipient genome is informationally unique in this model.

Therefore, the number of lateral gene transfers of segments up to length r from a donor to a recipient is $2rn_d n_r$, where n_d is the length of the donor genome and n_r is the length of the recipient genome.

We could make this calculation more precise by excluding certain short sequences. For instance, the number of possible sequences is 16 for doublets, 64 for triplets, and 256 for quadruplets. Presumably every donor genome has each one of these short sequences, and we already counted informationally equivalent mutations when we counted *de novo* insertions up to $r = 10$ nucleotides. However, the effect of this error will be small. If we count lateral transfers up to $r = 10,000$ nucleotides, removing events up to 10 nucleotides (in order to avoid double counting) is an effect of 1 part per 1000.

Finally, by specifying a distribution of donor genomes, we can enumerate the total number of possible lateral gene transfers from all possible donor genomes. For instance, suppose that the biosphere includes 10^{12} prokaryotic donor species (Locey and Lennon, 2016) and 10^7 eukaryotic donor species (Mora et al., 2011). Prokaryotic genome sizes are typically close to $n_d = 2 \times 10^6$ bp. Eukaryotic genomes are typically ten to 1000 times larger, but because prokaryotic species are 10^5 times more numerous, the size of the donor pool can be estimated from prokaryotes alone. If $D = 10^{12}$ is the number of prokaryotic donor species with genome size $n_d = 2 \times 10^6$, then the number of lateral gene transfers is $2rn_d n_r D = 8 \times 10^{28}$ given $r = 10^4$ for a prokaryotic recipient. Assuming that a typical eukaryotic genome is 1×10^9 bp, the same formula gives 4×10^{31} for a eukaryotic recipient.

B.8 Compound events

The recently named phenomenon of *kataegis* (Greek, "thunder") refers to clusters of point mutations representing regional domains of hypermutation. In the example shown previously (Section 2.8), the mutations are highly biased toward $C \to G$ and $C \to T$, and the affected regions are on the scale of kb (thousands of base-pairs).

Suppose that a domain of kataegis extends 10 kb, and involves 50 mutations. Suppose that these mutations happen only at cytosine (C) sites on one strand, thus given uniform nucleotide composition we expect $10,000/4 = 2,500$ C sites susceptible to mutation.

The number of possible configurations of mutated sites is a classic binomial "n choose k" problem, where $\binom{n}{k} = \frac{n!}{(n-k)!k!}$. As we have previously used the symbol n for the genome size, let us change n and k to d and e, for the domain of possible sites, and the number of implicated events, respectively. Given that each event must occur at a different site, we can compute the implicated sites using $\binom{d}{e} = \binom{2500}{50} \approx 10^{105}$. This is just the number of possibilities if we assume only one type of mutation in only one 10-kb block. If we consider two different mutational pathways, $C \to G$ and $C \to T$, we have to multiply this by a factor of $2^{50} \approx 10^{15}$. If we apply the same calculation, not just to a single 10-kb block, but to 200,000 blocks in an entire human genome, that adds another five orders of magnitude, or on the order of 10^{125} configurations, an inconceivably large number. By comparison, the number of atoms in the observable universe is estimated to be on the order of 10^{80}.

Though kataegis may be limited to eukaryotic cells under unusual conditions such as cancer, compound events involving multiple localized changes have been found in a much larger range of experimental and natural systems (see Section 2.8). A more general framework to address compound events would be to consider the length of a domain in which the events occur, the number of events, and the complexity of the types of events allowable. For instance, suppose that we want to compute complexity of compound events involving up to $e = 5$ point mutations in a domain of $d = 100$ bp. The choice of sites is a matter of $\binom{d}{e} = 7.5 \times 10^7$. Assuming $t = 6.7$ types of point mutations possible per site, the number of possible changes at those five sites is $t^e = 6.7^5 = 1.3 \times 10^4$, and thus the total number of possibilities in a single block is $\approx 1 \times 10^{12}$. Finally, the result of dividing the genome into nonoverlapping blocks is a set of $n/d = 2 \times 10^4$ blocks of 100 bp in a prokaryotic genome, thus the total number considering the whole genome is $\approx 2 \times 10^{16}$.

This yields a general formula for complex events of $\binom{d}{e} t^e n/d$. Note that this is very approximate, based on breaking the genome up into n/d nonoverlapping blocks. This underestimates the actual number of compound events because, for any way of breaking up the genome into blocks, legitimate compound events that span fewer than d nucleotides, but extend from one block into another, remain uncounted.

B.9 Summing up

Table B.1 illustrates the size of the universe of mutations assuming genome sizes of 2×10^6 bp and 10^9 bp for prokaryotes and eukaryotes, respectively. The maximum length of *de novo* insertions is assumed to be $i = 10$, and that for other rearrangements is assumed to be $r = 10^4$ bp. The numbers for lateral gene transfers (a transposition from one genome to another) are based on assuming $D = 10^{12}$ distinct donor species with genomes of an average length $n_d = 2 \times 10^6$.

Table B.1 A crude tally of the universe of mutations

Type	Formula	Prokaryotic	Eukaryotic
Point mutation	$6.7n$	1.3×10^7	6.7×10^9
Deletion	rn	2×10^{10}	10^{13}
Tandem duplication	rn	2×10^{10}	10^{13}
Inversion	rn	2×10^{10}	10^{13}
Insertion	$(4^{r+1}/3)n$	2.8×10^{12}	1.4×10^{15}
Transposition	$4rn^2$	1.6×10^{17}	4×10^{22}
Lateral gene transfer	$2rn_d n_r D$	8×10^{28}	4×10^{31}
Compound	$\binom{d}{e}t^e n/d$	?	?

The calculation for compound events is so variable, depending on the exact case, that a cautious estimate is not possible. Nevertheless, in the case of kataegis, compound events dwarf lateral gene transfers, and even in the case of more mundane compound events, such as small combinations of point mutations in a local region, the numbers quickly become larger than for most of the other categories.

If we ignore compound events, the universe of mutations is dominated by lateral gene transfers, with 8×10^{28} or 4×10^{31} possibilities for prokaryotic and eukaryotic recipients, respectively. If we ignore lateral gene transfers and compound events, the universe of mutations is dominated by transpositions.

In any case, the universe of mutations is enormously larger than the set of point mutations.

Note that the calculation of lateral gene transfers assumes that every segment of length r from a donor genome is unique. On the one hand, this overestimates lateral gene transfers to the extent that some short segments from highly conserved genes such as rDNA may be the same between different species. Presumably, this effect is small, because the larger segments outnumber the shorter ones, and they are likely to be distinct even for closely related species, e.g., if two sequences are 0.5 % different in sequence on average, then it follows from a Poisson assumption that 99 % of 1000-bp segments will be distinct.

On the other hand, the estimate of lateral gene transfers greatly underestimates the complexity of donor segments. First, we have underestimated the pool of distinct donor segments by ignoring the within-species diversity of cellular organisms, i.e., an individual from a particular species has rn segments to contribute, but when the segments are large, another individual from the same species will have many additional unique sequences to contribute. This is particularly true in prokaryotes, where members of the

same species often show > 1 % sequence difference. Second, we have underestimated the donor pool by ignoring the apparently enormous pool of noncellular (virus and phage) genomes (Casas and Rohwer, 2007). Thus, the calculation for lateral gene transfer presumably represents an underestimate.

B.10 Recurrences

Imagine briefly what it would be like if all these mutations happened at the same rate. In prokaryotes, the genomic mutation rate is typically something like $U = 0.003$ per generation (Drake et al., 1998), thus given 8×10^{28} types of mutations, the average rate for mutations is 4×10^{-32} per generation.

How often would we see the same mutation twice under these conditions? The "birthday problem" in statistics refers to a classic case of an intuitively surprising recurrence, namely, the chance of finding two individuals with the same birthday in a classroom full of students. In a class with just 23 students, the chance that two students share a birthday begins to become probable, i.e., begins to exceed 0.5. A simple approximation to the birthday problem is to take the square-root of the number of possibilities, e.g., $\sqrt{365} = 19.1$, which is not far from 23. The comparable estimate for the universe of mutations is $\sqrt{8 \times 10^{28}} = 2.8 \times 10^{14}$ mutations to achieve a 50 % chance of seeing a mutation twice. Given a genomic mutation rate of $U = 0.003$, this would require 9×10^{16} cell-generations. This is the number of mutations that would occur if we were to grow a flask with 10^{10} bacterial cells for nine million generations, or about 500 years, given 30 minutes per generation.

That is, if we had some way to monitor every mutation occurring in a prokaryotic population of 10^{10} cells, we would have to wait for 500 years for it to become likely to see the same mutation twice. Now, suppose that we observe mutations only via beneficial fixation events, assuming that 1 % of mutations are beneficial with a chance of fixation of 0.1 each. This 100-fold reduction in the universe of mutations results in a 10-fold reduction in the number of observations required to make a recurrence likely (via the square-root approximation). Yet, because the beneficial mutation rate is 100 times smaller, and only 10 % reach fixation, each observation requires 1000 times more generations. The net effect is a 100-fold increase in the size of the task, e.g., monitoring 100 populations (instead of just 1 population) for 500 years. Thus, if we ever see the same mutation twice in some laboratory experiment with prokaryotes, we know that mutation is nonuniform.

For typical large eukaryotic organisms, the number of organism-generations to require a recurrence is actually lower, because the effect of expanding the universe of mutations is more than offset by a higher genomic mutation rate of $U \approx 1$. However, it takes much longer to achieve a particular number of organism-generations, because these larger eukaryotic organisms have smaller populations and longer generations. A typical large organism has a population less than 10^7 in number, and a generation time on the order of a year. This means that, if we could somehow monitor every mutation, we would have to wait 630 million years to see the same mutation twice in a population of 10^7 organisms. If we aim to consider parallel adaptation in eukaryotes with these same assumptions, finding an exact parallelism would require us to track changes in thousands of populations for millions of years.

In short, seeing parallel mutations is proof that mutation is nonuniform. This confirms the intuition of Shull (1935) that "It strains one's faith in the laws of chance to imagine that identical changes should crop out again and again if the possibilities are endless and the probabilities equal" (p. 448).

APPENDIX C

Randomness quotations

"When *I* use a word," Humpty Dumpty said, in rather a scornful tone, "it means just what I choose it to mean—neither more nor less." "The question is," said Alice, "whether you can make words mean so many different things." "The question is," said Humpty Dumpty, "which is to be master—that's all."

Lewis Carroll, *Through the Looking-Glass*

C.1 Introduction

The collection of quotations below illustrates randomness claims from 1930 to the present. The main sources are major works of the eight individuals most often cited as founders of the Modern Synthesis: Wright, Fisher, Haldane, Huxley, Dobzhansky, Mayr, Simpson, and Stebbins. Additional sources include (1) other major figures identified with modern neo-Darwinism (Grant, Darlington, Gould, Maynard Smith) and (2) a small sample of orthodox sources including textbooks, reviews, and commentaries that address the issue of randomness. These quotations do not represent a proportional sample of all claims, but tend to focus on more interesting or significant claims.

The quotations are given in reverse chronological order, from the present back to the 1930s. In some cases, the citation gives a misleading date, e.g., Simpson (1984) is a reprint of his 1944 work. In these cases, the original date is provided in parentheses.

In some cases, the search for relevant statements is simplified and enriched by the availability of digitized texts, which greatly improve one's ability to assess the uniformity and frequency of statements on topics associated reliably with searchable keywords. For instance, "random" is used about 20 times in Mayr's 1942 *Systematics and the Origin of Species* (Mayr, 1999), about 40 times in Simpson's 1944 *Tempo and Mode in Evolution* (Simpson, 1984), and over 70 times in *Genetics of the Evolutionary Process* (Dobzhansky, 1970).

An important result of such systematic searches is that statements linking randomness to mutation are considerably *less frequent* than statements linking randomness to something else. As a rough approximation, I would estimate that (taken together) the canonical works of the Modern Synthesis apply "random" to mutation about as often as they apply it to each of the following: (1) ecological processes of mating, sampling, or dispersal; (2) population-genetic processes of gene loss, fixation, or fluctuations in frequencies (drift); and (3) genetic processes of assortment, recombination, or union of gametes. In addition, "random sampling" is used rarely, in regard to experimental design. Many of the references to randomness do not make a claim about the world, but only explain theoretical expectations (e.g., what is expected under random mating). *Typically these statements are not problematic and are consistent with "random" denoting a blind sample from the universe of possibilities.*

Statements invoking randomness in regard to mutation are more complex. They are the least revealing when "random" merely appears as as an unexplained epithet, e.g., "random mutation" occurs frequently without explanation in Mayr (1999). However, authors often invoke randomness in a more revealing way, or attempt to explain what they mean. This appendix tends to emphasize such statements, which are not evenly distributed among major authors, e.g., Simpson has quite a lot to say, and Mayr very little.

Finally, note that authors may make statements of fact about peculiarities or regularities of mutation, without ever entertaining any coherent evolutionary theory for a dispositional role of variation in evolution. Recognizing peculiarities or regularities of variation, and treating them as evolutionary dispositions, are two quite different things (namely, they are source laws and consequence laws per Chapter 7). Notably, Simpson, who has the most to say about peculiarities or regularities of mutation, also has the most interesting things to say about their possible role in evolution, though his comments never rise to the level of a coherent theory.

Mutation, Randomness, and Evolution. Arlin Stoltzfus, Oxford University Press (2021). © Arlin Stoltzfus. DOI: 10.1093/oso/9780198844457.001.0001

C.2 List of quotations

1. "A central tenet of the Modern Synthesis claims that genetic mutations occur by 'chance' or at 'random' with respect to adaptation, that is, with respect to the adaptive needs of the organisms concerned and the population to which they belong" (Merlin, 2010, p. 2, summarizing the default view based on her extensive literature review).

2. "The gist of the evolutionary notion of chance is that events are independent of an organism's need and of the directionality provided by natural selection in the process of adaptation. While for mutations evolutionary randomness is clear (the alternatives are Lamarckism and predetermination), for most other evolutionary phenomena this is not trivial, given the directionality implicit in the expectation of selection operating as a constructive, creative force toward adaptation over many generations (Mayr 1963; Gould, 1982). 'Directionality' is here materialistic and has no implication of teleology. Patterns are evolutionarily random whenever selection and adaptation are not directly involved" (Eble, 1999, summarizing the default view based on his conceptual analysis).

3. "Variation in the characteristics of organisms in a population originates through random mutation of DNA sequences (genes) that affect the characteristics. 'Random' here means that the mutations occur irrespective of their consequences for survival or reproduction" (from a white paper written by representatives of professional organizations: Futuyma and others, 2001, p. 4).

4. "Genetic variation occurs randomly, not in response to the needs of a population or organism . . . Genetic variation is random, but natural selection is not" (Working Group on Teaching Evolution, 1998).

5. "A central tenet of evolutionary theory is that mutation is random with respect to its adaptive consequences for individual organisms; that is, the production of variation precedes and does not cause adaptation" (Sniegowski and Lenski, 1995).

6. "A fundamental tenet of evolutionary biology is that mutations are random events. This tenet does not mean that mutation rates are unaffected by environmental factors or that all portions of the genome are equally susceptible to mutation. . . Rather, the randomness of mutation refers to the supposition that the likelihood of any particular mutational event is independent of its specific value to the organism" (Lenski and Mittler, 1993).

7. "We can now return to the original problem, of whether heredity alone can produce evolution. Because evolutionary innovations result from mutation and recombination, we should rephrase the question to ask whether mutation and recombination produce directed or undirected change. The answer is that, like pure Mendelian inheritance, they too are undirected. They do not consistently produce changes in any particular direction. Mutations affecting size are just as likely to produce taller organisms as shorter ones" (Ridley, 1985, p. 25).

8. "Furthermore, there is usually no reason to suppose that the developmental mechanisms in question evolved because of the particular phenotypes that they make readily accessible. In general, therefore, the direction of the resulting constraints (biases on the production of variant phenotypes) is 'accidental' or 'random' with respect to the demands of adaptive evolution" (Maynard Smith et al., 1985, p. 269).

9. "It has never been part of Darwinism to claim that phenotypic variation is random or unconstrained. In any given taxon, some variations occur commonly and others rarely or not at all. Darwin himself noted this phenomenon, and coined for it the term 'analogous variation'. There are, however, two senses in which a contemporary Darwinist would claim that variation is unconstrained. First, at the molecular level, we know of no constraints on the kinds of changes in DNA sequence that can arise by mutation. Secondly, the phenotypic changes resulting from a mutation do not in general adapt the organism to withstand the agent which caused the mutation. There is, therefore, nothing 'un-Darwinian' about the claim that there are developmental constraints. Our difficulty is that we can rarely predict the nature of those constraints from a knowledge of development" (Maynard Smith, 1983).

10. "By 'random' in this context, evolutionists mean only that variation is not inherently directed towards adaptation, not that all mutational changes are equally likely" (Gould, 1982, p. 386).

11. "The direction of mutations is entirely random with reference to their functional or adaptive value; there is no internal force or guiding principle that has directed the course of evolution from bacteria through amoeba, worms, fishes, reptiles and mammals to human beings. This is the aspect of chance" (Stebbins, 1982, p. 69, summarizing the arguments of Monod, 1971).

12. "Mutation is random in that the chance that a specific mutation will occur is not affected by how useful that mutation would be" (Futuyma, 1979, p. 249).

13. "Mutations are random changes because they occur independently of whether they are beneficial or harmful, and therefore they are a disordering process" (Dobzhansky et al., 1977, p. 66).

14. "At the level of the gene pool, mutations produce variations that are random with respect to the environ-

ment, and thus cannot by themselves direct evolution toward new adaptations" (Dobzhansky et al., 1977, p. 6).

15. "Transformation of the gene coding for hemoglobin into that coding for myoglobin . . . has a zero probability of occurring by a single mutation. Yet these transformations have in fact taken place in evolutionary history, by way of a sequence of mutations, presumably controlled by natural selection. While each mutation in this sequence was, if considered on the molecular level, an accident, the sequence as a whole is in no sense accidental or random" (Dobzhansky, 1970, p. 94).

16. "Mutations cannot be said to change the development in random directions because a single nucleotide substitution is rarely if ever sufficient for a gene to change its function radically . . . The successive mutational gene changes acquire a direction because natural selection controls the fitness of the resulting phenotypes and thus indirectly imposes a restriction on the randomness of the mutational events" (Dobzhansky, 1970, p. 93).

17. "The frequencies of spontaneous mutations of some genes have been ascertained in genetically well-studied organisms. Yet where, when, and in which individual a particular mutation will appear is unpredictable. Even the rather more specific chemical mutagens discussed in Chapter 2 rarely change 100 percent of the genes exposed to treatment. The mutational repertoire of the gene is great but not infinite; it is limited by the composition of the gene" (Dobzhansky, 1970, p. 92).

18. "Random mutations are the raw materials of the evolutionary process. Natural selection orders them in functionally coherent, adaptive systems. Mutations are often described as accidental, random, undirected, chance events. Just what do these epithets mean? Mutations are accidents, because the transmission of hereditary information normally involves precise copying. A mutant gene is, then, an imperfect copy of the ancestral gene. It would be absurd, however, to say that human genes are only distorted copies of the primeval genetic materials. The serviceability of human genes, or of those of any existing species, has been validated by natural selection. Mutations are undirected with respect to the adaptive needs of the species. They arise regardless of their actual or potential usefulness. It may seem a deplorable imperfection of nature that mutability is not restricted to changes that enhance the adaptedness of their carriers. However, only a vitalist Pangloss could imagine that the genes know how and when it is good for them to mutate" (Dobzhansky, 1970, p. 92).

19. "Reference was made in Chapter 2 to the distinction drawn by de Vries between mutations creating new species and Darwinian 'fluctuating' variability. This distinction is invalid. In point of fact, only chromosome doubling in interspecific hybrids (allopolyploidy) is a special kind of mutation that may lead directly to the emergence of new species (see Chapter 11). The process of mutation supplies only the building blocks, the raw materials, from which evolutionary changes, including species differences, are compounded by natural selection. Mutation is, then, the ultimate source of evolution, but there is more to evolution than mutation. It will be shown in the concluding pages of the present chapter that mutation is a random process with respect to the adaptive needs of the species. Therefore, mutation alone, uncontrolled by natural selection, would result in the breakdown and eventual extinction of life, not in adaptive or progressive evolution" (Dobzhansky, 1970, p. 65).

20. "The idea that evolution comes about from the interaction of a stochastic and a directed process was the essence of Darwin's theory. The stochastic process that he invoked was the occurrence of small random variations which he supposed, provided the raw material for natural selection, a process directed by the requirements of the environment and one that builds up, step by step, changes that would be inconceivably improbable at a single step . . . The meaning of 'random' . . . is that the variations are, as a group, not correlated with the course subsequently taken by evolution (which is determined by selection). The variations are, of course, severely limited in kind by the accumulated results of past evolution. Those that are seized upon by selection are ordinarily ones that are very slight phenotypically" (Wright, 1967, p. 117).

21. "Mutation . . . is at random with respect to the direction of adaptation" (Stebbins, 1966, p. 35).

22. "It is sometimes stated that the types of mutations arising in a population of organisms are at random. The randomness of mutations is not unrestricted however. Any genotype at any moment possesses an architecture which probably imposes physical limitations on the forms of gene alterations that can arise. The mutations in a gene are no doubt channelized along certain lines predetermined by the existing gene structure [here he cites Blum, 1955 and Waddington, 1957]. Among the array of mutant genes that are physically possible, however, there is no known tendency for the mutation process to produce preferentially those which fit the adaptive requirements of the organism. In fact the majority of the known mutations are detrimental in some degree to their carriers. The mutation process, though nonrandom in a physical sense, is

unoriented with reference to any standard of adaptedness, and hence is random in a biological sense" (Grant, 1963, pp. 163 to 164).

23. "A word of qualification is necessary in referring to the mutation process as random. Mutations are random in the sense that they may occur when they are not 'needed,' and may fail to occur when they are 'needed,' but their randomness has limits" (Moody, 1962).

24. "(1) Randomness of an event with respect to the significance of the event. Spontaneous mutation, caused by an 'error' in DNA replication, illustrates this cause for indeterminacy very well. The occurrence of a given mutation is in no way related to the evolutionary needs of the particular organism or of the population to which it belongs. The precise results of a given selection pressure are unpredictable because mutation, recombination, and developmental homeostasis are making indeterminate contributions to the response to this pressure. All the steps in the determination of the genetic contents of a zygote contain a large component of this type of randomness. What we have described for mutation is also true for crossing over, chromosomal segregation, gametic selection, mate selection, and early survival of the zygotes. Neither underlying molecular phenomena nor the mechanical motions responsible for this randomness are related to their biological effects" (Mayr, 1961).

25. "What do we mean by 20th century Darwinism, and what do we mean by the synthetic theory of evolution? I think its essence can be characterized by two postulates: (1) that all the events that lead to the production of new genotypes, such as mutation, recombination and fertilization, are essentially random and not in any way whatsoever finalistic, and (2) that the order in the organic world, manifested in the numerous adaptations of organisms to the physical and biotic environment, is due to the ordering effect of natural selection. Nothing has been discovered in the decades since these principles were first clearly stated that is in any way in conflict with these basic assumptions" (Mayr, 1959b, p. 4).

26. "It would be exceedingly difficult to visualize a mechanism by which the environment could induce directly a structural change in the DNA molecules that would result in the production of a superiorly adapted phenotype and more specifically in the appropriate response to a temporary need. Nor is there any evidence that this occurs. Indeed, there is no need for such an induction within the framework of the synthetic theory of evolution. Infinitely variable natural populations are of such evolutionary plasticity

that natural selection can mold them into almost any shape" (Mayr, 1959a, pp. 6 to 7).

27. "Variation is in some sense random, but natural selection picks out variations in one direction, and not in another" (Haldane, 1959, p. 147).

28. "It is frequently stated, particularly in popular scientific writings, that mutations are haphazard, chance, accidental, random, etc., changes of the genes. Such characterizations are misleading when given without qualification. For the only respect in which mutations are haphazard is that they arise regardless of the needs of the organism at a given time, and hence are far more likely to be deleterious than useful. But the kinds of mutations that a gene is capable of producing as well as the frequencies with which it produces them are far from indeterminate. They are controlled by the structure of the gene itself as well as by the genetic constitution of the organism" (Dobzhansky, 1955b, p. 58).

29. "Supposing that lack of horns was a factor in extinction of the latter, which is at least possible, we may say that they lost adaptation because they had no variation or mutation in the direction of horns [i.e., in the direction of growing horns]. We certainly cannot say why they did not have such variation, when their relative did have. In this and in the great number of similar cases, we can only clothe our ignorance in the words 'mutation is random'" (Simpson, 1953, p. 298, in a speculative analysis of rhinoceros evolution).

30. "The ultimate sources, mutations, are random with respect to selection" (Simpson, 1953, p. 180).

31. "That the mutations are spontaneous and random, at least in the special sense elsewhere defined, is a conclusion warranted and, with some restrictions, demanded by the experimental data. Calling them 'spontaneous' and 'random' means simply that they are not orderly in origin according to the demands of any one of the discarded theories. The incidence in time and the individuals affected seem to be random or nearly so in the same sense—that they do not agree with hypotheses that assume a more specific incidence. That the direction of mutations is entirely random is certainly not true; but neither is it true that mutations regularly occur in one direction only" (Simpson, 1953, p. 135).

32. "The conclusions are: (1) that mutations are not strictly random in the range of phenotypic effect or in relative frequency, (2) that given the limitations of range of effects and of characteristic rate, mutations are random in incidence, (3) that the nonrandom tendencies of effect and rate do not correlate with present adaptation or with past or present changes in adaptive type, so that mutation may also be said to be random from this special point of view, but (4) that mutations adap-

tive for changing or for different conditions do occur and are involved in adaptation" (Simpson, 1953, p. 88).

33. "On the other hand, the term 'randomness' as applied to mutation often refers to the lack of correspondence of phenotypic effect with the stimulus and with the actual or the adaptive direction of evolution. Heat-induced mutations do not produce phenotypic change related to heat tolerance [note 4: There remains a possibility that mutation induced chemically by reaction with the gene might produce physiological changes related to the inducing chemical. This, however, would only be the sort of exception that proves the rule. Its occurrence is still doubtful in the laboratory and unknown in nature.]. It is a well known fact, emphasized over and over in discussions of genetics and evolution, that the vast majority of known mutations are inadaptive" (Simpson, 1953, p. 87).

34. "There is, on one hand, a randomness as to where and when a mutation will occur. Mutations induced by hits by X-rays, hits that must be statistically random, tend to have about the same distribution as naturally occurring mutations although the frequencies are higher (Muller, 1947 and numerous references therein). This indicates a 'molar indeterminacy' (Muller), a randomness of energy supply or stimulus (whatever the source may be in spontaneous mutation), although still not a wholly random reaction, since different genes still mutate at greatly different rates" (Simpson, 1953, p. 86).

35. "This sort of limitation and the fact that different mutations may have widely and characteristically different rates of incidence show that mutations are not random in the full and usual sense of the word or in the way that some early Darwinists unrealistically considered as fully random the variation available for natural selection. I believe that the, in this sense, nonrandom nature of mutation has had a profound influence on the diversity of life and on the extent and character of adaptations. This influence is sometimes overlooked, probably because almost everyone speaks of mutations as random, which they are in other senses of the word" (Simpson, 1953, p. 86).

36. "Even within the limitations of a given reaction system, it is evident that not all conceivable mutations occur and those that do occur do so with very unequal frequencies . . . [in *Drosophila* and other cases] mutations are known affecting practically every part of the body and numerous different physiological characters, but effects do not necessarily take all directions" (Simpson, 1953, p. 85).

37. "As far as is known, however, such mutations are wholly random with respect to adaptation, and this makes the probability of their producing a viable

preadaptation so extremely small that it is hardly conceivable that they have any importance . . ." (Simpson, 1984, p. 211) (1944).

38. "Mutation is frequently directional in the sense that it occurs more frequently in one direction than in another, but it is usually random in the sense that this favored direction has no special tendency to coincide with advantageous modification or with the direction in which the group is really evolving" (Simpson, 1984, p. 178) (1944).

39. "Another extreme is seen in various theories that suppose adaptive characters to have arisen, not in response to external influences and individual needs or at random with respect to them, but in anticipation of them" (Simpson, 1984, p. 75) (1944).

40. "The virtual impossibility of the simultaneous appearance of a number of morphologically congruent random mutations and the obvious fact that different functionally related characters do evolve in unison underlie the revolt by some paleontologists and others against the belief that gene mutations as they occur in the laboratory have anything to do with evolution in any broad sense. In the first place, neither the experimental nor the observational data require or warrant the belief that mutations are completely random in their phenotypic effects, although they may be nearly random in incidence (i.e., have essentially random distribution in a population and in time). On the contrary, considered as samples of an infinite number of possible morphological changes, most possibilities of mutation appear to be rather rigidly limited. For many characters only two directions of change are possible, for instance, toward larger or toward smaller size, and mutations in one of these directions may be more frequent than in the other. If the chances are about even that a mutation will be toward or against an existing selection pressure, it may still be considered random from the point of view of adaptation and functional integration. It is, indeed, unnecessary to assume that mutations are normally random even in this limited sense. In a highly specialized and well-adapted organism the chances are that any one mutation will be disadvantageous, as in *Drosophila*. In a more poorly integrated and poorly adapted organism the chances of advantageous mutation are evidently much greater. In the second place, the problem of functional integration becomes much simpler if useful mutations are very small but numerous and may, indeed, be insuperable if mutations are few and large" (Simpson, 1984, pp. 55 to 56) (1944).

41. "Most of the mutations in the postulated situation could be only in one of two directions: an element of tooth structure can appear or disappear, become

larger or become smaller. For such characters half the mutations, if they were completely random, would be in the direction favored by selection or actually followed in the morphological progression of the phylum" (Simpson, 1984, p. 46) (1944).

42. "Finally, mutations, while they seem to occur more readily in certain directions than in others (Chapter 9), can be legitimately said to be random with regard to evolution. That is to say, the directions of the changes produced by them appear to be unrelated either to the direction of the evolutionary change to be observed in the type, or to the adaptive or functional needs of the organism. Evolutionary direction has to be imposed on random mutation through the sifting and therefore guiding action of selection" (Huxley, 1964, p. 54) (1942).

43. "The basic change factor is gene mutation, the occasional failure of precise duplication. Since the time of Lamarck, a school of biologists have held that the primary changes in hereditary constitution must be adaptive in direction in order to account for evolutionary advance. Unfortunately, the results of experimental study have given no support to this view. Instead, the characteristics of actually observed gene mutations seem about as unfavorable as could be imagined for adaptive evolution. In the first place, is their fortuitous occurrence. No correlation has been found between external conditions and direction of mutation, and those few agents which have been found to affect the rate (X-ray, radium, and to relatively unimportant extent, temperature) merely speed up the rate of random mutation. The great majority of mutations are either definitely injurious to the organism or produce such small effects as to be seemingly negligible. Muller has graphically compared the range of mutations to a spectrum in which the nonlethal conspicuous mutations form a narrow field between broad regions of individually inconspicuous mutations on the one hand and of sublethal and lethal mutations on the other" (Wright, 1931, p. 142).

Irrelevance quotations

D.1 Introduction

This appendix provides a variety of statements to the effect that the details of mutation are irrelevant to how evolution turns out, an argument that may take multiple forms, as explained in Chapter 6. Claims of this type have clearly diminished in scope and frequency over time. Contemporary sources tend to diminish the importance of mutation by calling it "random" and by saying that it provides only "raw materials." Earlier sources frequently made substantive and problematic claims, e.g., about the irrelevance of mutation rates to the rate of evolution, or the idea that selection is creative while mutation is not.

D.2 List of quotations

1. "Natural selection has shaped all the beautiful and extraordinary diverse phenotypes of living organisms,[1] while random mutations provide the raw material for this process.[2]" (Philippe et al., 2007, citing Darwin's *Origin of Species*, then Luria and Delbrück).

2. "If correct, Behe's calculations would at a stroke confound generations of mathematical geneticists, who have repeatedly shown that evolutionary rates are not limited by mutation. Single-handedly, Behe is taking on Ronald Fisher, Sewall Wright, J.B.S. Haldane, Theodosius Dobzhansky, Richard Lewontin, John Maynard Smith and hundreds of their talented co-workers and intellectual descendants. Notwithstanding the inconvenient existence of dogs, cabbages and pouter pigeons, the entire corpus of mathematical genetics, from 1930 to today, is flat wrong. Michael Behe, the disowned biochemist of Lehigh University, is the only one who has done his sums right. You think? The best way to find out is for Behe to submit a mathematical paper to The Journal of Theoretical Biology, say, or The American Naturalist, whose editors would send it to qualified referees" (from the *New York Times* book review of Behe's *The Edge of Evolution* by Dawkins, 2007).

3. "Since orthogenesis can only operate when mutation pressure becomes high enough to act as an agent of evolutionary change, empirical data on low mutation rates sound the death-knell of internalism" (Gould, 2002, p. 510, before citing the opposing pressures argument from Fisher in quotation D.36).

4. "Darwin knew nothing of this [i.e., genetics] but as it turned out, his ignorance was sublimely irrelevant to the problem he was really interested in tackling: evolution. This point was not fully grasped by biologists. Many early geneticists at the dawn of the 20th century, thought their discoveries of the fundamental principles of genetics somehow cast doubt [on], or rendered obsolete, the concept of natural selection. It took several decades of experimentation and theoretical (including mathematical) analysis to show not only that there was no conflict inherent between the emerging results of genetics and the older Darwinian notion of natural selection, but that the two operate in different domains. The principles of inheritance work within single organisms—two organisms, in the case of sexual reproduction. In contrast, natural selection involves differential reproductive success among large numbers of genetically varying organisms within a representative population" (Eldredge, 2001, p. 67).

5. "For instance, it is possible to say confidently that natural selection exerts so much stronger a force than mutation on many phenotypic characters that the direction and rate of evolution is ordinarily driven by selection even though mutation is ultimately necessary for any evolution to occur" (Futuyma and others, 2001, in a white paper written by representatives of various professional societies).

6. "In summary, mutation creates the genetic diversity that is the raw material for evolution. In principle, mutation can also produce changes in allele frequencies across generations. In practice, mutation's role as the source of genetic variation is usually more important than its role as a mechanism of evolution" (Freeman and Herron, 1998, p. 145).

7. "Those authors who thought that mutations alone supplied the variability on which selection can act, often called natural selection a chance theory. They

said that evolution had to wait for the lucky accident of a favorable mutation before natural selection could become active. This is now known to be completely wrong. Recombination provides in every generation abundant variation on which the selection of the relatively better adapted members of a population can work" (Mayr, 1994, p. 38).

8. "In practically all populations, however, the role of new mutations is not of immediate significance" (Strickberger, 1990, p. 464).

9. "The reason why I am disinclined to describe developmental constraints as directing evolution is the reason why I would not describe an artist's material as directing his handiwork" (Leigh, 1987, p. 229).

10. "Novelty does not arise because of unique mutations or other genetic changes that appear spontaneously and randomly in populations, regardless of their environment. Selection pressure for it is generated by the appearance of novel challenges presented by the environment and by the ability of certain populations to meet such challenges" (Stebbins, 1982, p. 160).

11. "The opinion expressed throughout this book, that mutations are most important as a way of replenishing the gene pool as it becomes depleted by natural selection, is now supported by a wealth of experimental and observational evidence. Nevertheless, the old idea that mutations direct evolutionary change is still held by many biologists and is expressed most often in popular accounts of evolution. Other misconceptions about mutations are also widespread. One misconception, which follows from the mistaken idea that mutations can establish the rate and direction of evolution, is that bursts of rapid evolutionary change are produced by increases in the rate of mutations" (Stebbins, 1982, p. 79).

12. "The term 'Darwinism' in the following discussions refers to the theory that selection is the only direction-giving factor in evolution" (Mayr, 1980, p. 1 footnote).

13. "Each unitary random variation is therefore of little consequence, and may be compared to random movements of molecules within a gas or liquid. Directional movements of air or water can be produced only by forces that act at a much broader level than the movements of individual molecules, e.g., differences in air pressure, which produce wind, or differences in slope, which produce stream currents. In an analogous fashion, the directional force of evolution, natural selection, acts on the basis of conditions existing at the broad level of the environment as it affects populations" (Dobzhansky et al., 1977, p. 6).

14. "The large number of variants arising in each generation by mutation represents only a small fraction of the total amount of genetic variability present in natural populations... It follows that rates of evolution are not likely to be closely correlated with rates of mutation... Even if mutation rates would increase by a factor of 10, newly induced mutations would represent only a very small fraction of the variation present at any one time in populations of outcrossing, sexually reproducing organisms" (Dobzhansky et al., 1977, p. 72).

15. "But why was natural selection compared to a composer by Dobzhansky; to a poet by Simpson; to a sculptor by Mayr; and to, of all people, Mr. Shakespeare by Julian Huxley? I won't defend the choice of metaphors, but I will uphold the intent, namely, to illustrate the essence of Darwinism—the creativity of natural selection. Natural selection has a place in all anti-Darwinian theories that I know. It is cast in a negative role as an executioner, a headsman for the unfit (while the fit arise by such non-Darwinian mechanisms as the inheritance of acquired characters or direct induction of favorable variation by the environment). The essence of Darwinism lies in its claim that natural selection creates the fit. Variation is ubiquitous and random in direction. It supplies the raw material only. Natural selection directs the course of evolutionary change. It preserves favorable variants and builds fitness gradually. In fact, since artists fashion their creations from the raw material of notes, words, and stone, the metaphors do not strike me as inappropriate." (Gould, 1977, p. 44).

16. "Mutation is of course originally responsible for the diversity of the genetic units. But living organisms are the product of evolution controlled not by mutation but by powerful selection" (Ford, 1971, p. 391).

17. "Furthermore, the fact that such powerful selection is normally operating reduces to negligible limits the effects of random drift upon changes in the frequency of widespread genetic qualities: the time when the operation of chance may indeed be critical is not then but when a mutant begins to spread or, as Darlington has pointed out, when an inversion occurs in a suitable place. Similarly, if ever it could have been thought that mutation is important in the control of evolution, it is impossible to think so now; for not only do we observe it to be so rare that it cannot compete with the forces of selection but we know this must inevitably be so... Situations in which evolution is controlled by mutation, and that envisaged in Lamarckism is one, are the more clearly impossible now that we appreciate the intensity of selection for advantageous qualities (its intensity against disadvantageous ones has been obvious since the time of Darwin)" (Ford, 1971, p. 391).

18. "That selection can work only with raw materials arisen ultimately by mutation is manifestly true. But it is also true that populations, particularly

those of diploid, outbreeding species, have stored in them a profusion of genetic variability. A temporary suppression of the mutation process, even if it could be brought about, would have no immediate effect on evolutionary plasticity" (Dobzhansky, 1970, p. 201).

19. "Adaptation has a known mechanism: natural selection acting on the genetics of populations ... In seeking the orienting factor in evolution we have seen that in some cases this must, by all reasonable inferences, be adaptation and in all, even the most doubtful, it could be adaptation. Thus we have a choice between a concrete factor with a known mechanism and the vagueness of inherent tendencies, vital urges, or cosmic goals, without known mechanism" (Simpson, 1967, p. 159).

20. "Our calculations lead us, therefore, to the conclusion which has been reached by most geneticists who are studying evolutionary processes. The chief limiting factor on the supply of variability for the action of natural selection is not the availability or rate of occurrence of mutations, but the restrictions on gene exchange and recombination which are imposed by the mating structure of populations and the structural patterns of chromosomes. Natural selection directs evolution not by accepting or rejecting mutations as they occur, but by sorting new adaptive combinations out of a gene pool of variability which has been built up through the combined action of mutation, gene recombination, and selection over many generations. . . Although mutations may occasionally play an important role in directing natural selection along a particular channel, most of them are important only as contributors to the gene pool. Consequently, the rate of mutation rarely if ever has an influence on the rate of evolution" (Stebbins, 1966, pp. 30–31).

21. "Mutations are rarely if ever the direct source of variation upon which evolutionary change is based. Instead, they replenish the supply of variability in the gene pool which is constantly being reduced by selective elimination of unfavorable variants. Because in any one generation the amount of variation contributed to a population by mutation is tiny compared to that brought about by recombination of pre-existing genetic differences, even a doubling or trebling of the mutation rate will have very little effect upon the amount of genetic variability available to the action of natural selection. Consequently, we should not expect to find any relationship between rate of mutation and rate of evolution. There is no evidence that such a relationship exists" (Stebbins, 1966, p. 29).

22. "An evolutionary line of organisms which is changing through eons of time can be likened to an automobile being driven along the highway. Mutation then corresponds to the gasoline in the tank. Since it is the only possible source of new genetic variation, it is essential for continued progress, but it is not the immediate source of motive power. This source is genetic recombination, through the shuffling of genes and chromosomes which goes on during the sexual cycle. Since this process provides the immediate source of variability upon which selection exerts its primary action it can be compared to the engine of the automobile. Natural selection, which directs genetic variability toward adaptation to the environment, can be compared to the driver of the vehicle" (Stebbins, 1966, p. 3).

23. "*Mutation as an evolutionary force.* In the early days of genetics it was believed that evolutionary trends are directed by mutation, or, as Dobzhansky (1959) recently phrased this view, 'that evolution is due to occasional lucky mutants which happen to be useful rather than harmful.' In contrast, it is held by contemporary geneticists that mutation pressure as such is of small immediate evolutionary consequence in sexual organisms, in view of the relatively far greater contribution of recombination and gene flow to the production of new genotypes and of the overwhelming role of selection in determining the change in the genetic composition of populations from generation to generation" (Mayr, 1963, p. 101).

24. "It is most important to clear up first some misconceptions still held by a few, not familiar with modern genetics: (1) Evolution is not primarily a genetic event. Mutation merely supplies the gene pool with genetic variation; it is selection that induces evolutionary change" (Mayr, 1963, p. 613).

25. "Darwin's contention is fully supported by modern genetics. If one had to rely on mutation pressure as the only evolutionary factor, one would need such a high rate of mutation that it would result in an enormous production of 'hopeful monsters'. All available evidence is opposed to such an assumption . . . The real function of mutation is to replenish the gene pool and to provide material for recombination as a source of individual variability in populations" (Mayr, 1960, p. 355).

26. "Second, mutation neither directs evolution, as the early mutationists believed, nor even serves as the immediate source of variability on which selection may act. It is, rather, a reserve or potential source of variability which serves to replenish the gene pool as it becomes depleted through the action of selection ... The factual evidence in support of these postulates, drawn from a wide variety of animals and plants, is now so extensive and firmly based upon observation and experiment that we who are familiar with it cannot imagine the appearance of new facts

which will either overthrow any of them or seriously limit their validity" (Stebbins, 1959, p. 305).

27. "Infinitely variable natural populations are of such evolutionary plasticity that natural selection can mold them into almost any shape . . . The study of evolutionary rates is a branch of the science of evolution that has been undeservedly neglected. One solution, though frequently proposed, is almost certainly wrong: that rates of evolution are controlled by rates of mutation . . . What is important is that the genetic constitution of [a] population, as analyzed by population genetics, reveals a genetic, and hence evolutionary, plasticity that provides almost unlimited evolutionary potentialities. If there is a selective premium on perfection of the eye, as among nocturnal birds of prey (owls), there is enough variability to supply material for the achievement of this perfection" (Mayr, 1959a, pp. 7 to 9).

28. "Darwin assumed in the *Origin of Species* that the evolution of living organisms depended on the origin of new forms which varied from old forms by continuous differences in no constant or predictable direction. Crossed together the new and the old showed blending inheritance. To these variations direction was given by a process of natural selection which, like artificial selection, preserved some while it destroyed others. A direction, an adaptive direction, was thus given to variations after they arose. This view was intended by Darwin to supplant the alternative view that direction was given to variations before they arose" (Darlington, 1958, p. 231).

29. "But the objection [that natural selection cannot be the guiding agent in evolution because it produces nothing new] became invalid in the light of modern biological knowledge . . . We should clearly distinguish the two basic evolutionary processes: that of the origin of the raw materials from which evolutionary changes can be constructed, and that of building and perfecting the organic form and function. Evolution can be compared to a factory: any factory needs a supply of raw materials to work with, but when the materials are available they must be transformed into a finished product by means of some manufacturing process" (Dobzhansky, 1955a, p. 131).

30. "The process of mutation supplies the raw materials of evolution, but the tempo of evolution is determined at the populational level by natural selection in conjunction with the ecology and the reproductive biology of the group of organisms" (Dobzhansky, 1955b, p. 282).

31. "There remains, nevertheless, the possibility of rare exceptions, cases in which mutation rate is so large and other factors, especially selection, so weak that mutation might control the direction of evolution in

spite of the evidence that this is not usually true. Wright (1940) has shown that the situation could arise in small populations. The systematic effect of mutation in such cases would be degenerative but there is some, although extremely small, chance that it might lead to progressive change" (Simpson, 1953, p. 114).

32. "Progressive rectilinearity in evolution may occur rather in spite of favored mutational directions than because of them, although there must be some mutation in the evolutionary direction, and this may be increased by natural selection . . . A real momentum effect could result from mutation predominantly in one direction, but only under unusual conditions; this has not been an important factor in evolution, and it is not definitely known that it has ever occurred" (Simpson, 1984, p. 178) (1944).

33. "For no rate of hereditary change hitherto observed in nature would have any evolutionary effect in the teeth of even the slightest degree of adverse selection. Either mutation-rates many times higher than any as yet detected must be sometimes operative, or else the observed results [apparent evolutionary trends] can be far better accounted for by selection" (Huxley, 1942, p. 56).

34. "In general, mutation is a necessary but not sufficient cause of evolution. Without mutation there would be no gene differences for natural selection to act upon. But the actual evolutionary trend would seem usually to be determined by selection, for the following reason.

A simple calculation shows that recurrent mutation (except of a gene so unstable as to be classifiable as multimutating) can not overcome selection of quite moderate intensity. Consider two phenotypes whose relative fitnesses are in the ratios 1 and $1 - k$, that is to say, that on the average one leaves $(1 - k)$ times as many progeny as the other [i.e., in Haldane's notation k is a selection coefficient]. Then, if p is the probability that a gene mutates to a less fit allelomorph in the course of a life cycle, it has been shown (Haldane, 1932) that when k is small, the mutant gene will only spread through a small fraction of the population unless p is about as large as k or larger. This is true whether the gene is dominant or recessive" (Haldane, 1933, p. 6).

35. [In this section, Haldane poses a question, then derives the mutation-selection balance— p/k in his notation and otherwise μ/s— , then concludes] "Under what conditions can mutation overcome selection? This is quite a simple problem. . . . Hence, unless k is so small as to be of the same order as p, the new type will not spread to any significant extent. . . . Thus until it has been shown that anywhere in nature conditions

produce a mutation rate considerably higher than this, we cannot regard mutation as a cause likely by itself to cause large changes in a species. But I am not suggesting for a moment that selection alone can have any effect at all. The material on which selection acts must be supplied by mutation" (Haldane, 1932, p. 109 to 110).

36. "For mutations to dominate the trend of evolution it is thus necessary to postulate mutation rates immensely greater than those which are known to occur, and of an order of magnitude which, in general, would be incompatible with particulate inheritance . . .

    For any evolutionary tendency which is supposed to act by favouring mutations in one direction rather than another, and a number of such mechanisms have from time to time been imagined, will lose its force many thousand-fold, when the particulate theory of inheritance, in any form, is accepted; wheras the directing power of Natural Selection, depending as it does on the amount of heritable variance maintained, is totally uninfluenced by any such change . . .

    The whole group of theories which ascribe to hypothetical physiological mechanisms, controlling the occurrence of mutations, a power of directing the course of evolution, must be set aside, once the blending theory of inheritance is abandoned. The sole surviving theory is that of Natural Selection" (Fisher, 1930b, Ch. 1).

37. "Mutation therefore determines the course of evolution as regards factors of negligible advantage or disadvantage to the species. It can only lead to results of importance when its frequency becomes large" (Haldane, 1927).

38. "Throughout this chapter and elsewhere I have spoken of selection as the paramount power, yet its action absolutely depends on what we in our ignorance call spontaneous or accidental variability. Let an architect be compelled to build an edifice with uncut stones, fallen from a precipice. The shape of each fragment may be called accidental; yet the shape of each has been determined by the force of gravity, the nature of the rock, and the slope of the precipice—events and circumstances, all of which depend on natural laws; but there is no relation between these laws and the purpose for which each fragment is used by the builder. In the same manner the variations of each creature are determined by fixed and immutable laws; but these bear no relation to the living structure which is slowly built up through the power of selection, whether this be natural or artificial selection" (Darwin, 1868, end of Ch. 21).

Bibliography

Aardema, Matthew, and Peter Andolfatto. "Phylogenetic Incongruence and the Evolutionary Origins of Cardenolide-Resistant Forms of Na(+) ,K(+) -ATPase in *Danaus* Butterflies." *Evolution* 70, no. 8 (2016): 1913–21.

Aardema, Matthew, Ying Zhen, and Peter Andolfatto. "The Evolution of Cardenolide-Resistant Forms of Na(+),K(+)-ATPase in *Danainae* Butterflies." *Molecular Ecology* 21, no. 2 (2012): 340–9.

Alberch, Pere. "Ontogenesis and Morphological Diversification." *American Zoologist* 20, no. 4 (1980): 653.

Alberch, Pere. "From Genes to Phenotype: Dynamical Systems and Evolvability." *Genetica* 84, no. 1 (1991): 5–11.

Alberch, Pere, and Emily Gale. "A Developmental Analysis of an Evolutionary Trend: Digital Reduction in Amphibians." *Evolution* 39, no. 1 (1985): 8–23.

Alexandrov, Ludmil B., Serena Nik-Zainal, David C. Wedge, Samuel A. Aparicio, Sam Behjati, Andrew V. Biankin, Graham R. Bignell, Niccolò Bolli, Ake Borg, Anne-Lise Børresen-Dale, Sandrine Boyault, Birgit Burkhardt, Adam P. Butler, Carlos Caldas, Helen R. Davies, Christine Desmedt, Roland Eils, Jórunn Erla Eyfjörd, John A. Foekens, Mel Greaves, Fumie Hosoda, Barbara Hutter, Tomislav Ilicic, Sandrine Imbeaud, Marcin Imielinski, Natalie Jäger, David T. W. Jones, David Jones, Stian Knappskog, Marcel Kool, Sunil R. Lakhani, Carlos López-Otín, Sancha Martin, Nikhil C. Munshi, Hiromi Nakamura, Paul A. Northcott, Marcin Pajic, Elli Papaemmanuil, Angelo Paradiso, John V. Pearson, Xose S. Puente, Keiran Raine, Manasa Ramakrishna, Andrea L. Richardson, Julia Richter, Philip Rosenstiel, Matthias Schlesner, Ton N. Schumacher, Paul N. Span, Jon W. Teague, Yasushi Totoki, Andrew N. Tutt, Rafael Valdés-Mas, Marit M van Buuren, Laura van't Veer, Anne Vincent-Salomon, Nicola Waddell, Lucy R. Yates, Australian Pancreatic Cancer Genome, Initiative, ICGC Breast Cancer Consortium, ICGC MMML-Seq Consortium, ICGC PedBrain, Jessica Zucman-Rossi, P. Andrew Futreal, Ultan McDermott, Peter Lichter, Matthew Meyerson, Sean M. Grimmond, Reiner Siebert, Elías Campo, Tatsuhiro Shibata, Stefan M. Pfister, Peter J. Campbell, and Michael R. Stratton. "Signatures of Mutational Processes in Human Cancer." *Nature* 500, no. 7463 (2013): 415–21.

Allen, Emily Graves, Sallie B. Freeman, Charlotte Druschel, Charlotte A. Hobbs, Leslie A. O'Leary, Paul A. Romitti, Marjorie H. Royle, Claudine P. Torfs, and Stephanie L. Sherman. "Maternal Age and Risk for Trisomy 21 Assessed by the Origin of Chromosome Nondisjunction: A Report from the Atlanta and National Down Syndrome Projects." *Human Genetics* 125, no. 1 (2009): 41–52.

Altenberg, Lee. "Evolution of the Genotype-Phenotype Map." In *Evolution and Biocomputation: Computational Models of Evolution*, edited by Wolfgang Banzhaf and Frank H. Eeckman, 205–59. Vol. 899, Springer-Verlag Lecture Notes in Computer Sciences. Berlin: Springer-Verlag, 1995.

Altshuler, Douglas L., and Robert Dudley. "The Physiology and Biomechanics of Avian Flight at High Altitude." *Integrative and Comparative Biology* 46, no. 1 (2006): 62–71.

Aminpour, Maral, Carlo Montemagno, and Jack A. Tuszynski. "An Overview of Molecular Modeling for Drug Discovery with Specific Illustrative Examples of Applications." *Molecules* 24, no. 9 (2019): 1693.

Amos, William. "Even Small SNP Clusters are Non-Randomly Distributed: Is This Evidence of Mutational Non-Independence?" *Proceedings of the Royal Society B: Biological Sciences* 277, no. 1686 (2010): 1443–9.

Amundson, Ronald. "Adaptation, Development, and the Quest for Common Ground." In *Adaptation and Optimality*, edited by Steven Hecht Orzack and Elliott Sober, 303–34. New York: Cambridge University Press, 2001.

Amundson, Ronald. *The Changing Role of the Embryo in Evolution.* Cambridge Studies in Philosophy and Biology. Cambridge, UK: Cambridge University Press, 2005.

Anfinsen, Christian. *The Molecular Basis of Evolution.* New York: Wiley and Sons, 1959.

Antonovics, Janis, and Peter H. van Tienderen. "Ontoecogenophyloconstraints? The Chaos of Constraint Terminology." *Trends in Ecology & Evolution* 6, no. 5 (1991): 166–8.

Arthur, Wallace. "Developmental Drive: An Important Determinant of the Direction of Phenotypic Evolution." *Evolution & Development* 3, no. 4 (2001): 271–8.

Arthur, Wallace. *Biased Embryos and Evolution*. Cambridge, UK: Cambridge University Press, 2004.

Arthur, Wallace, and Malcolm Farrow. "The Pattern of Variation in Centipede Segment Number as an Example of Developmental Constraint in Evolution." *Journal of Theoretical Biology* 200 no. 2 (1999): 183–91.

Asenjo, Ana B., Jeanne Rim, and Daniel D. Oprian. "Molecular Determinants of Human Red/Green Color Discrimination." *Neuron* 12, no. 5 (1994): 1131–8.

Ayala, Francisco J. "Teleological Explanations in Evolutionary Biology." *Philosophy of Science* 37, no. 1 (1970): 1–15.

Ayala, Francisco J., and Walter M. Fitch. "Genetics and the Origin of Species: An Introduction." *Proceedings of the National Academy of Sciences of the United States of America* 94, no. 15 (1997): 7691–7.

Azevedo, Ricardo B. R., Peter D. Keightley, Camilla Laurén-Määttä, Larissa L. Vassilieva, Michael Lynch, and Armand M Leroi. Spontaneous mutational variation for body size in *Caenorhabditis elegans*. *Genetics* 162 (2002): 755–765.

Balasubramanian, Bhavani, Wendy K. Pogozelski, and Thomas D. Tullius. "DNA Strand Breaking by the Hydroxyl Radical is Governed by the Accessible Surface Areas of the Hydrogen Atoms of the DNA Backbone." *Proceedings of the National Academy of Sciences of the United States of America* 95, no. 17 (1998): 9738–43.

Baldi, Christopher, Jeffrey Viviano, and Ronald E. Ellis. "A Bias Caused by Ectopic Development Produces Sexually Dimorphic Sperm in Nematodes." *Current Biology* 21, no. 16 (2011): 1416–20.

Ball, Edward V., Peter D. Stenson, Shaun S. Abeysinghe, Michael Krawczak, David N. Cooper, and Nadia A. Chuzhanova. "Microdeletions and Microinsertions Causing Human Genetic Disease: Common Mechanisms of Mutagenesis and the Role of Local DNA Sequence Complexity." *Human Mutation* 26, no. 3 (2005): 205–13.

Bankhead, Troy, and George Chaconas. "The Role of VlsE Antigenic Variation in the Lyme Disease Spirochete: Persistence Through a Mechanism that Differs from Other Pathogens." *Molecular Microbiology* 65, no. 6 (2007): 1547–58.

Bar-On, Yinon M., Rob Phillips, and Ron Milo. "The Biomass Distribution on Earth." *Proceedings of the National Academy of Sciences of the United States of America* 115, no. 25 (2018): 6506–11.

Barrangou, Rodolphe. "Diversity of CRISPR-Cas Immune Systems and Molecular Machines." *Genome Biol* 16 (2015): 247.

Barrangou, Rodolphe, Christophe Fremaux, Hélène Deveau, Melissa Richards, Patrick Boyaval, Sylvain Moineau, Dennis A. Romero, and Philippe Horvath. "CRISPR Provides Acquired Resistance Against Viruses in Prokaryotes." *Science* 315, no. 5819 (2007): 1709–12.

Barry, J. D., James P. Hall, and Lindsey Plenderleith. "Genome Hyperevolution and the Success of a Parasite." *Annals of the New York Academy of Sciences* 1267 (2012): 11–7.

Barton, Nick, and Linda Partridge. "Limits to Natural Selection." *Bioessays* 22, no. 12 (2000): 1075–84.

Bassing, Craig H., Wojciech Swat, and Frederick W. Alt. "The Mechanism and Regulation of Chromosomal V(D)J Recombination." *Cell* 109, Suppl (2002): S45–55.

Bateson, William. *Materials for the Study of Variation, Treated with Especial Regard to Discontinuity in the Origin of Species*. London: Macmillan, 1894.

Bateson, William. *Mendel's Principles of Heredity: A Defense*. Cambridge: Cambridge University Press, 1902.

Bateson, William. "Heredity and Variation in Modern Light." In *Darwin and Modern Science: Essays in Commemoration of the Centenary of the Birth of Charles Darwin and of the Fiftieth Anniversary of the Publication of the Origin of Species*, edited by A. Seward, 85–101. London: Cambridge University Press, 1909a.

Bateson, William. *Mendel's Principles of Heredity*. New York: Putnam's Sons, 1909b.

Bateson, William. "Inaugural address." *Nature* 93 (1914): 635–42.

Bateson, William, and Saunders, Edith. "Experimental Studies in the Physiology of Heredity." In *Reports to the Evolution Committee of the Royal Society*. London: Harrison and Sons, 1902.

Baucom, Regina S. "Evolutionary and Ecological Insights from Herbicide-Resistant Weeds: What Have We Learned about Plant Adaptation, and What is Left to Uncover?" *New Phytologist* 223 (2019): 68–82.

Beatty, John. "Reconsidering the Importance of Chance Variation." In *Evolution: The Extended Synthesis*, edited by Massimo Pigliucci and Gerd B. Müller, 21–44. Cambridge, Mass.: MIT Press, 2010.

Beatty, John. "Darwin's Cyclopean Architect." In *Evolutionary Biology: Conceptual, Ethical and Religious Issues*, edited by R. Paul Thompson and Denis M. Walsh, 175–92. Cambridge: Cambridge University Press, 2014.

Beatty, John. "The Creativity of Natural Selection? Part I: Darwin, Darwinism, and the Mutationists." *Journal of the History of Biology* 49, no. 4 (2016): 659–84.

Beckie, Hugh J., Suzanne I. Warwick, and Connie A. Sauder. "Basis for Herbicide Resistance in Canadian Populations of Wild Oat (*Avena fatua*)." *Weed Science* 60, no. 1 (2012): 10–18.

Bégin, Mattieu, and Derek A. Roff. "The Constancy of the G Matrix through Species Divergence and the Effects of Quantitative Genetic Constraints on Phenotypic Evolution: A Case Study in Crickets." *Evolution* 57, no. 5 (2003): 1107–20.

Beletskii, Anton, and Ashok S. Bhagwat. "Transcription-Induced Mutations: Increase in C to T Mutations in The Nontranscribed Strand During Transcription in *Escherichia coli.*" *Proceedings of the National Academy of Sciences of the United States of America* 93, no. 24 (1996): 13919–24.

Berg, Leo S. *Nomogenesis*. Cambridge, Mass.: MIT Press, 1969.

Bidmos, Fadil A., and Christopher D. Bayliss. "Genomic and Global Approaches to Unravelling how Hypermutable Sequences Influence Bacterial Pathogenesis." *Pathogens* 3, no. 1 (2014): 164–84.

Birch, Jonathan, and Samir Okasha. "Kin Selection and Its Critics." *BioScience* 65, no. 1 (2014): 22–32.

Bissler, John J. "DNA Inverted Repeats and Human Disease." *Frontiers in Bioscience* 3 (1998): d408–18.

Blake, Richard D., Samuel T. Hess, and Janice Nicholson-Tuell. "The Influence of Nearest Neighbors on the Rate and Pattern of Spontaneous Point Mutations." *Journal of Molecular Evolution* 34, no. 3 (1992): 189–200.

Blanquart, François, Guillaume Achaz, Thomas Bataillon, and Olivier Tenaillon. "Properties of Selected Mutations and Genotypic Landscapes under Fisher's Geometric Model." *Evolution* 68, no. 12 (2014): 3537–54.

Bloom, Jesse D. "An Experimentally Determined Evolutionary Model Dramatically Improves Phylogenetic Fit." *Molecular Biology and Evolution* 31, no. 8 (2014): 1956–78.

Blount, Zachary D., Christina Z. Borland, and Richard E. Lenski. "Historical Contingency and the Evolution of a Key Innovation in an Experimental Population of *Escherichia coli.*" *Proceedings of the National Academy of Sciences of the United States of America* 105, no. 23 (2008): 7899–906.

Boland, Michael J., and Judith K. Christman. "Characterization of Dnmt3b:Thymine-DNA Glycosylase Interaction and Stimulation of Thymine Glycosylase-Mediated Repair by DNA Methyltransferase(s) and RNA." *Journal of Molecular Biology* 379, no. 3 (2008): 492–504.

Bolstad, Geir H., Thomas F. Hansen, Christophe Pélabon, Mohsen Falahati-Anbaran, Rocío Pérez-Barrales, and W. Scott Armbruster. "Genetic Constraints Predict Evolutionary Divergence in *Dalechampia* Blossoms." *Philosophical Transactions of the Royal Society of London B: Biological Sciences* 369, no. 1649 (2014): 20130255.

Borst, Piet, and David R. Greaves. "Programmed Gene Rearrangements Altering Gene Expression." *Science* 235, no. 4789 (1987): 658–67.

Bowler, Peter. *The Non-Darwinian Revolution: Reinterpreting a Historical Myth*. Baltimore: Johns Hopkins University Press, 1988.

Braendle, Christian, Charles F. Baer, and Marie-Anne Félix. "Bias and Evolution of the Mutationally Accessible Phenotypic Space in a Developmental System." *PLoS Genetics* 6, no. 3 (2010): e1000 877.

Brent, Richard P. "Some Long-Period Random Number Generators Using Shifts and Xors." *ANZIAM Journal* 48 (2007): C188–C202.

Brigandt, Ingo. "From Developmental Constraint to Evolvability: How Concepts Figure in Explanation and Disciplinary Identity." In *Conceptual Change in Biology: Scientific and Philosophical Perspectives on Evolution and Development*, edited by Alan C. Love, 305–25. Dordrecht: Springer Netherlands, 2015.

Brisson, Dustin. "The Directed Mutation Controversy in an Evolutionary Context." *Critical Reviews in Microbiology* 29, no. 1 (2003): 25–35.

Brookfield, John F. Y. "Where Does Animal Diversity Come From?" *Current Biology* 15, no. 22 (2005): R908–10.

Brooks, Daniel R. "The Mastodon in the Room: How Darwinian is NeoDarwinism?" *Studies in History and Philosophy of Biological and Biomedical Sciences* 42, no. 1 (2011): 82–8.

Brown, Rachael L. "What Evolvability Really Is." *British Journal for the Philosophy of Science* 65 (2014): 549–72.

Bull, James J., Marty R. Badgett, Holly A. Wichman, John P. Huelsenbeck, David M. Hillis, Ashu Gulati, Clark Ho, and Ian J. Molineux. "Exceptional Convergent Evolution in a Virus." *Genetics* 147, no. 4 (1997): 1497–507.

Bull, James J. and Sarah P. Otto. "The First Steps in Adaptive Evolution." *Nature Genetics* 37, no. 4 (2005): 342–3.

Burgess, Rebecca C., Tom Misteli, and Phillip Oberdoerffer. "DNA Damage, Chromatin, and Transcription: The Trinity of Aging." *Current Opinion in Cell Biology* 24, no. 6 (2012):724–30.

Cairns, John, Julie Overbaugh, and Shelly Miller. "The Origin of Mutants." *Nature* 335, no. 6186 (1988): 142–5.

Campbell, Catarina D. and Evan E. Eichler. "Properties and Rates of Germline Mutations in Humans." *Trends in Genetics* 29, no. 10 (2013): 575–84.

Cannataro, Vincent L., Stephen G. Gaffney, and Jeffrey P. Townsend. "Effect Sizes of Somatic Mutations in Cancer." *Journal of the National Cancer Institute* 110, no. 11 (2018): 1171–7.

Caporale, Lynn. *Darwin in the Genome: Molecular Strategies in Biological Evolution*. New York: McGraw-Hill, 2003.

Carroll, Sean B. "Evo-Devo and an Expanding Evolutionary Synthesis: A Genetic Theory of Morphological Evolution." *Cell* 134, no. 1 (2008): 25–36.

Carter, Ashley J. R., Joachim Hermisson, and Thomas F. Hansen. "The Role of Epistatic Gene Interactions in the

Response to Selection and the Evolution of Evolvability." *Theor Popul Biol* 68, no. 3 (2005): 179–96.

Casas, Veronica, and Forest Rohwer. "Phage Metagenomics." *Methods in Enzymology* 421 (2007): 259–68.

Cavalli-Sforza, Luigi Luca, and Joshua Lederberg. "Isolation of Pre-Adaptive Mutants in Bacteria by Sib Selection." *Genetics* 41, no. 3 (1956): 367–81.

Cerdeño-Tárraga, Ana M., Sheila Patrick, Lisa C. Crossman, Garry Blakely, Val Abratt, Nicola Lennard, Ian Poxton, Brian Duerden, Barbara Harris, Mike A. Quail, Andrew Barron, Louise Clark, Craig Corton, Jonathan Doggett, Matthew T. G. Holden, Natasha Larke, Alexandra Line, Angela Lord, Halina Norbertczak, Doug Ormond, Claire Price, Ester Rabbinowitsch, John Woodward, Bart Barrell, and Julian Parkhill. "Extensive DNA Inversions in the *B. fragilis* Genome Control Variable Gene Expression." *Science* 307, no. 5714 (2005): 1463–5.

Chan, Kin, and Dmitry A. Gordenin. "Clusters of Multiple Mutations: Incidence and Molecular Mechanisms." *Annual Review of Genetics* 49 (2015): 243–67.

Charlesworth, Brian. "On the Origins of Novelty and Variation." *Science* 310, no. 5754 (2005): 1619–20.

Charlesworth, Brian, and Nicholas H. Barton. "Recombination Load Associated with Selection for Increased Recombination." *Genet Res* 67, no. 1 (1996): 27–41.

Charlesworth, Brian, and Deborah Charlesworth. "Darwin and Genetics." *Genetics* 183, no. 3 (2009): 757–66.

Charlesworth, Deborah, Nicholas H. Barton, and Brian Charlesworth. "The Sources of Adaptive Variation." *Proceedings of the Royal Society B: Biological Sciences* 284, no. 1855 (2017): 20162864.

Chetverikov, Sergei Sergeevich. *On Certain Aspects of the Evolutionary Process from the Standpoint of Modern Genetics*. Placitas, New Mexico: Genetics Heritage Press, 1997.

Cheverud, James M. "Quantitative Genetics and Developmental Constraints on Evolution." *Journal of Theoretical Biology* 110 (1984): 155–71.

Cheverud, James M. "A Comparison of Genotypic and Phenotypic Correlations." *Evolution* 42, no. 5 (1988): 958–68.

Chevin, Luis-Miguel, Guillaume Martin, and Thomas Lenormand. "Fisher's Model and the Genomics of Adaptation: Restricted Pleiotropy, Heterogenous Mutation, and Parallel Evolution." *Evolution* 64, no. 11 (2010): 3213–31.

Chopra-Dewasthaly, Rohini, Joachim Spergser, Martina Zimmermann, Christine Citti, Wolfgang Jechlinger, and Renate Rosengarten. "Vpma Phase Variation is Important for Survival and Persistence of *Mycoplasma agalactiae* in the Immunocompetent Host." *PLoS Pathogens* 13, no. 9 (2017): e1006 656.

Clegg, Steven, Janet Wilson, and Jeremiah Johnson. (2011). "More than One Way to Control Hair Growth: Regulatory Mechanisms in Enterobacteria That Affect Fimbriae Assembled by the Chaperone/Usher Pathway." *Journal of Bacteriology* 193, no. 9 (2011): 2081–8.

Colby, Chris. (1996). "Introduction to Evolutionary Biology. v.2," The TalkOrigins Archive, updated January 7, 1996, http://www.talkorigins.org/faqs/faq-intro-to-biology.html.

Colegrave, Nick, and Sinead Collins. "Experimental Evolution: Experimental Evolution and Evolvability." *Heredity (Edinburgh)* 100, no. 5 (2008): 464–70.

Conant, Gavin C., and Andreas Wagner. "The Rarity of Gene Shuffling in Conserved Genes." *Genome Biology* 6, no. 6 (2005): R50.

Conrad, Donald F., Christine Bird, Ben Blackburne, Sarah Lindsay, Lira Mamanova, Charles Lee, Daniel J Turner, and Matthew E Hurles. "Mutation Spectrum Revealed by Breakpoint Sequencing of Human Germline CNVs." *Nature Genetics* 42, no. 5 (2010): 385–91.

Conrad, Michael. "Evolution of the Adaptive Landscape." In *Theoretical Approaches to Complex Systems*, edited by R. Heim and G. Palm, 147–69. Berlin: Springer, 1978.

Conrad, Michael. "The Geometry of Evolution." *Biosystems* 24, no. 1 (1990): 61–81.

Conrad, Michael, and Michael V. Volkenstein. "Replaceability of Amino Acids and the Self-Facilitation of Evolution." *Journal of Theoretical Biology* 92, no. 3 (1981): 293–9.

Coombes, David, James W.B. Moir, Anthony M. Poole, Tim F. Cooper, and Renwick C.J. Dobson. "The Fitness Challenge of Studying Molecular Adaptation." *Biochemical Society Transactions* 47, no. 5 (2019): 1533–42.

Cooper, David N., Albino Bacolla, Claude Férec, Karen M. Vasquez, Hildegard Kehrer-Sawatzki, and Jian-Min Chen. "On the Sequence-Directed Nature of Human Gene Mutation: The Role of Genomic Architecture and the Local DNA Sequence Environment in Mediating Gene Mutations Underlying Human Inherited Disease." *Human Mutation* 32, no. 10 (2011): 1075–99.

Cordella, Daniela, Marina Muzza, Luisella Alberti, Paolo Colombo, Pietro Travaglini, Paolo Beck-Peccoz, Laura Fugazzola, and Luca Persani. "An In-Frame Complex Germline Mutation in the Juxtamembrane Intracellular Domain Causing RET Activation in Familial Medullary Thyroid Carcinoma." *Endocrine-Related Cancer* 13, no. 3 (2006): 945–53.

Couce, A., Alexandro Rodríguez-Rojas, and Jesús Blázquez. "Bypass of genetic constraints during mutator evolution to antibiotic resistance." *Proceedings of the Royal Society of London B: Biological Sciences* 282, no. 1804 (2015): 20142698.

Coutte, Loïc., Douglas J Botkin, Lihui Gao, and Steven J Norris. "Detailed Analysis of Sequence Changes Occurring During vlsE Antigenic Variation in the Mouse

Model of *Borrelia burgdorferi* Infection." *PLoS Pathogens* 5, no. 2 (2009): e1000 293.

Cowperthwaite, Matthew C., Evan P. Economo, William R. Harcombe, Eric L. Miller, and Lauren Ancel Meyers. "The Ascent of the Abundant: How Mutational Networks Constrain Evolution." *PLoS Computational Biology* 4, no. 7 (2008): e1000 110.

Coyne, Jerry A. *Why Evolution is True*. London: Penguin, 2010.

Crick, Francis H. H. "The Origin of the Genetic Code." *Journal of Molecular Biology* 38 (1968): 367–79.

Crill, Wayne D., Holly A. Wichman, and James J. Bull. "Evolutionary Reversals During Viral Adaptation to Alternating Hosts." *Genetics* 154, no. 1 (2000): 27–37.

Cromer, Deborah, Timothy E. Schlub, Redmond P. Smyth, Andrew J. Grimm, Abha Chopra, Simon Mallal, Miles P. Davenport, and Johnson Mak. "HIV-1 Mutation and Recombination Rates Are Different in Macrophages and T-cells." *Viruses* 8, no. 4 (2016): 118.

Cronin, H. *The Ant and the Peacock*. Cambridge, UK: Cambridge University Press, 1991.

Crow, James. "Genetic Loads and the Cost of Natural Selection." In *Mathematical Topics in Population Genetics*, edited by Ken-ichi Kojima, 128–77. New York: Springer, 1970.

Crow, James. and Kimura, M. *An Introduction to Population Genetics Theory*. Minneapolis: Burgess, 1970.

Crow, James. F. "Mid-Century Controversies in Population Genetics." *Annual Review of Genetics* 42 (2008): 1–16.

Crow, James F. "On Epistasis: Why it is Unimportant in Polygenic Directional Selection." *Philosophical Transactions of the Royal Society B: Biological Sciences* 365, no. 1544 (2010): 1241–4.

Croyle, Michelle L., Alison L. Woo, and Jerry B. Lingrel. "Extensive Random Mutagenesis Analysis of the Na+/K+-ATPase Alpha Subunit Identifies Known and Previously Unidentified Amino Acid Residues that Alter Ouabain Sensitivity–Implications for Ouabain Binding." *European Journal of Biochemistry* 248, no. 2 (1997): 488–95.

Cuénot, Lucien. "Recent Views of L. Cuénot on the Origin of Species by Mutation." *Science* 30, no. 778 (1909): 768–769.

Cunningham, Brian C., and James A. Wells. "High-Resolution Epitope Mapping of hGH-Receptor Interactions by Alanine-Scanning Mutagenesis." *Science* 244, no. 4908 (1989): 1081–5.

Czech, Benjamin, and Gregory J. Hannon. "One Loop to Rule Them All: The Ping-Pong Cycle and piRNA-Guided Silencing." *Trends in Biochemical Sciences* 41, no. 4 (2016): 324–337.

Dai, Qiyuan, Blanca I. Restrepo, Stephen F. Porcella, Sandra J. Raffel, Tom G. Schwan, and Alan G. Barbour.

"Antigenic Variation by *Borrelia hermsii* Occurs Through Recombination Between Extragenic Repetitive Elements on Linear Plasmids." *Molecular Microbiology* 60, no. 6 (2006): 1329–43.

Darlington, Cyril D. *The Evolution of Genetic Systems*, 2nd ed. New York: Basic Books, 1958.

Darwin, Charles. *Variation of Animals and Plants under Domestication*, vol. II. London: Murray, 1868.

Darwin, Charles. *On the Origin of Species*, 6th ed. London: John Murray, 1872.

Darwin, Charles. "Inherited Instinct." *Nature* 7, no. 172 (1873): 281–281.

Darwin, Charles. *Variation of Animals and Plants under Domestication*, vol. I, 2nd ed. New York: Appleton and Co., 1883.

Datta, Abhijit, and Sue Jinks-Robertson. "Association of Increased Spontaneous Mutation Rates with High Levels of Transcription in Yeast." *Science* 268, no. 5217 (1995): 1616–9.

Davenport, Charles B. (1909). "Mutation." In *Fifty Years of Darwinism: Modern Aspects of Evolution*, 160–81. Henry Holt and Company, New York.

Dawkins, Richard. *The Selfish Gene*. New York: Oxford University Press, 1976.

Dawkins, Richard. *The Blind Watchmaker*. New York: W.W. Norton and Company, 1987.

Dawkins, Richard. "The Evolution of Evolvability." In *Artificial Life*, vol. VI, edited by Christopher G. Langton, 201–20. Boston: Addison-Wesley, 1988.

Dawkins, Richard. *Climbing Mount Improbable*. New York: W.W. Norton and Company, 1996.

Dawkins, Richard. "Review: *The Edge of Evolution*." *The New York Times*, July 1, 2007.

Deitsch, Kirk W., Sheila A. Lukehart, and James R. Stringer. "Common Strategies for Antigenic Variation by Bacterial, Fungal and Protozoan Pathogens." *Nature Reviews Microbiology* 7, no. 7 (2009): 493–503.

del Re, Giuseppe. "Chance, Cause and the State-Space Approach." In *Probability in the Sciences*, edited by Evandro Agazzi, 89–101. Dordrect: Kluwer Academic Publishers, 1988.

Delage, Yves, and Marie Goldsmith. *The Theories of Evolution*. New York: Huebsch, 1913.

Desai, Michael M. and Daniel S. Fisher. "Beneficial Mutation Selection Balance and the Effect of Linkage on Positive Selection." *Genetics* 176, no. 3 (2007): 1759–98.

Di Noia, Javier Marcelo, and Michael S. Neuberger. "Molecular Mechanisms of Antibody Somatic Hypermutation." *Annual Review of Biochemistry* 76 (2007): 1–22.

Dickinson, W. Joe, and Jon Seger. "Cause and Effect in Evolution." *Nature*, 399, no. 6731 (1999): 30.

Dietrich, Michael R. "Paradox and Persuasion: Negotiating the Place of Molecular Evolution within

Evolutionary Biology." *Journal of the History of Biology* 31, no. 1 (1998): 85–111.

Dobzhansky, Theodosius. *Genetics and the Origin of Species*. New York: Columbia University Press, 1937.

Dobzhansky, Theodosius. *Evolution, Genetics and Man*. New York: Wiley and Sons, Inc., 1955a.

Dobzhansky, Theodosius. *Genetics and the Origin of Species*. New York: Wiley and Sons, Inc., 1955b.

Dobzhansky, Theodosius. *Genetics of the Evolutionary Process*. New York: Columbia University Press, 1970.

Dobzhansky, Theodosius. "Chance and Creativity in Evolution." In *Studies in the Philosophy of Biology: Reduction and Related Problems*, edited by Francisco J. Ayala and Theodosius Dobzhansky, 307–38. Berkeley: University of California Press, 1974.

Dobzhansky, Theodosius, Francisco J. Ayala, G. Ledyard Stebbins, and James W. Valentine. *Evolution*. San Francisco: W.H. Freeman and Co., 1977.

Domingo-Calap, Pilar, José M. Cuevas, Rafael Sanjuán. "The Fitness Effects of Random Mutations in Single-Stranded DNA and RNA Bacteriophages." *PLoS Genetics* 5, no. 11 (2009): e1000 742.

Doulatov, Sergei, Asher Hodes, Lixin Dai, Neeraj Mandhana, Minghsun Liu, Rajendar Deora, Robert W. Simons, Steven Zimmerly, and Jeff F. Miller. "Tropism Switching in *Bordetella* Bacteriophage Defines a Family of Diversity-Generating Retroelements." *Nature* 431, no. 7007 (2004): 476–81.

Drake, John W., Anna Bebenek, Grace E. Kissling, and Shyamal Peddada. "Clusters of Mutations from Transient Hypermutability." *Proceedings of the National Academy of Sciences of the United States of America* 102, no. 36 (2005): 12 849–54.

Drake, John W., Brian Charlesworth, Deborah Charlesworth, and James F. Crow. "Rates of Spontaneous Mutation." *Genetics* 148, no. 4 (1998): 1667–86.

Drobetsky, Elliot A., Andrew J. Grosovsky, and Barry W. Glickman. "The Specificity of UV-Induced Mutations at an Endogenous Locus in Mammalian Cells." *Proceedings of the National Academy of Sciences of the United States of America* 84, no. 24 (1987): 9103–7.

Duret, Laurent. (2008). "Neutral Theory: The Null Hypothesis of Molecular Evolution." *Nature Education* 1, no. 1 (2008): 218.

Duret, Laurent, and Peter F. Arndt. "The Impact of Recombination on Nucleotide Substitutions in the Human Genome." *PLoS Genetics* 4, no. 5 (2008): e1000 071.

Dutra, Bethany E. and Lovett, Susan T. "Cis and Trans-Acting Effects on a Mutational Hotspot Involving a Replication Template Switch." *Journal of Molecular Biology* 356, no. 2 (2006): 300–11.

Dworkin, Joel, and Martin J. Blaser. "Generation of *Campylobacter fetus* Slayer Protein Diversity Utilizes a Single Promoter on an Invertible DNA Segment." *Molecular Microbiology* 19, no. 6 (1996): 1241–53.

Eble, Gunther J. "On the Dual Nature of Chance in Evolutionary Biology and Paleobiology." *Paleobiology* 25, no. 1 (1999): 75–87.

Eck, Richard V., and Margaret Dayhoff. *Atlas of Protein Sequence and Structure*. Silver Spring, MD: National Biomedical Research Foundation, 1966.

Edwards, Anthony W. F. *Foundations of Mathematical Genetics*. New York: Cambridge University Press, 1977.

Eimer, Theodor. *On Orthogenesis; and The Impotence of Natural Selection in Species-Formation*, vol. 29. Chicago: Open Court Publishing Co., 1898.

Elard, Loïc, Ana M. Comes, and Jean-Francois Humbert. "Sequences of beta-tubulin cDNA from benzimidazole-susceptible and -resistant strains of *Teladorsagia circumcincta*, a nematode parasite of small ruminants." *Molecular Biochemical Parasitology* 79, no. 2 (1996): 249–53.

Eldredge, Niles. *The Triumph of Evolution and the Failure of Creationism*. New York: W H Freeman and Co., 2001.

Ellis, Nathan, and Jonathan Gallant. "An Estimate of the Global Error Frequency in Translation." *Molecular Gen Genetics* 188, no. 2 (1982): 169–72.

Elowitz, Michael B., Michael G. Surette, Pierre-Etienne Wolf, Jeffry B. Stock, and Stanislas Leibler. "Protein Mobility in the Cytoplasm of *Escherichia coli*." *Journal of Bacteriology* 181, no. 1 (1999): 197–203.

Emlen, Douglas J. (2000). "Integrating Development with Evolution: A Case Study with Beetle Horns." *BioScience* 50, no. 5 (2000): 403.

Eshel, Ilan, and Marcus W. Feldman. "Optimality and Evolutionary Stability under Short-Term and Long-Term Selection." In *Adaptationism and Optimality*, edited by Steven Hecht Orzack and Elliott Sober, 161–90. Cambridge, UK: Cambridge University Press, 2001.

Eyre-Walker, Adam, and Keightley, Peter D. "The distribution of fitness effects of new mutations." *Nature Reviews Genetics* 8, no. 8 (2007): 610–8.

Falcon, Andrea. (2019). "Aristotle on Causality." The Stanford Encyclopedia of Philosophy, updated March 7, 2019, https://plato.stanford.edu/entries/aristotle-causality/.

Farlow, Ashley, Eshwar Meduri, and Christian Schlötterer. "DNA Double-Strand Break Repair and the Evolution of Intron Density." *Trends in Genetics* 27, no. 1 (2011): 1–6.

Feldman, Chris R., Edmund D. Brodie Jr., Edmund D. Brodie III, and Michael E. Pfrender. "Constraint Shapes Convergence in Tetrodotoxin-Resistant Sodium Channels of Snakes." *Proceedings of the National Academy of Sciences of the United States of America* 109, no. 12 (2012): 4556–61.

Felsenstein, Joseph. "Maximum-Likelihood Estimation of Evolutionary Trees from Continuous Characters." *American Journal of Human Genetics* 25, no. 5 (1973): 471–92.

Felsenstein, Joseph. *Inferring Phylogenies*. Sunderland, Mass.: Sinauer, 2004.

Ferris, Martin T., Paul Joyce, and Christina L. Burch. "High Frequency of Mutations that Expand the Host Range of an RNA Virus." *Genetics* 176, no. 2 (2007): 1013–22.

ffrench-Constant, Richard H., Phillip J. Daborn, and Gaelle Le Goff. "The Genetics and Genomics of Insecticide Resistance." *Trends in Genetics* 20, no. 3 (2004): 163–70.

ffrench-Constant, Richard H., Jessica C. Steichen, Thomas A. Rocheleau, Kate Aronstein, and Richard T. Roush. "A Single-Amino Acid Substitution in a Gamma-Aminobutyric Acid Subtype A Receptor Locus is Associated with Cyclodiene Insecticide Resistance in *Drosophila* Populations." *Proceedings of the National Academy of Sciences of the United States of America* 90, no. 5 (1993): 1957–61.

Fineran, Peter C., and Emmanuelle Charpentier. "Memory of Viral Infections by CRISPR-Cas Adaptive Immune Systems: Acquisition of New Information." *Virology* 434, no. 2 (2012): 202–9.

Firnberg, Elad, Jason W. Labonte, Jeffrey J. Gray, and Marc Ostermeier. "A Comprehensive, High-Resolution Map of a Gene's Fitness Landscape." *Mol Biol Evol* 31, no. 6 (2014): 1581–92.

Fisher, Ronald. "The Correlation Between Relatives on the Supposition of Mendelian Inheritance." *Philosophical Transactions of the Royal Society of Edinburgh*, 52 (1918): 399–433.

Fisher, Ronald. "The Distribution of Gene Ratios for Rare Mutations." *Proceedings of the Royal Society of Edinburgh*, 50 (1930a): 205–20.

Fisher, Ronald. *The Genetical Theory of Natural Selection*. London: Oxford University Press, 1930b.

Fisher, Ronald. See comments in "Watson, discussion on the Theory of Natural Selection." *Proceedings of the Royal Society of London B: Biological Sciences* 121 (1936).

Foley, Janet. "Mini-Review: Strategies for Variation and Evolution of Bacterial Antigens." *Computational and Structural Biotechnology Journal* 13 (2015): 407–16.

Fontana, Walter. "Modelling 'Evo-Devo' with RNA." *Bioessays* 24, no. 12 (2002): 1164–77.

Force, Allen, Michael Lynch, F. Bryan Pickett, Angel Amores, Yi-lin Yan, and John Postlethwait. "Preservation of Duplicate Genes by Complementary, Degenerative Mutations." *Genetics* 151, no. 4 (1999): 1531–45.

Ford, Edmund B. *Ecological Genetics*, 3rd ed. London: Chapman and Hall, 1971.

Foster, Patricia L. "Adaptive Mutation: Has the Unicorn Landed?" *Genetics* 148, no. 4 (1998): 1453–9.

Fowler, Douglas M., and Stanley Fields. "Deep Mutational Scanning: A New Style of Protein Science." *Nature Methods* 11, no. 8 (2014): 801–7.

Fox, Ronald F. "Review of Stuart Kauffman, *The Origins of Order: Self-Organization and Selection in Evolution*." *Biophysical Journal* 65, no. 6 (1993): 2698–9.

Frank, Steven A. *Immunology and Evolution of Infectious Disease*. Princeton: Princeton University Press, 2002.

Frank, Steven A., and Alan G. Barbour. (2006). "Within-Host Dynamics of Antigenic Variation." *Infect Genet Evolution* 6, no. 2 (2006): 141–6.

Freeland, Stephen J. and Laurence D. Hurst. "The Genetic Code is One in a Million." *Journal of Molecular Evolution* 47, no. 3 (1998): 238–48.

Freeland, Stephen J., Tao Wu, and Nick Keulmann. "The Case for an Error-Minimizing Standard Genetic Code." *Origins of Life and Evolution of the Biosphere* 33, no. 4–5 (2003): 457–77.

Freeman, Scott, and Jon C Herron. *Evolutionary Analysis*. Upper Saddle River, NJ: Prentice-Hall, 1998.

Freese, Ernst. "On the Evolution of the Base Composition of DNA." *Journal of Theoretical Biology* 3 (1962): 82–101.

Friedberg, Errol C. "The Discovery of Enzymatic Photoreactivation and the Question of Priority: The Letters of Salvador Luria and Albert Kelner." *Biochimie* 81, no. 1–2 (1999): 7–13.

Friedberg, Errol C., Graham C. Walker, Wolfram Siede, Richard D. Wood, Roger A. Schultz, and, Tom Ellenberger. *DNA Repair and Mutagenesis*, 2nd ed. Washington, D.C.: ASM Press, 2006.

Fryxell, Karl J., and Emil Zuckerkandl. "Cytosine Deamination Plays a Primary Role in the Evolution of Mammalian Isochores." *Molecular Biology and Evolution* 17, no. 9 (2000): 1371–83.

Fudenberg, Drew, and Lorens A. Imhof. "Phenotype Switching and Mutations in Random Environments." *Bulletin of Mathematical Biology* 74, no. 2 (2012): 399–421.

Futuyma, Douglas. *Evolutionary Biology*. Sunderland, Mass.: Sinauer, 1979.

Futuyma, Douglas J. "Sturm und Drang and the Evolutionary Synthesis." *Evolution* 42, no. 2 (1988): 217–26.

Futuyma, Douglas J. "Evolution, Science, and Society." *The American Naturalist* 158, no. S4 (2001): S1–S46.

Futuyma, Douglas J. "Evolutionary Constraint and Ecological Consequences." *Evolution* 64, no. 7 (2010): 1865–84.

Futuyma, Douglas J. "Evolutionary Biology Today and the Call for an Extended Synthesis." *Interface Focus* 7, no. 5 (2017): 20160145.

Galen, Spencer C., Chandrasekhar Natarajan, Hideaki Moriyama, Roy E. Weber, Angela Fago, Phred

M. Benham, Andrea N. Chavez, Zachary A. Cheviron, Jay F. Storz, and Christopher C. Witt. "Contribution of a Mutational Hotspot to Hemoglobin Adaptation in High-Altitude Andean House Wrens." *Proceedings of the National Academy of Sciences of the United States of America* 112, no. 45 (2015): 13958–63.

Gally, David L., Joseph A. Bogan, Barry I. Eisenstein., and Ian C. Blomfield. "Environmental Regulation of the fim Switch Controlling Type 1 Fimbrial Phase Variation in *Escherichia coli* K-12: Effects of Temperature and Media." *Journal of Bacteriology* 175, no. 19 (1993): 6186–93.

Gao, Ziyue, Priya Moorjani, Thomas A. Sasani, Brent S. Pedersen, Aaron R. Quinlan, Lynn B. Jorde, Guy Amster, and Molly Przeworski. (2019). "Overlooked Roles of DNA Damage and Maternal Age in Generating Human Germline Mutations." *Proceedings of the National Academy of Sciences of the United States of America* 116, no. 19 (2019): 9491–9500.

Garson, Justin, Linton Wang, and Sahotra Sarkar. "How Development May Direct Evolution." *Biology and Philosophy* 18, no. 2 (2003): 353–70.

Gayon, Jean. *Darwinism's Struggle for Survival: Heredity and the Hypothesis of Natural Selection.* Cambridge, UK: Cambridge University Press, 1998.

Gerstein, Aleeza C., Jasmine Ono, Dara S. Lo, Marcus L. Campbell, Anastasia Kuzmin, and Sarah P. Otto. "Too Much of a Good Thing: The Unique and Repeated Paths Toward Copper Adaptation." *Genetics* 199, no. 2 (2015): 555–71.

Ghiselin, Michael T. (1994). "The Imaginary Lamarck: A Look at Bogus 'History' in Schoolbooks," published September 1994; accessed February 18, 2018, https://wseas-cscc.blogspot.com/2008/01/look-at-bogus-history-in-schoolbooks.html.

Gigerenzer, Gerd, Zeno Swijtink, Theodore Porter, Lorraine Daston, John Beatty, and Lorenz Kruger. *The Empire of Chance.* Cambridge, UK: Cambridge University Press, 1989.

Gilbert, Walter. "Why Genes in Pieces?" *Nature,* 271 (1978).

Gilbert, Walter. "The Exon Theory of Genes." *Cold Spring Harbor Symposia on Quantitative Biology* 52 (1987): 901–5.

Gillespie, John H. *The Causes of Molecular Evolution.* Oxford Series in Ecology and Evolution. New York: Oxford University Press, 1991.

Gillespie, John H. "A Simple Stochastic Gene Substitution Model." *Theoretical Population Biology* 23, no. 2 (1983a): 202–15.

Gillespie, John H. "Some Properties of Finite Populations Experiencing Strong Selection and Weak Mutation." *The American Naturalist,* 121, no. 5 (1983b): 691–708.

Gillespie, John H. "Molecular Evolution over the Mutational Landscape." *Evolution* 38, no. 5 (1984): 1116–29.

Gillespie, John H. "Why k=4Nus is Silly." In *The Evolution of Population Biology,* edited by Rama S. Singh and Marcy K. Uyenoyama, 178–92. Cambridge: Cambridge University Press, 2004.

Glover, Lucy, Sebastian Hutchinson, Sam Alsford, Richard McCulloch, Mark C. Field, and David Horn. "Antigenic Variation in African Trypanosomes: The Importance of Chromosomal and Nuclear Context in VSG Expression Control." *Cell Microbiology* 15, no. 12 (2013): 1984–93.

Godfrey-Smith, Peter. "Three Kinds of Adaptationism." In *Adaptationism and Optimality,* edited by Steven Hecht Orzack and Elliott Sober, 335–57. Cambridge: Cambridge University Press, 2001.

Goldberg, Gregory W., Wenyan Jiang, David Bikard, and Luciano A. Marraffini. "Conditional Tolerance of Temperate Phages via Transcription-Dependent CRISPRCas Targeting." *Nature* 514, no. 7524 (2014): 633–7.

Golding, G. Brian. "Nonrandom Patterns of Mutation are Reflected in Evolutionary Divergence and May Cause Some of the Unusual Patterns Observed in Sequences." In *Genetic Constraints on Adaptive Evolution,* edited by Volker Loeschcke, 151–72. Berlin: Springer-Verlag, 1987.

Gomez, Kevin, Jason Bertram, and Masel, Joanna. "Mutation Bias can Shape Adaptation in Large Asexual Populations Experiencing Clonal Interference." *bioRxiv* (2020): 2020.02.17.953265.

Goodsell, David S. "The Molecular Perspective: Ultraviolet Light and Pyrimidine Dimers." *Oncologist* 6, no. 3 (2001): 298–9.

Goodwin, Brian. *How the Leopard Changed Its Spots: The Evolution of Complexity.* New York: Charles Scribner's Sons, 1994.

Gordenin, Dmitry A., Kirill S. Lobachev, Natalya P. Degtyareva, Anna L. Malkova, Edward L. Perkins, and Michael A. Resnick. "Inverted DNA Repeats: A Source of Eukaryotic Genomic Instability." *Molecular and Cellular Biology* 13, no. 9 (1993): 5315–22.

Gotelli, Nicholas J., and Brian J. McGill. "Null Versus Neutral Models: What's the Difference?" *Ecography* 29, no. 5 (2006): 793–800.

Gould, Stephen J. *Ever Since Darwin.* New York: W.W. Norton and Co., 1977.

Gould, Stephen J. "Change in Developmental Timing as a Mechanism of Macroevolution." In *Evolution and Development,* edited by J. Bonner, 333–46. New York: Springer-Verlag, 1982.

Gould, Stephen J. *The Structure of Evolutionary Theory.* Cambridge, Mass.: Harvard University Press, 2002.

Gould, Stephen J. and Richard C. Lewontin. "The Spandrels of San Marco and the Panglossian Paradigm: A Critique of the Adaptationist Program." *Proceedings of the Royal Society of London B: Biological Sciences* 205 (1979): 581–98.

Grant, Verne. *The Origin of Adaptations*. New York: Columbia University Press, 1963.

Graves, Christopher J., Vera I. D. Ros, Brian Stevenson, Paul D. Sniegowski, and Dustin Brisson. "Natural Selection Promotes Antigenic Evolvability." *PLoS Pathogens* 9, no. 11 (2013): e1003 766.

Greenbaum, Jason A., Bo Pang, and Thomas D. Tullius. "Construction of a Genome-Scale Structural Map at Single-Nucleotide Resolution." *Genome Research* 17, no. 6 (2007): 947–53.

Gregory, T. Ryan. "Genome Size Evolution in Animals." In *The Evolution of the Genome*, edited by T. Ryan Gregory, 3–87. San Diego: Elsevier, 2005a.

Gregory, T. Ryan, ed. *The Evolution of the Genome*. San Diego: Elsevier, 2005b.

Griesemer, James R., and William C. Wimsatt. "Picturing Weismannism: A Case Study of Conceptual Evolution." In *What Philosophy of Biology Is. Essays dedicated to David Hull*, edited by Michael Ruse, 75–137. Dordrecht: Kluwer Academic Publishers, 1989.

Grodwohl, Jean-Baptiste. "'The Theory was Beautiful Indeed': Rise, Fall and Circulation of Maximizing Methods in Population Genetics (1930-1980)." *Journal of the History of Biology* 50, no. 3 (2016): 571–608.

Gu, Xun, David Hewett-Emmett, Wen-Hsiung Li. "Directional Mutational Pressure Affects the Amino Acid Composition and Hydrophobicity of Proteins in Bacteria." *Genetica* 103, no. 1–6 (1998): 383–91.

Guizetti, Julien, and Artur Scherf. (2013). "Silence, Activate, Poise and Switch! Mechanisms of Antigenic Variation in *Plasmodium falciparum*." *Cell Microbiology* 15(5): 718–26.

Guo, Huatao, Diego Arambula, Partho Ghosh, and Jeff F. Miller. (2014). "Diversity-Generating Retroelements in Phage and Bacterial Genomes." *Microbiology Spectrum* 2, no. 6 (2014): MDNA3-0029-2014.

Gutierrez, Arnaud, Luisa Laureti, Steve Crussard, Heni Abida, Alexandro Rodríguez-Rojas, Jesus Blázquez, Zeynep Baharoglu, Didier Mazel, Fabien Darfeuille, J. Vogel, and Ivan Matic. "*β*-Lactam Antibiotics Promote Bacterial Mutagenesis via an RpoS-Mediated Reduction in Replication Fidelity." *Nature Communication* 4 (2013): 1610.

Haber, James E. "Mating-Type Genes and MAT Switching in *Saccharomyces cerevisiae*." *Genetics* 191, no. 1 (2012): 33–64.

Haldane, John B. S. (1927). "A Mathematical Theory of Natural and Artificial Selection. V. Selection and Mutation." *Mathematical Proceedings of the Cambridge Philosophical Society* 26: 220–230.

Haldane, John B. S. *The Causes of Evolution*. New York: Longmans, Green and Co., 1932.

Haldane, John B. S. "The Part Played by Recurrent Mutation in Evolution." *American Naturalist* 67, no. 708 (1933): 5–19.

Haldane, John B. S. (1959). "Natural Selection." In *Darwin's Biological Work: Some Aspects Reconsidered*, edited by Peter R. Bell, 101–49. Cambridge: Cambridge University Press, 1959.

Hall, James P. J., Huanhuan Wang, and J. David Barry. "Mosaic VSGs and the Scale of *Trypanosoma brucei* Antigenic Variation." *PLoS Pathogens* 9, no. 7 (2013): e1003 502.

Hallet, Bernard. "Playing Dr Jekyll and Mr Hyde: Combined Mechanisms of Phase Variation in Bacteria." *Current Opinion in Microbiology* 4, no. 5 (2001): 570–81.

Halligan, Daniel L., and Peter D. Keightley. "Spontaneous Mutation Accumulation Studies in Evolutionary Genetics." *Annual Review of Ecology, Evolution, and Systematics* 40 (2009): 151–72.

Hamburger, Viktor. "Embryology and the Modern Synthesis in Evolutionary Theory." In *The Evolutionary Synthesis*, edited by Ernst Mayr and William B. Provine, 97–112. Cambridge, Mass.: Harvard University Press, 1980.

Hansen, Thomas F. "Is Modularity Necessary for Evolvability? Remarks on the Relationship Between Pleiotropy and Evolvability." *Biosystems* 69, no. 2–3 (2003): 83–94.

Hansen, Thomas F. "Why Epistasis is Important for Selection and Adaptation." *Evolution* 67, no. 12 (2013): 3501–11.

Hanson, Gavin, and Jeff Coller. "Codon Optimality, Bias and Usage in Translation and mRNA Decay." *Nat Rev Mol Cell Biol*, 19, no. 1 (2018): 20–30.

Hanson, Sara J., and Kenneth H. Wolfe. "An Evolutionary Perspective on Yeast Mating-Type Switching." *Genetics* 206, no. 1 (2017): 9–32.

Hartl, Daniel L., and Clifford H. Taubes. "Towards a Theory of Evolutionary Adaptation." *Genetica* 103, no. 1–6 (1998): 525–33.

Hawkes, Lucy A., Sivananinthaperumal Balachandran, Nyambayar Batbayar, Patrick J. Butler, Peter B. Frappell, William K. Milsom, Natsagdorj Tseveenmyadag, Scott H. Newman, Graham R. Scott, Ponnusamy Sathiyaselvam, John Y. Takekawa, Martin Wikelski, and Charles M. Bishop. "The Trans-Himalayan Flights of Bar-Headed Geese (*Anser indicus*)." *Proceedings of the National Academy of Sciences of the United States of America* 108, no. 23 (2011): 9516–9.

Hayes, Jeffrey J., Thomas D. Tullius, and Alan P. Wolffe. "The Structure of DNA in a Nucleosome." *Proceedings of the National Academy of Sciences of the United States of America* 87, no. 19 (1990): 7405–9.

Henderson, Ian R., Peter Owen, and James P. Nataro. "Molecular Switches: The ON and OFF of Bacterial

Phase Variation." *Molecular Microbiology* 33, no. 5 (1999): 919–32.

Hendrikse, Jesse Love, Trish Elizabeth Parsons, and Benedikt Hallgrímsson. "Evolvability as the Proper Focus of Evolutionary Developmental Biology." *Evolution and Development* 9, no. 4 (2007): 393–401.

Hershberg, Ruth, and Dmitri A. Petrov. (2010). "Evidence that Mutation is Universally Biased Towards AT in Bacteria." *PLoS Genet,* 6(9): e1001 115.

Hicks, James B., and Ira Herskowitz. "Interconversion of Yeast Mating Types II. Restoration of Mating Ability to Sterile Mutants in Homothallic and Heterothallic Strains." *Genetics* 85, no. 3 (1977): 373–93.

Higgins, Brian P., Chandra D. Carpenter, and Anna C. Karls. "Chromosomal Context Directs High-Frequency Precise Excision of IS492 in *Pseudoalteromonas atlantica*." *Proceedings of the National Academy of Sciences of the United States of America* 104, no. 6 (2007): 1901–6.

Hill, Kathleen A., Jicheng Wang, Kelly D. Farwell, and Steve S. Sommer. "Spontaneous Tandem-Base Mutations (TBM) Show Dramatic Tissue, Age, Pattern and Spectrum Specificity." *Mutation Research* 534, no. 1–2 (2003): 173–86.

Hodgkinson, Alan, and Adam Eyre-Walker. "Variation in the Mutation Rate Across Mammalian Genomes." *Nature Reviews Genetics* 12, no. 11 (2011): 756–66.

Hodgkinson, Alan, Emmanuel Ladoukakis, and Adam Eyre-Walker. "Cryptic variation in the human mutation rate." *PLoS Biology* 7, no. 2 (2009): e1000 027.

Hoekstra, Hopi E., and Jerry A. Coyne. "The Locus of Evolution: Evo Devo and the Genetics of Adaptation." *Evolution* 61, no. 5 (2007): 995–1016.

Holden, Nicola, Ian C Blomfield, Bernt-Eric Uhlin, Makrina Totsika, Don Hemantha Kulasekara, and David L Gally. "Comparative Analysis of FimB and FimE Recombinase Activity." *Microbiology* 153, Pt 12 (2007): 4138–49.

Holzinger, Ferdinand, and Michael Wink. "Mediation of Cardiac Glycoside Insensitivity in the Monarch Butterfly (*Danaus plexippus*): Role of an Amino Acid Substitution in the Ouabain Binding Site of Na+,K+-ATPase." *Journal of Chemical Ecology* 22, no. 10 (1996): 1921–37.

Hook, Ernest B., Philip K. Cross, and Dina M. Schreinemachers. "Chromosomal Abnormality Rates at Amniocentesis and In Live-Born Infants." *Journal of the American Medical Association* 249, no. 15 (1983): 2034–8.

Horn, David. "Antigenic variation in African trypanosomes." *Molecular Biochemical Parasitology* 195, no. 2 (2014): 123–9.

Houle, David. "Comparing evolvability and variability of quantitative traits." *Genetics* 130, no. 1 (1992): 195–204.

Houle, David. "How Should We Explain Variation in the Genetic Variance of Traits?" *Genetica* 102–3, no. 1–6 (1998): 241–53.

Houle, David, Geir H. Bolstad, Kim van der Linde, and Thomas F. Hansen. "Mutation Predicts 40 Million Years of Fly Wing Evolution." *Nature* 548, no. 7668 (2017): 447–50.

Hua, Xia, and Lindell Bromham. "Darwinism for the Genomic Age: Connecting Mutation to Diversification." *Frontiers in Genetics* 8 (2017): 12.

Hurst, Laurence D., and Alexa R. Merchant. "High Guanine-Cytosine Content is Not an Adaptation to High Temperature: A Comparative Analysis Amongst Prokaryotes." *Proceedings of the Royal Society B: Biological Sciences* 268, no. 1466 (2001): 493–7.

Huxley, Julian. *Evolution: The Modern Synthesis.* London: George Allen and Unwin, 1942.

Huxley, Julian. *Evolution: The Modern Synthesis.* New York: Wiley and Sons, Inc., 1964.

Huxley, Julian, Alfred E. Emerson, Theodosius Dobzhansky, E. B. Ford, Ernst Mayr, A. J. Nicholson, Everett C. Olson, C. Ladd Prosser, G. Ledyard Stebbins, and Sewell Wright. "Panel Two: The Evolution of Life." In *Evolution After Darwin: Issues in Evolution,* edited by Sol Tax and Charles Callender, vol. III. Chicago: University of Chicago Press, 1960.

Hwang, Dick G., and Phil Green. "Bayesian Markov Chain Monte Carlo Sequence Analysis Reveals Varying Neutral Substitution Patterns in Mammalian Evolution." *Proceedings of the National Academy of Sciences of the United States of America* 101, no. 39 (2004): 13 994–4001.

Hynes, Alexander P., Geneviève M. Rousseau, Daniel Agudelo, Adeline Goulet, Beatrice Amigues, Jeremy Loehr, Dennis A. Romero, Christophe Fremaux, Philippe Horvath, Yannick Doyon, Christian Cambillau, and Sylvain Moineau. "Widespread Anti-CRISPR Proteins in Virulent Bacteriophages Inhibit a Range of Cas9 Proteins." *Nature Communications* 9, no. 1 (2018): 2919.

Hynes, Alexander P., Manuela Villion, and Sylvain Moineau. "Adaptation in Bacterial CRISPR-Cas Immunity Can Be Driven by Defective Phages." *Nature Communications* 5 (2014): 4399.

Iida, Shigeru, Hans E. Huber, Rosemarie Hiestand-Nauer, Jürg Meyer, Thomas A. Bickle, and Werner Arber. "The Bacteriophage P1 Site-Specific Recombinase cin: Recombination Events and DNA Recognition Sequences." *Cold Spring Harbor Symposium of Quantitative Biology* 49 (1984): 769–77.

Jablonka, Eva, and Marion J. Lamb. *Evolution in Four Dimensions: Genetic, Epigenetic, Behavioral, and Symbolic Variation in the History of Life.* Cambridge, Mass.: MIT Press, 2005.

Jain, Kavita, and Sarada Seetharaman. "Multiple Adaptive Substitutions During Evolution in Novel Environments." *Genetics* 189, no. 3 (2011): 1029–43.

Jaynes, Edwin T. *Probability Theory: The Logic of Science.* Cambridge, UK: Cambridge University Press, 2003.

Johnson, Philip L. F., and Ines Hellmann. "Mutation Rate Distribution Inferred from Coincident SNPs and Coincident Substitutions." *Genome Biology and Evolution* 3 (2011): 842–50.

Jones, Adam G., Arnold, Stevan J., and Reinhard Burger. "The Mutation Matrix and the Evolution of Evolvability." *Evolution* 61, no. 4 (2007): 727–45.

Jost, Manda Clair, David M. Hillis, Ying Lu, John W. Kyle, Harry A. Fozzard, and Harold H. Zakon. "Toxin-Resistant Sodium Channels: Parallel Adaptive Evolution Across a Complete Gene Family." *Molecular Biology and Evolution* 25, no. 6 (2008): 1016–24.

Jovelin, Richard, and Asher D. Cutter. "Fine-Scale Signatures of Molecular Evolution Reconcile Models of Indel-Associated Mutation." *Genome Biol Evolution* 5, no. 5 (2013): 978–86.

Jung, David, and Frederick W. Alt. "Unraveling V(D)J Recombination; Insights into Gene Regulation." *Cell* 116, no. 2 (2004): 299–311.

Katju, Vaishali, and Ulfar Bergthorsson. "Old Trade, New Tricks: Insights into the Spontaneous Mutation Process from the Partnering of Classical Mutation Accumulation Experiments with High-Throughput Genomic Approaches." *Genome Biology and Evolution* 11, no. 1 (2019): 136–65.

Kauffman, Stuart A. *The Origins of Order: Self-Organization and Evolution.* New York: Oxford University Press, 1993.

Kauffman, Stuart A. "Requirements for Evolvability in Complex Systems: Orderly Dynamics and Frozen Components." *Physica D: Nonlinear Phenomena* 42, no. 1 (1990): 135–52.

Kavanagh, Kathryn D., Oren Shoval, Benjamin B. Winslow, Uri Alon, Brian P. Leary, Akinori Kan, and Clifford J. Tabin. "Developmental Bias in the Evolution of Phalanges." *Proceedings of the National Academy of Sciences of the United States of America* 110, no. 45 (2013): 18 190–5.

Keightley, Peter D., and Adam Eyre-Walker. "What Can We Learn about the Distribution of Fitness Effects of New Mutations from DNA Sequence Data?" *Philosophical Transactions of the Royal Society of London B: Biological Sciences* 365, no. 1544 (2010): 1187–93.

Keith, Nathan, Abraham E. Tucker, Craig E. Jackson, Way Sung, José Ignacio Lucas Lledó, Daniel R. Schrider, Sarah Schaack, Jeffry L. Dudycha, Matthew Ackerman, Andrew J. Younge, Joseph R. Shaw, and Michael Lynch. "High Mutational Rates of Largescale Duplication and Deletion in *Daphnia pulex*." *Genome Research* 26, no. 1 (2016): 60–9.

Keller, Irene, Douda Bensasson, and Richard A. Nichols "Transition-Transversion Bias is not Universal: A Counter Example from Grasshopper Pseudogenes." *PLoS Genetics* 3, no. 2 (2007): e22.

Kenigsberg, Ephraim, Yishai Yehuda, Lisette Marjavaara, Andrea Keszthelyi, Andrei Chabes, Amos Tanay, and Itamar Simon. "The Mutation Spectrum in Genomic Late Replication Domains Shapes Mammalian GC Content." *Nucleic Acids Research* 44, no. 9 (2016): 4222–32.

Keulen, Wilco, Nicole K. T. Back, Albert van Wijk, Charles A. B. Boucher, and Ben Berkhout. "Initial Appearance of the 184Ile Variant in Lamivudine-Treated Patients is Caused by the Mutational Bias of Human Immunodeficiency Virus Type 1 Reverse Transcriptase." *Journal of Virology* 71, no. 4 (1997): 3346–50.

Kidd, Jeffrey M., Tina Graves, Tera L. Newman, Robert Fulton, Hillary S. Hayden, Maika Malig, Joelle Kallicki, Rajinder Kaul, Richard K. Wilson, and Evan E. Eichler. "A Human Genome Structural Variation Sequencing Resource Reveals Insights into Mutational Mechanisms." *Cell* 143, no. 5 (2010): 837–47.

Kim, Nayun, Jang-Eun Cho, Yue C. Li, and Sue Jinks-Robertson. (2013). "RNA:DNA Hybrids Initiate Quasi-Palindrome-Associated Mutations in Highly Transcribed Yeast DNA." *PLoS Genetics* 9, no. 11 (2013): e1003 924.

Kim, Nayun, and Sue Jinks-Robertson. "Transcription as a Source of Genome Instability." *Nature Reviews Genetics* 13, no. 3 (2012): 204–14.

Kimura, Motoo. "A simple method for estimating evolutionary rates of base substitutions through comparative studies of nucleotide sequences." *Journal of Molecular Evolution* 16, no. 2 (1980): 111–20.

Kimura, Motoo. *The Neutral Theory of Molecular Evolution.* Cambridge, UK: Cambridge University Press, 1983.

Kimura, Motoo, and Takeo Maruyama. "The Substitutional Load in a Finite Population." *Heredity (Edinburgh)* 24, no. 1 (1969): 101–14.

King, Jack L. "The Influence of the Genetic Code on Protein Evolution." In *Biochemical Evolution and the Origin of Life*, edited by E Schoffeniels, 3–13. Viers: North-Holland Publishing Company, 1971.

King, Jack L. "The Role of Mutation in Evolution." In *Sixth Berkeley Symposium on Mathematical Statistics and Probability*, vol. V, edited by L. Le Cam, J. Neyman, and E. Scott, 69–88. Los Angeles: University of California Press, 1972.

King, Jack L., and Thomas H. Jukes. "Non-Darwinian Evolution." *Science* 164 (1969): 788–97.

Kirschner, Marc, and John K. Gerhart. "Evolvability." *Proceedings of the National Academy of Sciences of the United States of America* 95 (1988): 8420–7.

Kirschner, Marc, and John K. Gerhart. *The Plausibility of Life: Resolving Darwin's Dilemma.* New Haven, Conn.: Yale University Press, 2005.

Klapacz, Joanna, and Ashok S. Bhagwat. "Transcription-Dependent Increase in Multiple Classes of Base Substitution Mutations in *Escherichia coli*." *Journal of Bacteriology* 184, no. 24 (2002): 6866–72.

Klar, Amar J. "The Yeast Mating-Type Switching Mechanism: A Memoir." *Genetics* 186, no. 2 (2010): 443–9.

Kleina, Lynn G., and Jeffrey H. Miller. "Genetic Studies of the *lac* Repressor. XIII. Extensive Amino Acid Replacements Generated by the Use of Natural and Synthetic Nonsense Suppressors." *Journal of Molecular Biology* 212, no. 2 (1990): 295–318.

Klimasauskas, Casimir C. "Not Knowing Your Random Number Generator Could Be Costly." *PC AI Magazine* (2002).

Knight, Rob D., Stephen J. Freeland, and Laura F. Landweber. "A Simple Model Based on Mutation and Selection Explains Trends in Codon and Amino-Acid Usage and GC Composition Within and Across Genomes." *Genome Biology* 2, no. 4 (2001): research0010.1– research0010.13.

Koenraadt, Harrie, Shauna C. Somerville, and Al Jones. "Characterization of Mutations in the Beta-Tubulin Gene of Benomyl-Resistant Field Strains of *Venturia inaequalis* and Other Plant Pathogenic Fungi." *Phytopathology*, 82 (1992): 1348–54.

Kolesnikov, Alexander I., George F. Reiter, Narayani Choudhury, Timothy R. Prisk, Eugene Mamontov, Andrey Podlesnyak, George Ehlers, Andrew G. Seel, David J. Wesolowski, and Lawrence M. Anovitz. "Quantum Tunneling of Water in Beryl: A New State of the Water Molecule." *Physics Review Letters* 116, no. 16 (2016): 167802.

Komano, Teruya. "Shufflons: Multiple Inversion Systems and Integrons." *Annual Review of Genetics* 33 (1999): 171–91.

Kondrashov, Alexey. S., and Igor B. Rogozin. "Context of Deletions and Insertions in Human Coding Sequences." *Human Mutation* 23, no. 2 (2004): 177–85.

Koonin, Eugene V. *The Logic of Chance*. Upper Saddle River, NJ: FT Press, 2011.

Koonin, Eugene V. "Evolution of RNA- and DNA-Guided Antivirus Defense Systems in Prokaryotes and Eukaryotes: Common Ancestry vs Convergence." *Biology Direct* 12, no. 1 (2017): 5.

Koonin, Eugene V. "Lamarckian or Not, CRISPR-Cas is an Elaborate Engine of Directed Evolution." *Biology and Philosophy* 34, no. 1 (2019): 17.

Koonin, Eugene V., and Yuri I. Wolf. "Genomics of Bacteria and Archaea: The Emerging Dynamic View of the Prokaryotic World." *Nucleic Acids Research* 36, no. 21 (2008): 6688–719.

Koonin, Eugene V., and Yuri I. Wolf. "Is Evolution Darwinian or/and Lamarckian?" *Biology Direct* 4 (2009): 42.

Krawczak, Michael, Edward V. Ball, and David N. Cooper. "Neighboring Nucleotide Effects on the Rates of Germ-Line Single-Base-Pair Substitution in Human Genes." *American Journal of Human Genetics* 63, no. 2 (1998): 474–88.

Krašovec, Roc, Huw Richards, Danna R. Gifford, Charlie Hatcher, Katy J. Faulkner, Roman V. Belavkin, Alastair Channon, Elizabeth Aston, Andrew J. McBain, and Christopher G. Knight. "Spontaneous Mutation Rate is a Plastic Trait Associated with Population Density Across Domains of Life." *PLoS Biology* 15, no. 8 (2017): e2002731.

Kreitman, Martin. "The Neutral Theory is Dead. Long Live the Neutral Theory." *Bioessays* 18, no. 8 (1996): 678–83.

Kumar, Dinesh, Amy L. Abdulovic, Jörgen Viberg, Anna Karin Nilsson, Thomas A. Kunkel, and Andrei Chabes. "Mechanisms of Mutagenesis *In Vivo* Due to Imbalanced dNTP Pools." *Nucleic Acids Research* 39, no. 4 (2011): 1360–71.

Kussell, Edo, and Stanislaus Leibler. "Phenotypic Diversity, Population Growth, and Information in Fluctuating Environments." *Science* 309, no. 5743 (2005): 2075–8.

Kuwahara, Tomomi, Atsushi Yamashita, Hideki Hirakawa, Haruyuki Nakayama, Hidehiro Toh, Natsumi Okada, Satoru Kuhara, Masahira Hattori, Tetsuya Hayashi, and Yoshinari Ohnishi. "Genomic Analysis of *Bacteroides fragilis* Reveals Extensive DNA Inversions Regulating Cell Surface Adaptation." *Proceedings of the National Academy of Sciences of the United States of America* 101, no. 41 (2004): 14919–24.

Lafay, Bénédicte, Andrew T. Lloyd, Michael J. McLean, Kevin M. Devine, Paul M. Sharp, and Kenneth H. Wolfe. "Proteome Composition and Codon Usage in Spirochaetes: Species-Specific and DNA Strand-Specific Mutational Biases." *Nucleic Acids Research* 27, no. 7 (1999): 1642–9.

Laland, Kevin, Tobias Uller, Marc Feldman, Kim Sterelny, Gerd B. Müller, Armin Moczek, Eva Jablonka, John Odling-Smee, Gregory A. Wray, Hopi E. Hoekstra, Douglas J. Futuyma, Richard E. Lenski, Trudy F. C. Mackay, Dolph Schluter, and Joan E. Strassmann. "Does Evolutionary Theory Need a Rethink?" *Nature* 514, no. 7521 (2014): 161–4.

Lande, Russell, and Stevan J. Arnold. "The Measurement of Selection on Correlated Characters." *Evolution* 37, no. 6 (1983): 1210–26.

Lane, M. Chelsea, Xin Li, Melanie M. Pearson, Amy N. Simms, and Harry L. T. Mobley. "Oxygen-Limiting Conditions Enrich for Fimbriate Cells of Uropathogenic *Proteus mirabilis* and *Escherichia coli*." *Journal of Bacteriology* 191, no. 5 (2009): 1382–92.

Lang, Gregory I., and Andrew W. Murray. "Estimating the Per-Base-Pair Mutation Rate in the Yeast *Saccharomyces cerevisiae.*" *Genetics* 178, no. 1 (2008): 67–82.

Lange, Axel, Hans L. Nemeschkal, and Gerd B. Müller. "Biased Polyphenism in Polydactylous Cats Carrying a Single Point Mutation: The Hemingway Model for Digit Novelty." *Evolutionary Biology* 41, no. 2 (2014): 262–75.

Laplace, Pierre Simon. *A Philosophical Essay on Probabilities,* 6th ed. New York: Dover Publications, 1951.

Laureti, Luisa, Ivan Matic, and Arnaud Gutierrez. "Bacterial Responses and Genome Instability Induced by Subinhibitory Concentrations of Antibiotics." *Antibiotics* 2, no. 1 (2013): 100–14.

Lawrence, Jeffrey G., and John R. Roth. "Selfish Operons: Horizontal Transfer May Drive the Evolution of Gene Clusters." *Genetics* 143 (1996): 1843–60.

L'Ecuyer, Pierre. "Random Number Generation with Multiple Streams for Sequential and Parallel Computing." In *Proceedings of the 2015 Winter Simulation Conference,* edited by Levent Yilmaz, Wai Kin Victor Chan, Il-Chul Moon, Teresa M. K. Roeder, Charles Macal, and Manuel D. Rossetti, 31–44. Hoboken, NJ: IEEE, 2015.

Le Rouzic, Arnaud, and José M. Álvarez-Castro. "Epistasis-Induced Evolutionary Plateaus in Selection Responses." *Am Nature* 188, no. 6 (2016): E134–E150.

Leigh, Egbert G. Jr. "Ronald Fisher and the Development of Evolutionary Theory. II. Influences of New Variation on Evolutionary Process." *Oxford Surveys in Evolutionary Biology* 4 (1987): 212–63.

Leighow, Scott M., ChuanLiu, Haider Inam, Boyang Zhao, and Justin R. Pritchard. "Multi-Scale Predictions of Drug Resistance Epidemiology Identify Design Principles for Rational Drug Design." *Cell Reports* 30, no. 12 (2020): 3951–63 e4.

Lenski, R. E. and Mittler, J. E. "The Directed Mutation Controversy and Neo-Darwinism." *Science* 259, no. 5092 (1993): 188–94.

Levins, Richard. "The Strategy of Model Building in Population Biology." *American Scientist,* 54, no. 4 (1966): 421–31.

Levy, Asaf, Moran G. Goren, Ido Yosef, Oren Auster, Miriam Manor, Gil Amitai, Rotem Edgar, Udi Qimron, and Rotem Sorek. "CRISPR Adaptation Biases Explain Preference for Acquisition of Foreign DNA." *Nature* 520, no. 7548 (2015): 505–10.

Lewontin, Richard C. *The Genetic Basis of Evolutionary Change.* New York: Columbia University Press, 1974.

Lewontin, Richard C. "The Organism as the Subject and Object of Evolution." In *The Dialectical Biologist,* edited by R. Levins and Richard C. Lewontin, 85–97. Cambridge, Mass.: Harvard University Press, 1985.

Li, Wen-Hsiung, Chi-Cheng Luo, and Chun-I Wu. "Evolution of DNA Sequences." In *Molecular Evolutionary Genetics,* edited by Ross J. MacIntyre, New York: Plenum, 1985.

Liao, Hans, Tim McKenzie, and Robert Hageman. "Isolation of a Thermostable Enzyme Variant by Cloning and Selection in a Thermophile." *Proceedings of the National Academy of Sciences of the United States of America* 83, no. 3 (1986): 576–80.

Lind, Peter A., Eric Libby, Jenny Herzog, and Paul B. Rainey. "Predicting Mutational Routes to New Adaptive Phenotypes." *eLife,* 8 (2019): e38 822.

Liu, Wenjie, Dion K. Harrison, Dominika Chalupska, Piotr Gornicki, Chris C. O'Donnell, Steve W. Adkins, Robert Haselkorn, and Richard R. Williams. "Single-Site Mutations in the Carboxyltransferase Domain of Plastid Acetyl-CoA Carboxylase Confer Resistance to Grass-Specific Herbicides." *Proceedings of the National Academy of Sciences of the United States of America* 104, no. 9 (2007): 3627–32.

Liu, Zhen, Qi, Fei-Yan Qi, Xin Zhou, Hai-Qing Ren, and Peng Shi. "Parallel Sites Implicate Functional Convergence of the Hearing Gene Prestin Among Echolocating Mammals." *Molecular Biology Evolution* 31, no. 9 (2014): 2415–24.

Locey, Kenneth J., and Jay T. Lennon. "Scaling Laws Predict Global Microbial Diversity." *Proceedings of the National Academy of Sciences of the United States of America* 113, no. 21 (2016): 5970–5.

Long, Hongan, Way Sung, Sibel Kucukyildirim, Emily Williams, Samuel F. Miller, Wanfeng Guo, Caitlyn Patterson, Colin Gregory, Chloe Strauss, Casey Stone, Cécile Berne, David Kysela, William R. Shoemaker, Mario E. Muscarella, Haiwei Luo, Jay T. Lennon, Yves V. Brun, and Michael Lynch. "Evolutionary Determinants of Genome-Wide Nucleotide Composition." *Nat Ecol Evolution* 2, no. 2 (2018): 237–40.

Losos, Jonathan B., Stevan J. Arnold, Gill Bejerano, E. D. Brodie III, David Hibbett, Hopi E. Hoekstra, David P. Mindell, Antónia Monteiro, Craig Moritz, H. Allen Orr, Dmitri A. Petrov, Susanne S. Renner, Robert E. Ricklefs, Pamela S. Soltis, and Thomas L. Turner. "Evolutionary Biology for the 21st Century." *PLoS Biology* 11, no. 1 (2013): e1001466.

Love, Donald R., Sarah B. England, Astrid Speer, Rosalind F. Marsden, J. F. Bloomfield, Anya L. Roche, Gareth S. Cross, Roger C. Mountford, Terry J. Smith, and Kay E. Davies. "Sequences of Junction Fragments in the Deletion-Prone Region of the Dystrophin Gene." *Genomics* 10, no. 1 (1991): 57–67.

Løvtrup, Søren. *Darwinism: The Refutation of a Myth.* New York: Croom Helm, 1987.

Lu, Sheng Dong, Deru Lu, and Max Gottesman. "Stimulation of IS1 Excision by Bacteriophage P1 ref Function." *Journal of Bacteriology* 171, no. 6 (1989): 3427–32.

Lynch, Michael. "The Frailty of Adaptive Hypotheses for the Origins of Organismal Complexity." *Proceedings of the National Academy of Sciences of the United States of America* 104, Suppl 1 (2007a): 8597–604.

Lynch, Michael. *The Origins of Genome Architecture*. Sunderland, Mass.: Sinauer Associates, Inc, 2007b.

Lynch, Michael. "Rate, Molecular Spectrum, and Consequences of Human Mutation." *Proceedings of the National Academy of Sciences of the United States of America* 107, no. 3 (2010): 961–8.

Maclean, Craig, Gabriel G. Perron, and Andy Gardner. "Diminishing Returns from Beneficial Mutations and Pervasive Epistasis Shape the Fitness Landscape for Rifampicin Resistance in *Pseudomonas aeruginosa*." *Genetics* 186, no. 4 (2010): 1345–54.

Maekawa, Hiromi, and Yoshinobu Kaneko. "Inversion of the Chromosomal Region between Two Mating Type Loci Switches the Mating Type in *Hansenula polymorpha*." *PLoS Genetics* 10, no. 11 (2014): e1004796.

Makarova, Kira S., Yuri I. Wolf, Omer S. Alkhnbashi, Fabrizio Costa, Shiraz A. Shah, Sita J. Saunders, Rodolphe Barrangou, Stan J. J. Brouns, Emmanuelle Charpentier, Daniel H. Haft, Philippe Horvath, Sylvain Moineau, Francisco J. M. Mojica, Rebecca M. Terns, Michael P. Terns, Malcolm F. White, Alexander F. Yakunin, Roger A. Garrett, John van der Oost, Rolf Backofen, and Eugene V. Koonin. "An Updated Evolutionary Classification of CRISPR-Cas Systems." *Nature Reviews Microbiology* 13, no. 11 (2015): 722–36.

Maki, Hisaji. "Origins of Spontaneous Mutations: Specificity and Directionality of Base-Substitution, Frameshift, and Sequence-Substitution Mutageneses." *Annual Review of Genetics* 36 (2002): 279–303.

Margoliash, Emanuel. "Primary Structure and Evolution of Cytochrome C." *Proceedings of the National Academy of Sciences of the United States of America* 50 (1963): 672–9.

Marston, Hilary D., Dennis M. Dixon, Jane M. Knisely, Tara N. Palmore, and Anthony S. Fauci. "Antimicrobial Resistance." *Journal of the American Medical Association* 316, no. 11 (2016): 1193–1204.

Masel, Joanna. "Evolutionary Capacitance May Be Favored by Natural Selection." *Genetics* 170, no. 3 (2005): 1359–71.

Mathews, Christopher K. "Deoxyribonucleotides as Genetic and Metabolic Regulators." *The FASEB Journal* 28, no. 9 (2014): 3832–40.

Matuszewski, Sebastian, Joachim Hermisson, and Michael Kopp. "Fisher's Geometric Model with a Moving Optimum." *Evolution* 68, no. 9 (2014): 2571–88.

Maynard Smith, John. "Natural Selection and the Concept of a Protein Space." *Nature* 225, no. 5232 (1970): 563–4.

Maynard Smith, John. *The Theory of Evolution*, 3rd ed. Cambridge, UK: Cambridge University Press, 1975.

Maynard Smith, John. "Evolution and Development." In *Development and Evolution*, edited by Brian Goodwin, Nigel Holder, and Christopher C. Wylie, 33–46. New York: Cambridge University Press, 1983.

Maynard Smith, John, Richard Burian, Stuart A. Kauffman, Pere Alberch, John H. Campbell, Brian Goodwin, Russell Lande, David M. Raup, and Lewis Wolpert. "Developmental Constraints and Evolution: A Perspective from the Mountain Lake Conference on Development and Evolution." *Quarterly Review of Biology* 60, no. 3 (1985): 265–87.

Mayr, Ernst. "Darwin and the Evolutionary Theory in Biology." In *Evolution and Anthropology: A Centennial Appraisal*, 1–10. Washington, D.C.: Anthropological Society, 1959a.

Mayr, Ernst. "Where Are We?" *Cold Spring Harbor Symposium of Quantitative Biology* 24 (1959b): 1–14.

Mayr, Ernst. "The Emergence of Evolutionary Novelties." In *Evolution After Darwin: The University of Chicago Centennial*, vol. I, edited by Sol Tax and Charles Callender, 349–80. Chicago: University of Chicago Press, 1960.

Mayr, Ernst. "Cause and Effect in Biology." *Science* 134 (1961): 1501–6.

Mayr, Ernst. *Animal Species and Evolution*. Cambridge, Mass.: Harvard University Press, 1963.

Mayr, Ernst. "Some Thoughts on the History of the Evolutionary Synthesis." In *The Evolutionary Synthesis: Perspectives on the Unification of Biology*, edited by Ernst Mayr and William B. Provine, 1–48. Cambridge, Mass.: Harvard University Press, 1980.

Mayr, Ernst. *The Growth of Biological Thought: Diversity, Evolution, and Inheritance*. Cambridge, Mass.: Harvard University Press, 1982.

Mayr, Ernst. "How to Carry Out the Adaptationist Program?" *The American Naturalist* 121 (1983): 324–34.

Mayr, Ernst. "Response to John Beatty." *Biology and Philosophy*, 9 (1994): 357–8.

Mayr, Ernst. *Systematics and the Origin of Species*. Cambridge, Mass.: Harvard University Press, 1999.

Mayr, Ernst. *What Evolution Is*. New York: Basic Books, 2001.

McCandlish, David M., and Arlin Stoltzfus. "Modeling evolution using the probability of fixation: history and implications." *Quarterly Review of Biology* 89, no. 3 (2014): 225–52.

McCracken, Kevin G., Christopher P. Barger, Mariana Bulgarella, Kevin P. Johnson, Sarah A. Sonsthagen, Jorge Trucco, Thomas H. Valqui, Robert E. Wilson, Kevin

Winker, and Michael D. Sorenson. "Parallel Evolution in the Major Haemoglobin Genes of Eight Species of Andean Waterfowl." *Molecular Ecology* 18, no. 19 (2009): 3992–4005.

McDowell, John V., Shian-Ying Sung, Linden T. Hu, and Richard T. Marconi. "Evidence That the Variable Regions of the Central Domain of VlsE Are Antigenic During Infection with Lyme Disease Spirochetes." *Infect Immun* 70, no. 8 (2002): 4196–203.

McGuigan, Katrina. "Studying Phenotypic Evolution Using Multivariate Quantitative Genetics." *Molecular Ecology* 15, no. 4 (2006): 883–96.

McLaughlin, Richard N. Jr, and Harmit S. Malik. "Genetic Conflicts: The Usual Suspects and Beyond." *Journal of Experimental Biology*, 220, Pt 1 (2017): 6–17.

Medawar, Peter B. *The Art of the Soluble*. London: Methuen and Co., 1967.

Mell, Joshua Chang, and Rosemary J. Redfield. "Natural Competence and the Evolution of DNA Uptake Specificity." *Journal of Bacteriology* 196, no. 8 (2014): 1471–83.

Melo, Diogo, Arthur Porto, James M. Cheverud, and Gabriel Marroig. "Modularity: Genes, Development and Evolution." *Annual Review of Ecology, Evolution, and Systematics* 47 (2016): 463–86.

Merlin, Francesca. "Evolutionary Chance Mutation: A Defense of the Modern Synthesis' Consensus View." *Philosophy and Theory in Biology*, 2 (2010).

Metzgar, David, and Christopher Wills. "Evidence for the Adaptive Evolution of Mutation Rates." *Cell* 101, no. 6 (2000): 581–4.

Meyer, Justin R., Devin T. Dobias, Joshua S. Weitz, Jeffrey E. Barrick, Ryan T. Quick, and Richard E. Lenski. "Repeatability and Contingency in the Evolution of a Key Innovation in Phage Lambda." *Science* 335, no. 6067 (2012): 428–32.

Michod, Richard E. "The Theory of Kin Selection." *Annual Review of Ecology and Systematics* 13, no. 1 (1982): 23–55.

Mill, John Stuart. *On Liberty*, 4th ed. London: Longman, Roberts and Green, 1869.

Miller, Craig R., Paul Joyce, and Holly A. Wichman. "Mutational Effects and Population Dynamics During Viral Adaptation Challenge Current Models." *Genetics* 187, no. 1 (2011): 185–202.

Millstein, Roberta L. "The Chances of Evolution: An Analysis of the Roles of Chance in Microevolution and Macroevolution." PhD diss., University of Minnesota, 1997. https://www.researchgate.net/publication/236670398_The_Chances_of_Evolution_An_Analysis_of_the_Roles_of_Chance_in_Microevolution_and_Macroevolution/link/0046353419bdf5bb99000000/download.

Mivart, St George J. *On the Genesis of Species*. London: R. Clay, Son and Taylor, 1871.

Molla, Akhteruzzaman, Marina Korneyeva, Qing Gao, Sudthida Vasavanonda, Paaline J. Schipper, Hong-Mei Mo, Martin Markowitz, Tatyana Chernyavskiy, Ping Niu, Nicholas Lyons, Ann Hsu, G. Richard Granneman, David D. Ho, Charles A.B. Boucher, John M. Leonard, Daniel W. Norbeck, and Dale J. Kempf. "Ordered Accumulation Of Mutations in HIV Protease Confers Resistance to Ritonavir." *Nature Medicine* 2, no. 7 (1996): 760–6.

Monod, Jacques. *Chance and Necessity: An Essay on the Natural Philosophy of Modern Biology*. New York: Vintage Books, 1972.

Moody, Paul. *Introduction to Evolution*, 2nd ed. New York: Harper and Brothers, 1962.

Mooers, Arne Ø., and Edward C. Holmes. "The Evolution of Base Composition and Phylogenetic Inference." *Trends in Ecology and Evolution* 15, 9 (2000): 365–9.

Mora, Camilo, Derek P. Tittensor, Sina Adl, Alastair G. B. Simpson, and Boris Worm. "How Many Species are There on Earth and in the Ocean?" *PLoS Biology* 9, no. 8 (2011): e1001127.

Morgan, Thomas H. *Evolution and Adaptation*. New York: Macmillan, 1903.

Morgan, Thomas H. "The Origin of Species through Selection Contrasted with their Origin through the Appearance of Definite Variations." *Popular Science Monthly* (1904): 54–65.

Morgan, Thomas H. "For Darwin." *Popular Science Monthly* 74 (1909): 367–80.

Morgan, Thomas H. "The American Society of Naturalists: Chance or Purpose in the Origin and Evolution of Adaptation." *Science* 31, no. 789 (1910): 201–10.

Morgan, Thomas H. *A Critique of the Theory of Evolution*. Princeton, NJ: Princeton University Press, 1916.

Morgan, Thomas H. *The Physical Basis of Heredity*. Philadelphia: J.B. Lippincott, 1919.

Morgan, Thomas H. "The Bearing of Mendelism on the Origin of Species." *The Scientific Monthly* 16 (1923): 237–47.

Morgan, Thomas H. *Evolution and Genetics*, 2nd ed. Princeton, NJ: Princeton University Press, 1925.

Morris, Tracy, and John Thacker. "Formation of Large Deletions by Illegitimate Recombination in the HPRT Gene of Primary Human Fibroblasts." *Proceedings of the National Academy of Sciences of the United States of America* 90, no. 4 (1993): 1392–6.

Moxon, E. Richard, Paul B. Rainey, Martin A. Nowak, and Richard E. Lenski. "Adaptive Evolution of Highly Mutable Loci in Pathogenic Bacteria." *Current Biology* 4, no. 1 (1994): 24–33.

Moxon, E. Richard, Chris Bayliss, and Derek Hood. "Bacterial Contingency Loci: The Role of Simple Sequence DNA Repeats in Bacterial Adaptation." *Annual Reviews Genetics* 40 (2006): 307–33.

Mugal, Carina F., Peter F. Arndt, Lena Holm, and Hans Ellegren. "Evolutionary Consequences of DNA Methylation on the GC Content in Vertebrate Genomes." *G3 (Bethesda)* 5, no. 3 (2015): 441–7.

Mugnier, Monica R., Erec C. Stebbins, and F. Nina Papavasiliou. "Masters of Disguise: Antigenic Variation and the VSG Coat in Trypanosoma brucei." *PLoS Pathogen* 12, no. 9 (2016): e1005 784.

Mukai, Terumi. "The Genetic Structure of Natural Populations of *Drosophila melanogaster*. I. Spontaneous Mutation Rate of Polygenes Controlling Viability." *Genetics* 50, no. 1 (1964): 1–19.

Muller, Fritz, and Heinz Tobler. "Chromatin Diminution in the Parasitic Nematodes *Ascaris suum* and *Parascaris univalens*." *International Journal of Parasitology* 30, no. 4 (2000): 391–9.

Natarajan, Chandrasekhar, Joana Projecto-Garcia, Hideaki Moriyama, Roy E. Weber, Violeta Muñoz-Fuentes, Andy J. Green, Cecilia Kopuchian, Pablo L. Tubaro, Luis Alza, Mariana Bulgarella, Matthew M. Smith, Robert E. Wilson, Angela Fago, Kevin G. McCracken, and Jay F. Storz. "Convergent Evolution of Hemoglobin Function in High-Altitude Andean Waterfowl Involves Limited Parallelism at the Molecular Sequence Level." *PLoS Genetics* 11, no. 12 (2015): e1005681.

National Academy of Sciences. "The Role of Theory in Advancing 21st Century Biology: Catalyzing Transformative Research." Washington, D.C.: National Academy of Sciences, 2007.

Nei, Masatoshi. *Mutation-Driven Evolution*. Oxford: Oxford University Press, 2013.

Nevarez, P. Andrew, Christopher M. DeBoever, Benjamin J. Freeland, Marissa A. Quitt, and Eliot C. Bush. "Context Dependent Substitution Biases Vary within the Human Genome." *BMC Bioinformatics* 11 (2010): 462.

Nguyen, Hoa Anh, Toshito Tomita, Morihiko Hirota, Jun Kaneko, Tetsuya Hayashi, and Yoshiyuki Kamio. "DNA Inversion in the Tail Fiber Gene Alters the Host Range Specificity of *carotovoricin Er*, a Phage-Tail-Like Bacteriocin of Phytopathogenic *Erwinia carotovora* subsp. *carotovora Er*." *Journal of Bacteriology* 183, no, 21 (2001): 6274–81.

Nieuwenhuis, Bart P. S., Sergio Tusso, Pernilla Bjerling, Josefine Stångberg, Jochen B. W. Wolf, and Simone Immler. "Repeated Evolution of Self-Compatibility for Reproductive Assurance." *Nature Communications* 9, no. 1 (2018): 1639.

Niggli, Hugo J., and Peter A. Cerutti. "Nucleosomal Distribution of Thymine Photodimers Following Far- and Near-Ultraviolet Irradiation." *Biochemical and Biophysical Research Communications* 105, no. 3 (1982): 1215–23.

Noble, Denis. "Evolution Beyond Neo-Darwinism: A New Conceptual Framework." *Journal of Experimental Biology* 218, Pt 8 (2015): 1273.

Nordenskiöld, Erik. *The History of Biology: A Survey*. London: Kegan Paul, Trench, Trubner and Co., 1929.

Norris, Steven J. "Antigenic Variation with a Twist–The *Borrelia* Story." *Molecular Microbiology* 60, no. 6 (2006): 1319–22.

Noto, Tomoko, and Kazufumi Mochizuki. "Whats, Hows and Whys of Programmed DNA Elimination in Tetrahymena." *Open Biology* 7, no. 10 (2017): https://doi.org/10.1098/rsob.170172.

Nuño de la Rosa, Laura. "Computing the Extended Synthesis: Mapping the Dynamics and Conceptual Structure of the Evolvability Research Front." *Journal of Experimental Zoology Part B: Molecular and Developmental Evolution* 328, no. 5 (2017): 395–411.

Odell, Ian D., Susan S. Wallace, and David S. Pederson. "Rules of Engagement for Base Excision Repair in Chromatin." *Journal of Cell Physiology* 228, no. 2 (2013): 258–66.

Orr, H. Allen "The Evolutionary Genetics of Adaptation: A Simulation Study." *Genetics Research* 74, no. 3 (1999): 207–14.

Orr, H. Allen. "The Genetics of Species Differences." *Trends in Ecology & Evolution* 16, no. 7 (2001): 343–350.

Orr, H. Allen. "The Population Genetics of Adaptation: The Adaptation of DNA Sequences." *Evolution* 56, no. 7 (2002): 1317–30.

Orr, H. Allen. (2005a). "The Genetic Theory of Adaptation: A Brief History." *Nat Rev Genet*, 6(2): 119–27.

Orr, H. Allen. "Theories of Adaptation: What They Do and Don't Say." *Genetica* 123, no. 1–2 (2005b): 3–13.

Orr, H. Allen. "Turned On." *The New Yorker*, October 24, 2005c.

Orr, H. Allen. and Coyne, Jerry A. "The Genetics of Adaptation: A Reassessment." *American Naturalist* 140, no. 5 (1992): 725–42.

Ortmann, Arnold E. "Facts and Interpretations in the Mutation Theory." *Science* 25, no. 631 (1907): 185–90.

Orzack, Steven Hecht, and Elliott Sober. "Introduction." In *Adaptationism and Optimality*, edited by Steven Hecht Orzack and Elliott Sober, 1–23. Cambridge, UK: Cambridge University Press, 2001.

Osawa, Syozo, and Thomas H. Jukes. "Codon Reassignment (Codon Capture) in Evolution." *Journal of Molecular Evolution* 28 (1989): 271–278.

Padian, Kevin. "Correcting Some Common Misrepresentations of Evolution in Textbooks and the Media." *Evolution: Education and Outreach* 6, no. 1 (2013): 11.

Palmer, Guy H., Troy Bankhead, and H. Steven Seifert. "Antigenic Variation in Bacterial Pathogens." *Microbiology Spectrum* 4, no. 1 (2016). 10.1128/microbiolspec.VMBF-0005-2015.

Palmer, Mary E., Marc Lipsitch, E. Richard Moxon, and Christopher D. Bayliss. "Broad Conditions Favor the Evolution of Phase-Variable Loci." *mBio* 4, no. 1 (2013): e00 430–12.

Palopoli, Michael F., and Nipam H. Patel. "Neo-Darwinian Developmental Evolution: Can We Bridge the Gap Between Pattern and Process?" *Current Opinion in Genetics & Development* 6, no. 4 (1996): 502–8.

Park, Stephen K., and Keith Willam Miller. "Random Number Generators: Good Ones Are Hard to Find." *Communications of the ACM* 31, no. 10 (1988): 1192–1201.

Park, Su-Chan, Damien Simon, and Joachim Krug. "The Speed of Evolution in Large Asexual Populations." *Journal of Statistical Physics* 138, no. 1–3 (2010): 381–410.

Parkinson, John S., Gerald L.Hazelbauer, and Joseph J. Falke. "Signaling and Sensory Adaptation in *Escherichia coli* Chemoreceptors: 2015 Update." *Trends in Microbiology* 23, no. 5 (2015): 257–66.

Patra, Pintu, and Stefan Klumpp. "Emergence of Phenotype Switching Through Continuous and Discontinuous Evolutionary Transitions." *Physical Biology* 12, no. 4 (2015): 046004.

Patterson, Adrian G., Mariya S. Yevstigneyeva, and Peter C. Fineran. "Regulation of CRISPR-Cas Adaptive Immune Systems." *Current Opinion in Microbiology* 37 (2017): 1–7.

Paul, Blair G., David Burstein, Cindy J. Castelle, Sumit Handa, Diego Arambula, Elizabeth Czornyj, Brian C. Thomas, Partho Ghosh, Jeff F. Miller, Jillian F. Banfield, and David L. Valentine. "Retroelement-Guided Protein Diversification Abounds in Vast Lineages of Bacteria and Archaea." *Nature Microbiology* 2 (2017): 17045.

Payne, Joshua L., Fabrizio Menardo, Andrej Trauner, Sonia Borrell, Sebastian M. Gygli, Chloe Loiseau, Sebastien Gagneux, and Alex R. Hall. "Transition Bias Influences the Evolution of Antibiotic Resistance in *Mycobacterium tuberculosis*." *PLoS Biology* 17, no. 5 (2019): e3000265.

Payne, Joshua L., and Andreas Wagner. "The Causes of Evolvability and Their Evolution." *Nature Reviews Genetics* 20, no. 1 (2019): 24–38.

Peacock, Andrew J. "ABC of Oxygen: Oxygen at High Altitude." *British Medical Journal* 317, no. 7165 (1998): 1063–6.

Penny, David. "Epigenetics, Darwin, and Lamarck." *Genome Biology and Evolution* 7, no. 6 (2015): 1758–60.

Peris, Joan Baptista, Paulina Davis, José M Cuevas, Miguel R. Nebot, and Rafael Sanjuán. "Distribution of Fitness Effects Caused by Single-Nucleotide Substitutions in Bacteriophage f1." *Genetics* 185, no. 2 (2010): 603–9.

Phear, Geraldine, Josephine Nalbantoglu, and Mark Meuth. "Next-Nucleotide Effects in Mutations Driven by DNA Precursor Pool Imbalances at the *aprt* Locus of Chinese Hamster Ovary Cells." *Proceedings of the National Academy of Sciences of the United States of America* 84, no. 13 (1987): 4450–4.

Philippe, Nadège, Estelle Crozat, Richard E. Lenski, and Dominique Schneider. "Evolution of Global Regulatory Networks During a Long-Term Experiment with *Escherichia coli*." *Bioessays* 29, no. 9 (2007): 846–60.

Pigliucci, Massimo. "Is Evolvability Evolvable?" *Nature Reviews Genetics* 9, no. 1 (2008): 75–82.

Plasterk, Ronald H. "Genetic Switches: Mechanism and Function." *Trends in Genetics* 8, no. 12 (1992): 403–6.

Poulton, Edward Bagnell. *Essays on Evolution, 1889–1907.* Oxford: Clarendon Press, 1908.

Poulton, Edward Bagnell. "Fifty Years of Darwinism." In *Fifty Years of Darwinism: Modern Aspects of Evolution,* 8–56. New York: Henry Holt and Company, 1909.

Price, Elmer M. and Jerry B. Lingrel. "Structure-Function Relationships in the Na,K-ATPase Alpha Subunit: Site-Directed Mutagenesis of Glutamine-111 to Arginine and Asparagine-122 to Aspartic Acid Generates a Ouabain-Resistant Enzyme." *Biochemistry* 27, no. 22 (1988): 8400–8.

Projecto-Garcia, Joana, Chandrasekhar Natarajan, Hideaki Moriyama, Roy E. Weber, Angela Fago, Zachary A. Cheviron, Robert Dudley, Jimmy A. McGuire, Christopher C. Witt, and Jay F. Storz. "Repeated Elevational Transitions in Hemoglobin Function During the Evolution of Andean Hummingbirds." *Proceedings of the National Academy of Sciences of the United States of America* 110, no. 51 (2013): 20669–74.

Prosperi, Mattia C. F, Roberto D'Autilia, Francesca Incardona, Andrea De Luca, Maurizio Zazzi, and Giovanni Ulivi. "Stochastic Modelling of Genotypic Drug-Resistance for Human Immunodeficiency Virus Towards Long-Term Combination Therapy Optimization." *Bioinformatics* 25, no. 8 (2009): 1040–7.

Provine, William B. *The Origins of Theoretical Population Genetics.* Chicago: University of Chicago Press, 1971.

Provine, William B. "The Role of Mathematical Population Geneticists in the Evolutionary Synthesis of the 1930s and 1940s." *Studies in History of Biology* 2 (1978): 167–92.

Provine, William B. *The Origins of Theoretical Population Genetics,* 2nd ed. Chicago: University of Chicago Press, 2001.

Psujek, Sean. and Randall D. Beer. "Developmental Bias in Evolution: Evolutionary Accessibility of Phenotypes in a Model Evo-Devo System." *Evolution & Development* 10, no. 3 (2008): 375–90.

Punnett, Reginald C. *Mendelism.* London: MacMillan and Bowes, 1905.

Punnett, Reginald C. *Mendelism,* 3rd ed. MacMillan, 1911.

Punnett, Reginald C. "More Mendelism and Mimicry." *Bedrock* 2 (1913): 496.

Punnett, Reginald C. *Mimicry in Butterflies.* London: Cambridge University Press, 1915.

Qiu, Li Yan, Elmar Krieger, Gijs Schaftenaar, Herman G. P. Swarts, Peter H. G. M. Willems, Jan Joep H. H. M.

De Pont, and Jan B. Koenderink. "Reconstruction of the Complete Ouabain-Binding Pocket of Na,K-ATPase in Gastric H,K-ATPase by Substitution of Only Seven Amino Acids." *Journal of Biological Chemistry* 280, no. 37 (2005): 32349–55.

Ram, Yoav, and Lilach Hadany. "The Probability of Improvement in Fisher's Geometric Model: A Probabilistic Approach." *Theoretical Population Biology* 99 (2015): 1–6.

Rastogi, Rajesh P., Sami Richa, Ashok Kumar, Madhu B. Tyagi, and Rajeshwar P. Sinha. "Molecular Mechanisms of Ultraviolet Radiation-Induced DNA Damage and Repair." *Journal of Nucleic Acids* 2010 (2010): 592980.

Razeto-Barry, Pablo, and Ramiro Frick. "Probabilistic Causation and the Explanatory Role Of Natural Selection." *Studies in History and Philosophy of Biological and Biomedical Sciences* 42, no. 3 (2011): 344–55.

Razeto-Barry, Pablo, and Davide Vecchi. "Mutational Randomness as Conditional Independence and the Experimental Vindication of Mutational Lamarckism." *Biological Reviews of the Cambridge Philosophical Society* 92, no. 2 (2016): 673–83.

Recker, Mario, Caroline O. Buckee, Andrew Serazin, Sue Kyes, Robert Pinches, Zóe Christodoulou, Amy L. Springer, Sunetra Gupta, and Chris I. Newbold. "Antigenic Variation in *Plasmodium falciparum* Malaria Involves a Highly Structured Switching Pattern." *PLoS Pathogens* 7, no. 3 (2011): e1001 306.

Reeve, Hudson Kern, and Paul W. Sherman. "Adaptation and the Goals of Evolutionary Research." *Quarterly Review of Biology* 68, no. 1 (1993): 1–32.

Reid, Thomas M., and Lawrence A. Loeb. "Tandem Double CC–>TT Mutations Are Produced by Reactive Oxygen Species." *Proceedings of the National Academy of Sciences of the United States of America* 90, no. 9 (1993): 3904–7.

Richardson, Michael K., and Ariel D. Chipman. "Developmental Constraints in a Comparative Framework: A Test Case Using Variations in Phalanx Number During Amniote Evolution." *Journal of Experimental Zoology B: Molecular and Developmental Evolution* 296, no. 1 (2003): 8–22.

Ridley, Mark. *The Problems of Evolution.* Oxford: Oxford University Press, 1985.

Ridley, Mark. *Evolution.* Cambridge, Mass.: Blackwell, 1993.

Ridley, Mark. "Natural Selection: An Overview." In *Encyclopedia of Evolution*, vol. 2, edited by M Pagel, 797–804. New York: Oxford University Press, 2002.

Rodrigue, Nicolas. and Hervé Philippe. "Mechanistic Revisions of Phenomenological Modeling Strategies in Molecular Evolution." *Trends in Genetics* 26, no. 6 (2010): 248–52.

Rokas, Antonis. "What Determines the Direction of Evolutionary Change?" *Trends in Ecology and Evolution* 19(6 (2004): 287–8.

Rokyta, Darin R., Paul Joyce, S. Brian Caudle, and Holly A. Wichman. "An Empirical Test of the Mutational Landscape Model of Adaptation Using a Single-Stranded DNA Virus." *Nature Genetics* 37, no. 4 (2005): 441–4.

Roll-Hansen, Nils. "The Crucial Experiment of Wilhelm Johannsen." *Biology and Philosophy* 4, no. 3 (1989): 303–329.

Rosenberg, Alex. "Discussion Note: Indeterminism, Probability, and Randomness in Evolutionary Theory." *Philosophy of Science* 68, no. 4 (2001): 536.

Rosenberg, Michael S., Sankar Subramanian, and Sudhir Kumar. "Patterns of Transitional Mutation Biases Within and Among Mammalian Genomes." *Molecular Biology and Evolution* 20, no. 6 (2003): 988–93.

Rosenberg, Susan M., and Philip J. Hastings. "Microbiology and Evolution. Modulating Mutation Rates in the Wild." *Science* 300, no. 5624 (2003): 1382–3.

Rosenberg, Susan M., and Philip J. Hastings. "Adaptive Point Mutation and Adaptive Amplification Pathways in the *Escherichia coli* Lac System: Stress Responses Producing Genetic Change." *Journal of Bacteriology* 186, no. 15 (2004): 4838–43.

Roth, John R., Elisabeth Kugelberg, Andrew B. Reams, Eric Kofoid, and Dan I. Andersson. "Origin of Mutations under Selection: The Adaptive Mutation Controversy." *Annual Review of Microbiology* 60 (2006): 477–501.

Sackman, Andrew M., Lindsey W. McGee, Anneliese J. Morrison, Jessica Pierce, Jeremy Anisman, Hunter Hamilton, Stephanie Sanderbeck, Cayla Newman, and Darin R. Rokyta. "Mutation-Driven Parallel Evolution During Viral Adaptation." *Molecular Biology and Evolution* 34, no. 12 (2017):

Sanjuan, Rafael, Andrés Moya, and Santiago F. Elena. "The Distribution of Fitness Effects Caused by Single-Nucleotide Substitutions in an RNA Virus." *Proceedings of the National Academy of Sciences of the United States of America* 101, no. 22 (2004): 8396–401.

Sarkar, Sahotra. "Lamarck Contre Darwin, Reduction Versus Statistics: Conceptual Issues in the Controversy Over Directed Mutagenesis in Bacteria." In *Organism and the Origins of Self*, edited by A. I. Tauber, 235–71. Dordrecht: Kluwer Academic Publishers, 1991.

Sarkar, Sahotra. "The Genomic Challenge to Adaptationism." *The British Journal for the Philosophy of Science*, 66 (2014): 505–536.

Schader, Susan M., Maureen Oliveira, Ruxandra-Ilinca Ibanescu, Daniela Moisi, Susan P. Colby-Germinario, and Mark A. Wainberg. "*In Vitro* Resistance Profile of the

Candidate HIV-1 Microbicide Drug Dapivirine." *Antimicrobial Agents and Chemotherapy* 56, no. 2 (2012): 751–6.

Schaerli, Yolanda, Alba Jiménez, José M Duarte, Ljiljana Mihajlovic, Julien Renggli, Mark Isalan, James Sharpe, and Andreas Wagner. "Synthetic Circuits Reveal How Mechanisms of Gene Regulatory Networks Constrain Evolution." *Mol Syst Biol* 14, no. 9 (2018): e8102.

Schenk, Martijn F., Ivan G. Szendro, Joachim Krug, and J. Arjan G. M. de Visser. "Quantifying the Adaptive Potential of an Antibiotic Resistance Enzyme." *PLoS Genetics* 8, no. 6 (2012): e1002783.

Schluter, Dolph. "Adaptive Radiation Along Genetic Lines of Least Resistance." *Evolution* 50, no. 5 (1966): 1766–74.

Scholl, Raphael, and Massimo Pigliucci. "The Proximate–Ultimate Distinction and Evolutionary Developmental Biology: Causal Irrelevance Versus Explanatory Abstraction." *Biology and Philosophy*, 30, no. 5 (2015): 653–70.

Schrempf, Dominik, Bui Quang Minh, Arndt von Haeseler, and Carolin Kosiol. "Polymorphism-Aware Species Trees with Advanced Mutation Models, Bootstrap, and Rate Heterogeneity." *Molecular Biology and Evolution* 36, no. 6 (2019): 1294–1301.

Schroeder, Jeremy W., William G. Hirst, Gabriella A. Szewczyk, and Lyle A. Simmons. "The Effect of Local Sequence Context on Mutational Bias of Genes Encoded on the Leading and Lagging Strands." *Current Biology* 26, no. 5 (2016): 692–7.

Schroeder, Jeremy W., Ponlkrit Yeesin, Lyle A. Simmons, and Jue D. Wang. "Sources of Spontaneous Mutagenesis in Bacteria." *Critical Reviews in Biochemistry and Molecular Biology* 53, no. 1 (2018): 29–48.

Schrödinger, Erwin. *What Is Life? Mind and Matter.* Cambridge, UK: Cambridge University Press, 1944.

Schultheis, Patrick J., Earl T. Wallick, and Jerry B. Lingrel. "Kinetic Analysis of Ouabain Binding to Native and Mutated Forms of Na,K-ATPase and Identification of a New Region Involved in Cardiac Glycoside Interactions." *Journal of Biological Chemistry* 268, no. 30 (1993): 22 686–94.

Sears, Karen E. "Quantifying the Impact of Development on Phenotypic Variation and Evolution." *Journal of Experimental Zoology B: Molecular and Developmental Evolution* 322, no. 8 (2014): 643–53.

Segerstråle, Ullica. "Neo-Darwinism." In *Encyclopedia of Evolution*, vol. 2, edited by M. Pagel, 807–10. New York: Oxford University Press, 2002.

Sekizuka, Tsuyoshki, Michiko Kawanishi, Mamoru Ohnishi, Ayaka Shima, Kengo Kato, Akifumi Yamashita, Mari Matsui, Satowa Suzuki, and Makoto Kuroda. "Elucidation of Quantitative Structural Diversity of Remarkable Rearrangement Regions, Shufflons, in IncI2 Plasmids." *Scientific Reports* 7, no. 1 (2017): 928.

Shah, Premal, and Michael A. Gilchrist. "Explaining Complex Codon Usage Patterns with Selection for Translational Efficiency, Mutation Bias, and Genetic Drift." *Proceedings of the National Academy of Sciences of the United States of America* 108, no. 25 (2011): 10 231–6.

Shapiro, James. *Evolution: A View From the 21st Century.* New York: FT Press, 2011.

Shen, Jiang-Cheng, William M. Rideout 3rd, and Peter A. Jones. "The Rate of Hydrolytic Deamination of 5-Methylcytosine in Double-Stranded DNA." *Nucleic Acids Research* 22, no. 6 (1994): 972–6.

Shi, Xinbai, Li Lu, Zijian Qiu, Wei He, and Joseph Frankel. "Microsurgically Generated Discontinuities Provoke Heritable Changes in Cellular Handedness of a Ciliate, *Stylonychia mytilus*." *Development* 111, no. 2 (1991): 337–56.

Shull, A. Franklin. "Weismann and Haeckel: One Hundred Years." *Science* 81, no. 2106 (1935): 443–51.

Shull, A. Franklin. *Evolution.* New York: McGraw-Hill, 1936.

Shull, George H. "Importance of the Mutation Theory in Practical Breeding." *Proceedings of the American Breeders' Association*, 3 (1907): 60–7.

Shyue, Song-Kun, David Hewett-Emmett, Harry G. Sperling, David M. Hunt, James K. Bowmaker, John D. Mollon, and Wen-Hsiung Li. "Adaptive Evolution of Color Vision Genes in Higher Primates." *Science* 269, no. 5228 (1995): 1265–7.

Simpson, George Gaylord. *The Meaning of Evolution*, 2nd ed. New Haven, Conn.: Yale University Press, 1967.

Simpson, George Gaylord. *The Major Features of Evolution.* New York: Simon and Schuster, 1953.

Simpson, George Gaylord. "Organisms and Molecules in Evolution." *Science* 146, no. 3651 (1964): 1535–8.

Simpson, George Gaylord. *Tempo and Mode in Evolution.* New York: Columbia University Press, 1944/1984.

Singer, G. A. and D A Hickey. "Nucleotide Bias Causes a Genome-wide Bias in the Amino Acid Composition of Proteins." *Mol Biol Evol* 17, no. 11 (2000): 1581–8.

Smith, Michael W, Da-Fei Feng, and Russell F. Doolittle. "Evolution by Acquisition: The Case for Horizontal Gene Transfers." *Trends in Biochemical Sciences* 17, no. 12 (1992): 489–93.

Smocovitis, Vassiliki Betty. *Unifying Biology: The Evolutionary Synthesis and Evolutionary Biology.* Princeton, NJ: Princeton University Press, 1996.

Sniegowski, Paul D., and Richard Lenski. "Mutation and Adaptation: The Directed Mutation Controversy in Evolutionary Perspective." *Annual Review of Ecology and Systematics* 26, no. 1 (1995): 553–578.

Sniegowski, Paul D., and Helen A. Murphy. "Evolvability." *Current Biology* 16, no. 19 (2006): R831–4.

Sober, Elliott. *The Nature of Selection: Evolutionary Theory in Philosophical Focus.* Cambridge, Mass.: MIT Press, 1984.

Sober, Elliott. "What is Adaptationism?" In *The Latest on the Best,* edited by J. Dupré, 105–18. Cambridge, Mass.: MIT Press, 1987.

Sober, Elliott. "Evolutionary Theory, Causal Completeness, and Theism: The Case of 'Guided' Mutation." In *Evolutionary Biology: Conceptual, Ethical, and Religious Issues,* edited by D. Walsh and P. Thompson, 31–44. Cambridge: Cambridge University Press, 2014.

Soderlund, David M. *Sodium Channels. Comprehensive Molecular Insect Science Series,* vol 5. New York: Elsevier, 2005.

Sonneborn, Tracy M. "Degeneracy of the Genetic Code: Extent, Nature, and Genetic Implications." In *Evolving Genes and Proteins,* edited by V. Bryson and H. J. Vogel, 377–97. New York: Academic Press, 1965.

Sousa, Ana, Catarina Bourgard, Lindi M. Wahl, and Isabel Gordo. "Rates of Transposition in *Escherichia coli.*" *Biol Lett* 9, no. 6 (2013): 20130 838.

Stamos, David N. "Quantum Indeterminism, Mutation, Natural Selection, and the Meaning of Life." In *Quantum Biochemistry,* edited by C. Matta, 837–72. New York: John Wiley and Sons, 2010.

Stamos, David N. "Quantum Indeterminism and Evolutionary Biology." *Philosophy of Science* 68, no. 2 (2001): 164.

Stearns, Stephen, and Rolf Hoekstra. *Evolution: An Introduction,* 2nd ed. New York: Oxford University Press, 2005.

Stebbins, G. Ledyard. "The Synthetic Approach to Problems of Organic Evolution." *Cold Spring Harbor Symposium of Quantitative Biology* 24 (1959): 305–11.

Stebbins, G. Ledyard. *Processes of Organic Evolution.* Englewood Cliffs, NJ: Prentice Hall, 1966.

Stebbins, G. Ledyard. *Darwin to DNA, Molecules to Humanity.* San Francisco: W.H. Freeman and Company, 1982.

Stephens, Christopher. "Selection, Drift, and the 'Forces' of Evolution." *Philosophy of Science* 71, no. 4 (2004): 550–70.

Steppan, Scott J., Patrick C. Phillips, and David Houle. "Comparative Quantitative Genetics: Evolution of the G Matrix." *Trends in Ecology and Evolution* 17, no. 7 (2002): 320–7.

Stern, David L., and Virginie Orgogozo. "The Loci of Evolution: How Predictable is Genetic Evolution?" *Evolution* 62, no. 9 (2008): 2155–77.

Stewart, Caro-Beth, James W. Schilling, and Allan C. Wilson. "Adaptive Evolution in the Stomach Lysozymes of Foregut Fermenters." *Nature* 330, no. 6146 (1987): 401–4.

Stiffler, Michael A, Doeke R. Hekstra, and Rama Ranganathan. "Evolvability as a Function of Purifying Selection in TEM-1 β-Lactamase." *Cell* 160, no. 5 (2015): 882–92.

Stocker, Bruce A. D. "Measurements of Rate of Mutation of Flagellar Antigenic Phase in *Salmonella typhimurium.*" *Journal of Hygiene (London)* 47, no. 4 (1949): 398–413.

Stoltzfus, Arlin. "On the Possibility of Constructive Neutral Evolution." *Journal of Molecular Evolution* 49, no. 2 (1999): 169–81.

Stoltzfus, Arlin. "Mutation-Biased Adaptation in a Protein NK Model." *Molecular Biology and Evolution* 23, no. 10 (2006a): 1852–62.

Stoltzfus, Arlin. "Mutationism and the Dual Causation of Evolutionary Change." *Evolution & Development* 8, no. 3 (2006b): 304–17.

Stoltzfus, Arlin. "Evidence for a Predominant role of Oxidative Damage in Germline Mutation in Mammals." *Mutation Research* 644, no. 1–2 (2008): 71–3.

Stoltzfus, Arlin. "Constructive Neutral Evolution: Exploring Evolutionary Theory's Curious Disconnect." *Biology Direct* 7, no. 1 (2012): 35.

Stoltzfus, Arlin. "In Search of Mutation-Driven Evolution." *Evolution & Development* 16, no. 1 (2014): 57–9.

Stoltzfus, Arlin. "Why We Don't Want Another 'Synthesis'." *Biology Direct* 12, no. 1 (2017): 23.

Stoltzfus, Arlin. "Understanding Bias in the Introduction of Variation as an Evolutionary Cause." In *Evolutionary Causation: Biological and Philosophical Reflections,* edited by Tobias Uller and Kevin Laland, 29–62. Vienna Series in Theoretical Biology. Cambridge, Mass.: MIT Press, 2019.

Stoltzfus, Arlin, and Kele Cable. "Mendelian-Mutationism: The Forgotten Evolutionary Synthesis." *Journal of the History of Biology* 47, no. 4 (2014): 501–46.

Stoltzfus, Arlin, and David M. McCandlish. "Mutation-Biased Adaptation in Andean House Wrens." *Proceedings of the National Academy of Sciences of the United States of America* 112, no. 45 (2015): 13 753–4.

Stoltzfus, Arlin, and David M. McCandlish. "Mutational Biases Influence Parallel Adaptation." *Mol Biol Evol* 34, no. 9 (2017): 2163–72.

Stoltzfus, Arlin, and David M. McCandlish. "Amino Acid Exchangeability from Deep Mutational Scanning Data." *in progress*

Stoltzfus, Arlin, and Ryan W. Norris. "On the Causes of Evolutionary Transition:Transversion Bias." *Mol Biol Evol* 33, no. 3 (2016): 595–602.

Stoltzfus, Arlin, David F. Spencer, Michael Zuker, John M. Logsdon, and W. Ford Doolittle. "Testing the Exon Theory of Genes: The Evidence from Protein Structure [see comments]." *Science* 265, no. 5169 (1994): 202–7.

Stoltzfus, Arlin, and Lev Y. Yampolsky. "Amino Acid Exchangeability and the Adaptive Code Hypothesis." *J Mol Evol* 65, no. 4 (2007): 456–62.

Stoltzfus, Arlin, and Lev Y. Yampolsky. "Climbing Mount Probable: Mutation as a Cause of Nonrandomness in Evolution." *Journal of Heredity* 100, no. 5 (2009): 637–47.

Storz, Jay F, Chandrasekhar Natarajan, Anthony V. Signore, Christopher C. Witt, David M. McCandlish, and Arlin Stoltzfus. "The Role of Mutation Bias in Adaptive Molecular Evolution: Insights from Convergent Changes in Protein Function." *Philosophical Transactions of the Royal Society of London B: Biological Sciences* 374, no. 1777 (2019): 20180238.

Streisfeld, Matthew A., and Mark D. Rausher. "Population Genetics, Pleiotropy, and the Preferential Fixation of Mutations During Adaptive Evolution." *Evolution* 65, no. 3 (2011): 629–42.

Strickberger, Monroe. *Evolution*. Boston: Jones and Bartlett Publishers, 1990.

Strotskaya, Alexandra, Ekaterina Savitskaya, Anastasia Metlitskaya, Natalia Morozova, Kirill A. Datsenko, Ekaterina Semenova, and Konstantin Severinov. "The Action of *Escherichia coli* CRISPR-Cas system on Lytic Bacteriophages with Different Lifestyles and Development Strategies." *Nucleic Acids Research* 45, no. 4 (2017): 1946–57.

Sturtevant, Alfred H. "An Interpretation of Orthogenesis." *Science* 59, no. 1539 (1924): 579–80.

Sueoka, Noboru. "On the Genetic Basis of Variation and Heterogeneity of DNA Base Composition." *Proceedings of the National Academy of Sciences of the United States of America* 48 (1962): 582–92.

Sueoka, Noboru. "Directional Mutation Pressure and Neutral Molecular Evolution." *Proceedings of the National Academy of Sciences of the United States of America* 85, no. 8 (1988): 2653–7.

Sutcliffe, Michael J., and Nigel S. Scrutton. "Enzymology Takes a Quantum Leap Forward." *Philosophical Transactions of the Royal Society A: Mathematical, Physical, and Engineering Sciences* 358, no. 1766 (2000): 367–86.

Svensson, Erik I. "On Reciprocal Causation in the Evolutionary Process." *Evolutionary Biology* 45, no. 1 (2018): 1–14.

Svensson, Erik I, and David Berger. "The Role of Mutation Bias in Adaptive Evolution." *Trends in Ecology & Evolution* 34, no. 5 (2019): 422–34.

Szöllösi, Gergely. J, Bastien Boussau, Sophie S. Abby, Eric Tannier, and Vincent Daubin. "Phylogenetic Modeling of Lateral Gene Transfer Reconstructs the Pattern and Relative Timing of Speciations." *Proceedings of the National Academy of Sciences of the United States of America* 109, no. 43 (2012): 17 513–8.

Tang, Hua, Gerald J. Wyckoff, Jian Lu, and Chung-I Wu. "A Universal Evolutionary Index for Amino Acid Changes." *Mol Biol Evol* 21, no. 8 (2004): 1548–56.

Tax, Sol, and Charles Callender. *Evolution After Darwin: The University of Chicago Centennial*, vol. I. Chicago: University of Chicago Press, 1960.

Tenaillon, Olivier. "The Utility of Fisher's Geometric Model in Evolutionary Genetics." *Annual Review of Ecology, Evolution, and Systematics* 45 (2014): 179–201.

Tenaillon, Olivier, Alejandra Rodríguez-Verdugo, Rebecca L. Gaut, Pamela McDonald, Albert F. Bennett, Anthony D. Long, and Brandon S. Gaut. "The Molecular Diversity of Adaptive Convergence." *Science* 335, no. 6067 (2012): 457–61.

Thompson, D'Arcy. *On Growth and Form*. Cambridge: Cambridge University Press, 1917.

Thomson, Keith. "Essay Review: The Relationship Between Development and Evolution." In *Oxford Surveys in Evolutionary Biology*, vol. 2, edited by Richard Dawkins and Mark Ridley, 220–33. Oxford: Oxford University Press, 1985.

Tillier, Elisabeth R, and Richard A. Collins. "High Apparent Rate of Simultaneous Compensatory Base-Pair Substitutions in Ribosomal RNA." *Genetics* 148, no. 4 (1998): 1993–2002.

Ujvari, Beata, Nicholas R. Casewell, Kartik Sunagar, Kevin Arbuckle, Wolfgang Wüster, Nathan Lo, Denis O'Meally, Christa Beckmann, Glenn F. King, Evelyne Deplazes, and Thomas Madsen. "Widespread Convergence in Toxin Resistance by Predictable Molecular Evolution." *Proceedings of the National Academy of Sciences of the United States of America* 112, no. 38 (2015): 11911–16.

Ujvari, Beata, Hee-chang Mun, Arthur D. Conigrave, Alessandra Bray, Jens Osterkamp, Petter Halling, and Thomas Madsen. "Isolation Breeds Naivety: Island Living Robs Australian Varanid Lizards of Toad-Toxin Immunity via Four-Base-Pair Mutation." *Evolution* 67, no. 1 (2013): 289–94.

Uller, Tobias, Armin P. Moczek, Richard A. Watson, Paul M. Brakefield, and Kevin N. Laland. "Developmental Bias and Evolution: A Regulatory Network Perspective." *Genetics* 209, no. 4 (2018): 949–66.

University of California Museum of Paleontology (2008). "Understanding Evolution," accessed June 26, 2020, http://evolution.berkeley.edu.

Valouev, Anton, Steven M. Johnson, Scott D. Boyd, Cheryl L. Smith, Andrew Z. Fire, and Arend Sidow. "Determinants of Nucleosome Organization in Primary Human Cells." *Nature* 474, no. 7352 (2011): 516–20.

van der Woude, Marjan W., and Andreas J. Bäumler. "Phase and Antigenic Variation in Bacteria." *Clin Microbiol Rev* 17, no. 3 (2004): 581–611.

van Ommen, Gert-Jan B. "Frequency of New Copy Number Variation in Humans." *Nature Genetics* 37, no. 4 (2005): 333–4.

Vandecraen, Joachim, Michael Chandler, Abram Aertsen, and Rob Van Houdt. "The Impact of Insertion Sequences on Bacterial Genome Plasticity and Adaptability." *Critical Reviews in Microbiology* 43, no. 6 (2017): 709–30.

Vavilov, Nikolai I. "The Law of Homologous Series in Variation." *Journal of Heredity* 12 (1922): 47–89.

Venkat, Aarti, Matthew W. Hahn, and Joseph W. Thornton. "Multinucleotide Mutations Cause False Inferences of Lineage-Specific Positive Selection." *Nat Ecol Evol* 2, no. 8 (2018): 1280–8.

Vink, Cornelis, Gloria Rudenko, and H. Steven Seifert. "Microbial Antigenic Variation Mediated by Homologous DNA Recombination." *FEMS Microbiology Reviews* 36, no. 5 (2012): 917–48.

Wagner, Andreas. "The Role of Randomness in Darwinian Evolution." *Philosophy of Science* 79, no. 1 (2012): 95–119.

Wagner, Andreas. *Arrival of the Fittest: Solving Evolution's Greatest Puzzle.* New York: Penguin Random House, 2014.

Wagner, Günter P., and Lee Altenberg. "Complex Adaptations and the Evolution of Evolvability." *Evolution* 50, no. 3 (1996): 967–76.

Wagner, Günter P., and Jianzhi Zhang. "The Pleiotropic Structure of the Genotype-Phenotype Map: The Evolvability of Complex Organisms." *Nature Reviews Genetics* 12, no. 3 (2011): 204–13.

Waisertreiger, Irina S., Victoria G. Liston, Miriam R. Menezes, Hyun Min Kim, Kirill S. Lobachev, Elena I. Stepchenkova, Tahir H. Tahirov, Igor B. Rogozin, and Youri I. Pavlov. "Modulation of Mutagenesis in Eukaryotes by DNA Replication Fork Dynamics and Quality of Nucleotide Pools." *Environmental and Molecular Mutagenesis* 53, no. 9 (2012): 699–724.

Waitt, Damon E., and Donald A. Levin. "Genetic and Phenotypic Correlations in Plants: A Botanical Test of Cheverud's Conjecture." *Heredity* 80 (1998): 310–19.

Wakeley, John. "The Excess of Transitions Among Nucleotide Substitutions: New Methods of Estimating Transition Bias Underscore its Significance." *Trends in Ecology and Evolution* 11, no. 4 (1996): 158–62.

Wallace, Bruce. "Can Embryologists Contribute to an Understanding of Evolutionary Mechanisms?" In *Integrating Scientific Disciplines*, edited by Bechtel, 149–63. Dordrecht: Martinus Nijhoff Publishers, 1986.

Wang, Yongge, and Tony Nicol. "On Statistical Distance-Based Testing of Pseudo Random Sequences and Experiments with PHP and Debian OpenSSL." *Computers and Security* 53 (2015): 44–64.

Watt, Danielle L., Robert J. Buckland, Scott A. Lujan, Thomas A. Kunkel, and Andrei Chabes. "Genome-Wide Analysis of the Specificity and Mechanisms of Replication Infidelity Driven by Imbalanced dNTP Pools." *Nucleic Acids Research* 44, no. 4 (2015): 1669–80.

Waxman, David. "A Unified Treatment of the Probability of Fixation when Population Size and the Strength of Selection Change over Time." *Genetics* 188, no. 4 (2011): 907–13.

Weber, James L., and Carmen Wong. "Mutation of Human Short Tandem Repeats." *Human Molecular Genetics* 2, no. 8 (1993): 1123–8.

Weill, Mylène, Georges Lutfalla, Knud Mogensen, Fabrice Chandre, Arnaud Berthomieu, Claire Berticat, Nicole Pasteur, Alexandre Philips, Philippe Fort, and Michel Raymond. "Comparative Genomics: Insecticide Resistance in Mosquito Vectors." *Nature* 423, no. 6936 (2003): 136–7.

Weinreich, Daniel M., Nigel F. Delaney, Mark A. Depristo, and Daniel L. Hartl. "Darwinian Evolution Can Follow Only Very Few Mutational Paths to Fitter Proteins." *Science* 312, no. 5770 (2006): 111–4.

Weinreich, Daniel M., and Jennifer L. Knies. "Fisher's Geometric Model of Adaptation Meets the Functional Synthesis: Data on Pairwise Epistasis for Fitness Yields Insights into the Shape and Size of Phenotype Space." *Evolution* 67, no. 10 (2013): 2957–72.

Weiss, Adam. "Lamarckian Illusions." *Trends in Ecology & Evolution* 30, no. 10 (2015): 566–8.

Weissman, Daniel B., Michael M. Desai, Daniel S. Fisher, and Marcus W. Feldman. "The Rate at which Asexual Populations Cross Fitness Valleys." *Theoretical Population Biology* 75, no. 4 (2009): 286–300.

Weissman, Jake L., Arlin Stoltzfus, Edze R. Westra, and Philip L. F. Johnson. "Avoidance of Self during CRISPR Immunization." *Trends in Microbiology* 28 no. 7 (2020): P543–53.

Wen, D Dingqiao, Yun Yu, and Luay Nakhleh. "Bayesian Inference of Reticulate Phylogenies under the Multispecies Network Coalescent." *PLoS Genetics* 12, no. 5 (2016): e1006006.

West-Eberhard, Mary Jane. *Developmental Plasticity and Evolution.* New York: Oxford University Press, 2003.

Westra, Edze R, Ümit Pul, Nadja Heidrich, Matthijs M. Jore, Magnus Lundgren, Thomas Stratmann, Reinhild Wurm, Amanda Raine, Melina Mescher, Luc Van Heereveld, Marieke Mastop, E. Gerhart H. Wagner, Karin Schnetz, John Van Der Oost, Rolf Wagner, and Stan J. J. Brouns. "H-NS-Mediated Repression of CRISPR-Based Immunity in *Escherichia coli* K12 Can Be Relieved by the Transcription Activator LeuO." *Molecular Microbiology* 77, no. 6 (2010): 1380–93.

Whitman, William B., David C. Coleman, and William J. Wiebe. "Prokaryotes: The Unseen Majority." *Proceedings of the National Academy of Sciences of the United States of America* 95, no. 12 (1998): 6578–83.

Wicken, Jeffrey S. *Evolution, Thermodynamics, and Information*. New York: Oxford University Press, 1987.

Wilkins, Adam S. "Evolutionary Developmental Biology: Where is it Going?" *BioEssays* 20, no. 10 (1998): 783–784.

Willems, Thomas, Melissa Gymrek, Gareth Highnam, The 1000 Genomes Project Consortium, David Mittelman, and Yaniv Erlich. "The Landscape of Human STR Variation." *Genome Research* 24, no. 11 (2014): 1894–904.

Williams, George C. *Natural Selection: Domains, Levels and Challenges*. Oxford Series in Ecology and Evolution. New York: Oxford University Press, 1992.

Winsor, Mary P. "The Creation of the Essentialism Story: An Exercise in Metahistory." *History and Philosophy of the Life Sciences* 28, no. 2 (2006): 149–74.

Winther, Rasmus Grønfeldt. "August Weismann on Germ-Plasm Variation." *Journal of the History of Biology* 34 (2001): 517–55.

Winther, Rasmus Grønfeldt. "Darwin on Variation and Heredity." *Journal of the History of Biology* 33 (2000): 425–55.

Wisniewski-Dyé, Florence, and Ludovic Vial. "Phase and Antigenic Variation Mediated by Genome Modifications." *Antonie Van Leeuwenhoek* 94, no. 4 (2008): 493–515.

Woese, Carl R. "On the Evolution of the Genetic Code." *Proceedings of the National Academy of Sciences of the United States of America* 54, no. 6 (1965): 1546–52.

Wojcik, Ewelina A., Anna Brzostek, Albino Bacolla, Pawel Mackiewicz, Karen M. Vasquez, Malgorzata Korycka-Machala, Adam Jaworski, and Jaroslaw Dziadek. "Direct and Inverted Repeats Elicit Genetic Instability by Both Exploiting and Eluding DNA Double-Strand Break Repair Systems in Mycobacteria." *PLoS One* 7, no. 12 (2012): e51 064.

Wolfe, Kenneth H. "Mammalian DNA Replication: Mutation Biases and the Mutation Rate." *Journal of Theoretical Biology* 149, no. 4 (1991): 441–51.

Working Group on Teaching Evolution. "Teaching About Evolution and the Nature of Science," accessed June 26, 2020, http://fermat.nap.edu/books/0309063647/html/index.html. Report 0-309-06364-7. New York: National Academy Press, 1998.

Wray, Gregory A., Matthew W. Hahn, Ehab Abouheif, James P. Balhoff, Margaret Pizer, Matthew V. Rockman, and Laura A. Romano. "The Evolution of Transcriptional Regulation in Eukaryotes." *Mol Biol Evol* 20, no. 9 (2003): 1377–419.

Wright, Barbara E. "Does Selective Gene Activation Direct Evolution?" *FEBS Letters* 402, no. 1 (1997): 4–8.

Wright, Barbara E. "A Biochemical Mechanism for Nonrandom Mutations and Evolution." *Journal of Bacteriology* 182, no. 11 (2000): 2993–3001.

Wright, Barbara E., Angelika Longacre, and Jacqueline M. Reimers. (1999). "Hypermutation in Derepressed Operons of *Escherichia coli* K12." *Proceedings of the National Academy of Sciences of the United States of America* 96, no. 9 (1999): 5089–94.

Wright, Sewall. "Evolution in Mendelian Populations." *Genetics* 16 (1931): 97–159.

Wright, Sewall. "Comments on the Preliminary Working Papers of Eden and Waddington." In *Mathematical Challenges to the Neo-Darwinian Interpretation of Evolution*, edited by P. Moorehead and M. Kaplan, 117–20. Philadelphia: Wistar Institutional Press, 1967.

Wright, Stephen I. "Mutationism 2.0: Viewing Evolution through Mutation's Lens." *Evolution* 68 (2014): 1225–7.

Wu, Li, Mari Gingery, Michael Abebe, Diego Arambula, Elizabeth Czornyj, Sumit Handa, Hamza Khan, Minghsun Liu, Mechthild Pohlschroder, Kharissa L. Shaw, Amy Du, Huatao Guo, Partho Ghosh, Jeff F. Miller, and Steven Zimmerly. "Diversity-Generating Retroelements: Natural Variation, Classification and Evolution Inferred from a Large-Scale Genomic Survey." *Nucleic Acids Research* 46, no. 1 (2018): 11–24.

Xue, Julian Z., André Costopoulos, and Frédéric Guichard. "A Trait-Based Framework for Mutation Bias as a Driver of Long-Term Evolutionary Trends." *Complexity* 21 (2015): 331–45.

Yampolsky, Lev. Y., and Arlin Stoltzfus. "Bias in the Introduction of Variation as an Orienting Factor in Evolution." *Evol Dev* 3, no. 2 (2001): 73–83.

Yampolsky, Lev. Y., and Arlin Stoltzfus. "The Exchangeability of Amino Acids in Proteins." *Genetics* 170, no. 4 (2005): 1459–72.

Yampolsky, Lev. Y., and Arlin Stoltzfus. "Mutational Bias." In *Encyclopedia of Life Sciences*. Chichester: John Wiley and Sons, 2008.

Yedid, Gabriel, and Graham Bell. "Macroevolution Simulated with Autonomously Replicating Computer Programs." *Nature* 420, no. 6917 (200): 810–2.

Yokoyama, Shozo, and F. Bernhard Radlwimmer. "The Molecular Genetics and Evolution of Red and Green Color Vision in Vertebrates." *Genetics* 158, no. 4, (2001): 1697–710.

Yu, Chuanhe, Michael J. Bonaduce, and Amar J. S. Klar. "Going in the Right Direction: Mating-Type Switching of *Schizosaccharomyces pombe* Is Controlled by Judicious Expression of Two Different *swi2* Transcripts." *Genetics* 190, no. 3 (2012): 977–87.

Yu, Li, Xiao-yan Wang, Wei Jin, Peng-tao Luan, Nelson Ting, and Ya-ping Zhang. "Adaptive Evolution of Digestive RNASE1 Genes in Leaf-Eating Monkeys Revisited: New Insights from Ten Additional Colobines." *Molecular Biology and Evolution* 27, no. 1 (2010): 121–31.

Zhang, Jianzhi. "Parallel Adaptive Origins of Digestive RNases in Asian and African Leaf Monkeys." *Nature Genetics* 38, no. 7 (2006): 819–23.

Zhen, Ying, M L Aardema, Matthew L. Aardema, Edgar M. Medina, Molly Schumer, and Peter Andolfatto. "Parallel Molecular Evolution in an Herbivore Community." *Science* 337, no. 6102 (2012): 1634–7.

Zhong, Weihao, and Nicholas K. Priest. "Stress-Induced Recombination and the Mechanism of Evolvability." *Behavioral Ecology and Sociobiology* 65, no. 3 (2011): 493–502.

Zhu, Yuan O., Mark L. Siegal, David W. Hall, and Dmitri A. Petrov. "Precise Estimates of Mutation Rate and Spectrum in Yeast." *Proceedings of the National Academy of Sciences of the United States of America* 111, no. 22 (2014): E2310–8.

Zieg, Janine, Michael R. Silverman, Marcia Hilmen, and Melvin Simon. "Recombinational Switch for Gene Expression." *Science* 196, no. 4286 (1977): 170–2.

Zirkle, Conway. "The Inheritance of Acquired Characters and the Provisional Hypothesis of Pangenesis." *The American Naturalist* 69, no. 724 (1935): 417–45.

Zuckerkandl, Emil, and Linus Pauling. "Molecular Disease, Evolution, and Genetic Heterogeneity." In *Horizons in Biochemistry*, edited by M. Kasha and B. Pullman, 189–225. New York: Academic Press, 1962.

Zuckerkandl, Emil, and Linus Pauling. "Evolutionary Divergence and Convergence in Proteins." In *Evolving Genes and Proteins*, edited by Vernon Bryson and Henry J. Vogel, 97–166. New York: Academic Press, 1965.

Zufall, Rebecca A., Tessa Robinson, and Laura A. Katz. "Evolution of Developmentally Regulated Genome Rearrangements in Eukaryotes." *Journal of Experimental Zoology B: Molecular and Developmental Evolution* 304, no. 5 (2005): 448–55.

Index

adaptation **123**
  altitude 43, 176, 177, 181, 194
  by avoidance 77
  physiological 49
  pre-adaptation 10, **123**, 124
adaptationism 39, 103, 130, 131
  empirical **39**, 98, 99, 108, 120,
    199, 204
  explanatory **39**, 105
  methodological **39**
adaptationist program 4, 39, 204
adaptive immunity 27, **69**, 78
adaptive responsiveness,
    see adaptation, physiological 49
antigenic variation 27, 68, **74**, 83, 84
architect analogy 59, 96, **106**, 108,
    120, 241
AT bias, *see* bias, GC 167

bias
  developmental 4, 52, 81, 97, 126,
    155, 156, **186**, 187, 195, 203
  directional 88, 109, 140, 143, **153**,
    153, 240, 241
  GC 128, 130, 143, **153**, 154, 167
  transition-transversion 11, 19, 41,
    44, 56, **153**, 166, 168, 171, 172, 177,
    180, 182, 187, 192
blending inheritance 51, 98, 104, 115,
    124, 240, 241
buffet regime 13, 147, **148**, 149,
    163, 164

cassette shuffling 11, **76**
causal completeness argument 120,
    125, 139
cause
  efficient **59**, 96, 157
  final **59**
  formal **59**, 157
  material **59**, 96, 107, 126, 157
  proximate 9, 97, 104, 125, 151, 157,
    185, 202
  ultimate 89, 96, 97, 111, 113, 125,
    198, 199, 205
chemotaxis 139, 140

Cheverud's conjecture 117, **186**
ciliate cortex 23
codon usage 90, 126, 128, 165, 173
conditional independence **60**, 60–62,
    64, 65, 85, 86
consequence laws 12, 113, **114**,
    115–117, 125, 126, 133, 140, 231
constraints 5, 9, 96, 124, 148, 151, 156,
    158, 200, 205, 207, 232, 238
contingency 5, 205, 206
contingency loci 87
creativity **95**, 96, 99, 100, 104, 120,
    147, 150, 200, 203, 237, 238
creativity-improbability 122
CRISPR-Cas 11, 26, 28, 70, 90, 199, 202

Darwin's architect, *see* architect
    analogy 59
deep mutational scanning 55, 57, 166,
    168, 221
developmental bias **155**, *see* bias,
    developmental 186
DFE, *see* distribution of fitness
    effects 55
directional bias, *see* bias 153
distribution of fitness effects 55, 172
diversity-generating retroelements
    (DGRs) **28**, 70, **71**
DNA damage 5, 6, 15, 22, 26, 29, 32,
    58, 59, **218**, 218–220
DNA precursor pools 31, 32, 59, 218
Down's syndrome 41

echolocation **181**, 195
equilibrium explanation 4, 204
error-prone repair 22, 26, 29, 58,
    218, 220
evo-devo 4, 5, 9, 12, 87, 97, 118, **124**,
    125, 156, 157, 183, 184, 201
evolvability 5, 9, **81**, 86, 90, 116, 124
  as *explanandum* (E3) **81**, 82,
    86–88, 90
  as *explanans* (E2) **81**, 87–90, 124
  as fact (E1) **81**, 87, 88
  research front 81, 88, 90
explanandum 39, **81**, 124, 203, 204

explanans **81**
Extended Evolutionary Synthesis
    (EES) 10, 212
extrapolationist doctrine **3**, 104,
    152, 200

Fisher's geometric model **120**, 121,
    122, 146
fitness landscape 119, **140**, 142, 143,
    159, 207
fluctuation 51, *see* indefinite
    variability 104, 115, 117,
    124, 205
fluctuation test **25**, 34, 35, 48,
    55, 199
forces theory 7, 97, **149**, 150, 151, 169,
    208–210
foregut fermentation 182
foresight **16**, 34, 35

G matrix 89, **119**, 207
gene conversion 29, 71, 76, 128
gene pool 10, 100, **102**, 102–104, 107,
    119, 122, 129, 147, 148, 151, 163,
    164, 172, 200, 204, 232, 238, 239
gene scrambling 153, 154
genetic code 4, 57, 82, 87, 88, 126, **143**,
    144, 155, 157, 166, 167, 203
genome composition 13, 90, 109, 126,
    128, 130, 131
gradualism **120**, 122, 123

Hardy–Weinberg equilibrium
    114, 184
horizontal gene transfer, *see* lateral
    gene transfer 21
hypermutation 29, 74

immune evasion 11, 27, 70, 74, **78**, 84,
    89, 91
indefinite variability 51, **98**, 116,
    124, 205
indeterminacy **21**, 21, 22, 34, 35, 45,
    185, 205, 234, 235

kataegis **29**, 30, 228, 229